Radar for Space Observation: Systems, Methods and Applications

Radar for Space Observation: Systems, Methods and Applications

Guest Editor

Vassilis Karamanavis

MDPI

Basel • Beijing • Wuhan • Barcelona • Belgrade • Novi Sad • Cluj • Manchester

Guest Editor
Vassilis Karamanavis
German Space Agency
at DLR
Bonn
Germany

Editorial Office
MDPI AG
Grosspeteranlage 5
4052 Basel, Switzerland

This is a reprint of the Special Issue, published open access by the journal *Remote Sensing* (ISSN 2072-4292), freely accessible at: https://www.mdpi.com/journal/remotesensing/special_issues/7C43H2J2V2.

For citation purposes, cite each article independently as indicated on the article page online and as indicated below:

Lastname, A.A.; Lastname, B.B. Article Title. *Journal Name* **Year**, *Volume Number*, Page Range.

ISBN 978-3-7258-6970-1 (Hbk)
ISBN 978-3-7258-6971-8 (PDF)
https://doi.org/10.3390/books978-3-7258-6971-8

Cover image courtesy of Vassilis Karamanavis

Contents

About the Editor

Vassilis Karamanavis

Vassilis Karamanavis is a Projects Officer at the German Space Agency at DLR. Previously, he was a Staff Scientist and Project Manager at Fraunhofer FHR, specializing in space domain awareness with radar. His expertise includes satellite tracking, inverse synthetic aperture radar (ISAR) imaging, the characterization of resident space objects, and radar-based analysis for space safety. He has also conducted research in radio astronomy at the Max Planck Institute for Radio Astronomy, focusing on very-long-baseline interferometry (VLBI) and multi-wavelength studies of cosmic radio sources. He holds a Ph.D. in Natural Sciences from the University of Cologne, Germany, and a degree in Physics from the Aristotle University of Thessaloniki, Greece.

remote sensing

MDPI

Editorial

Radar for Space Observation: Systems, Methods and Applications

Vassilis Karamanavis

Fraunhofer Institute for High Frequency Physics and Radar Techniques FHR, Fraunhoferstraße 20, 53343 Wachtberg, Germany; v.karamanavis.work@outlook.com

1. Context and Motivation

In today's world, where near-Earth space constitutes an increasingly congested and contested theater of economic, political, and military activities for the growing number of space-faring nations, the significance of actionable situational awareness cannot be overstated. The proliferation of objects in orbit results in an increased risk of collisions, requiring robust methods of debris detection and surveillance for the timely resolution of conjunctions. Radar systems, renowned for their versatility and precision, have emerged as indispensable tools for addressing these challenges. From tracking the trajectories of satellites and space debris to protect critical infrastructure, to imaging celestial bodies, they provide crucial data that underpin space domain awareness (SDA), space safety, and planetary defense. This Special Issue, titled "Radar for Space Observation: Systems, Methods and Applications", was established to serve as a platform for disseminating innovative research and fostering collaboration within this vital field, underscoring how cutting-edge radar technology contributes to safer, more effective operations in increasingly crowded orbits.

2. Overview of Contributions

Contributions to this Special Issue are divided into two thematic areas: Existing and Upcoming Radar Systems for Space Observation (Section 2.1) and Advanced Signal Processing, Models and Simulations (Section 2.2). In the following paragraphs, a brief overview of the contributions is given.

2.1. Existing and Upcoming Radar Systems for Space Observation

Very-high-frequency (VHF) wind profiling radars are a sensor class capable of observing resident space objects (RSOs) in low-Earth orbit. Holdsworth et al. [1] detail the use of the Buckland Park VHF radar for space surveillance. Their study, "Space Domain Awareness Observations Using the Buckland Park VHF Radar", presents a new beam scheduling and signal processing functionality of the radar system. The results reveal a three-fold increase in the detection rate over previous single-beam direction observations and the capability to detect objects in medium-Earth orbit. It is also shown that it is possible to investigate the rotational behavior of satellites and to monitor ionospheric plasma waves and instabilities.

In their contribution, "The Future of Radar Space Observation in Europe—Major Upgrade of the Tracking and Imaging Radar (TIRA)", Klare et al. [2] present the modernization effort centered around the TIRA system, a cornerstone of European space observation. This major upgrade will equip TIRA with advanced capabilities for detecting, tracking, and imaging objects in orbit. This work highlights the improvement of TIRA's tracking precision, its increased imaging resolution for monitoring the status and integrity of RSOs, and its enhanced data processing to address the challenges arising from the ever-growing space traffic. The article delves into the project's technical innovations and their critical role in supporting Europe's space sustainability initiatives.

Pandeirada et al. [3] present the recent progress of the ATLAS tracking radar, which is part of Portugal's contribution to EU SST (EU Space Surveillance and Tracking) and a promising test bed for space surveillance technologies. Their paper, titled "ATLAS: Latest Advancements and First, Observations", focuses on the activities of calibration of the pointing system and initial observations of RSOs. ATLAS aims to not only be a standalone radar system but rather a full-featured space surveillance and tracking ground station with the capacity to predict the orbit of RSOs, generate observing schedules, and maintain an internal objects' catalog.

Space-based radars for SDA have also been proposed and are the subject of ongoing research efforts by various space actors. In "Joint Power and Bandwidth Allocation with RCS Fluctuation Characteristic for Space Target Tracking", Yang et al. [4] present a method for optimal allocation of resources for a space-based radar aimed at tracking targets in orbit. The method takes into account the variability of the target's radar cross-section. This way, the power and the bandwidth required for an observation can be adjusted for optimal tracking accuracy.

This section is concluded with a fascinating review article, titled "Overview of High-Power and Wideband Radar Technology Development at MIT Lincoln Laboratory". Here, MacDonald et al. [5] present over 60 years of radar history and groundbreaking technology development at the MIT Lincoln Laboratory. This captivating review focuses on space observations from early research on satellite tracking and planetary radar to present capabilities for centimeter-resolution imaging of RSOs and future plans to extend this to the geosynchronous orbit domain.

2.2. Advanced Signal Processing, Models and Simulations

In "Inverse Synthetic Aperture Radar Imaging of Space Targets Using Wideband Pseudo-Noise Signals with Low Peak-to-Average Power Ratio", Anger et al. [6] investigate the application of pseudo-noise signals for imaging in the context of inverse synthetic aperture radar (ISAR) imaging of RSOs. The use of pseudo-noise signals provides several advantages, including a reduction in RF interference with other radar and communication systems operating within the same frequency band. Additionally, these signals facilitate the concurrent use of multiple channels and exhibit a low probability of intercept, thereby decreasing vulnerability to electronic support measures (ESM), electronic intelligence (ELINT) systems, and jamming. The paper details the implementation of pseudo-noise signals within the IoSiS radar system, developed by the Microwaves and Radar Institute of DLR.

The paper "The Radar Signal Processor of the First, Romanian Space Surveillance Radar" by Bîră et al. [7] focuses on the signal processing chain of the Cheia space surveillance radar. It offers an in-depth look at the signal processor's architecture, alongside a comprehensive review of state-of-the-art signal processing techniques for space radars. The study also includes a comparison between the signal processor's outputs and the results produced by benchmarking processing algorithms. This work represents a significant contribution to the field, providing valuable insights into the core radar processing functions of the Cheia radar, a highly capable space surveillance asset with substantial potential.

In their contribution "Micro-Doppler Signature Analysis for Space Domain Awareness Using VHF Radar", Heading et al. [8] characterize defunct satellites by analyzing their micro-Doppler signatures, which are linked to their rotational motion. They conducted this study using the Buckland Park Stratosphere–Troposphere VHF radar in South Australia. Computational electromagnetic simulations were used to provide a basis for comparison with the observed radar data, helping to characterize three RSOs. By comparing the simulated and measured micro-Doppler signatures, they were able to estimate key parameters, such as spin axis, attitude, number of solar panels, rotation rate, and approximate size.

In "Range-Doppler-Time Tensor Processing for Deep-Space Satellite Characterization Using Narrowband Radar", Serrano et al. [9] address the challenge of characterizing satellites in geostationary orbit (geosynchronous equatorial orbit, GEO), a domain where existing wideband radars, typically used for low-Earth Orbit satellites, lack the sensitivity required for the long ranges involved (approximately 36,000 km or greater). The authors propose a novel range–Doppler–time (RDT) tensor processing technique tailored for GEO satellite characterization using a narrowband radar. This method generates fully-resolved two-dimensional images of rotating GEO satellites under low-bandwidth observational conditions. In scenarios where the change in the aspect angle is limited, the technique provides enhanced Doppler information, improving the monitoring of a satellite's status. The efficacy of the technique is evaluated using simulated radar data and radar data collected in a compact range on a scaled satellite model.

In the context of space-based SDA, Zheng et al. [10] explore a novel algorithm for tracking objects in orbit, from orbit. In their article, titled "Space Target Tracking with the HRRP Characteristic-Aided Filter via Space-Based Radar", they utilize high-resolution range profiles to improve the radar system's tracking performance. This approach is also put to the test, and its performance is evaluated.

Radars play a significant role in the observation of planetary bodies and asteroids. The contribution by Anne Virkki [11], "Numerical Evaluation of Planetary Radar Backscatter Models for Self-Affine Fractal Surfaces", presents a valuable collection of open source Python codes for generating self-affine fractal surfaces and evaluating radar scattering laws, thus offering a versatile tool for researchers in remote sensing, electromagnetic simulations, and material sciences. The synthetic surfaces successfully reproduce measured rms slopes for natural terrains, reinforcing the model's validity. Key findings include the confirmation of the Gaussian law as an optimal scattering model for surfaces with rms slope angles up to 40°, along with insights into the role of roll-off length in surface roughness characteristics. Additionally, the article highlights the impact of multiple scattering on backscattered power and identifies key challenges in modeling wavelength-scale interactions on complex surfaces. By providing a computational framework for analyzing radar scattering, this work paves the way for future advancements in planetary and terrestrial surface characterization.

3. Knowledge Gaps and Future Research Directions

Recent advances in radar technology for space observation have focused on broader coverage, improved tracking performance, higher-resolution imaging, and rudimentary integration with AI-driven data analysis. Nonetheless, challenges persist, notably in the areas of real-time data fusion, autonomous tracking of RSOs, and radar-assisted space debris removal. Although this Special Issue has addressed several key topics, open questions remain. Among other directions, future research should explore the following:

- **AI-enhanced radar techniques:** Leveraging the full potential of AI to improve automation, increase efficiency, and enhance the characterization of RSOs.
- **Distributed sensor networks:** Strengthening international cooperation in space surveillance to achieve comprehensive coverage and multisensor data fusion.
- **Radar-driven investigation of critical size regimes:** Developing systems that can peer into the critical size regime of $\leqslant$10 cm, where cubesats and space debris proliferate and will continue to do so.

– **Radars as enablers of efficient space traffic management (STM):** Advancing the use of ground- and space-based radar technology and methods towards efficient, operational STM.
– **Radar-assisted space sustainability:** Investigating debris mitigation strategies and active removal technologies supported by radar monitoring.

4. Concluding Remarks

Surveillance of the near-Earth space environment is undergoing a profound transformation, driven by the increasing importance of monitoring space assets and debris in orbit. This Special Issue demonstrates the remarkable progress in radar technology applied to space observations and the vital role radar systems play in ensuring the sustainable and safe use of space. I am confident that the insights and innovations presented here will inspire further research and collaboration, driving the field forward towards meeting current and future challenges.

Funding: This research received no external funding.

Acknowledgments: I would like to express my deepest gratitude to the authors for their valuable and inspiring contributions. I am indebted to the reviewers for their meticulous and insightful feedback and wholeheartedly thank the editorial team of *Remote Sensing* for the invitation to edit this Special Issue and for their unwavering support throughout this effort. The success of this Special Issue is a testament to the collective efforts of a vibrant and dedicated research community.

Conflicts of Interest: The author declares no conflicts of interest.

References

1. Holdsworth, D.A.; Spargo, A.J.; Reid, I.M.; Adami, C.L. Space Domain Awareness Observations Using the Buckland Park VHF Radar. *Remote Sens.* **2024**, *16*, 1252. [CrossRef]
2. Klare, J.; Behner, F.; Carloni, C.; Cerutti-Maori, D.; Fuhrmann, L.; Hoppenau, C.; Karamanavis, V.; Laubach, M.; Marek, A.; Perkuhn, R.; et al. The Future of Radar Space Observation in Europe—Major Upgrade of the Tracking and Imaging Radar (TIRA). *Remote Sens.* **2024**, *16*, 4197. [CrossRef]
3. Pandeirada, J.; Bergano, M.; Marques, P.; Coelho, B.; Barbosa, D.; Figueiredo, M. ATLAS: Latest Advancements and First Observations. *Remote Sens.* **2024**, *16*, 704. [CrossRef]
4. Yang, Q.; Jiang, L.; Zheng, S.; Zhao, Y.; Wang, Z. Joint Power and Bandwidth Allocation with RCS Fluctuation Characteristic for Space Target Tracking. *Remote Sens.* **2023**, *15*, 3971. [CrossRef]
5. MacDonald, M.; Abouzahra, M.; Stambaugh, J. Overview of High-Power and Wideband Radar Technology Development at MIT Lincoln Laboratory. *Remote Sens.* **2024**, *16*, 1530. [CrossRef]
6. Anger, S.; Jirousek, M.; Dill, S.; Peichl, M. Inverse Synthetic Aperture Radar Imaging of Space Targets Using Wideband Pseudo-Noise Signals with Low Peak-to-Average Power Ratio. *Remote Sens.* **2024**, *16*, 1809. [CrossRef]
7. Bîră, C.; Ionescu, L.; Rusu-Casandra, A. The Radar Signal Processor of the First Romanian Space Surveillance Radar. *Remote Sens.* **2023**, *15*, 3630. [CrossRef]
8. Heading, E.; Nguyen, S.T.; Holdsworth, D.; Reid, I.M. Micro-Doppler Signature Analysis for Space Domain Awareness Using VHF Radar. *Remote Sens.* **2024**, *16*, 1354. [CrossRef]
9. Serrano, A.; Capper, J.; Morrison, R.L.; Abouzahra, M.D. Range-Doppler-Time Tensor Processing for Deep-Space Satellite Characterization Using Narrowband Radar. *Remote Sens.* **2024**, *16*, 1374. [CrossRef]
10. Zheng, S.; Jiang, L.; Yang, Q.; Zhao, Y.; Wang, Z. Space Target Tracking with the HRRP Characteristic-Aided Filter via Space-Based Radar. *Remote Sens.* **2023**, *15*, 4808. [CrossRef]
11. Virkki, A. Numerical Evaluation of Planetary Radar Backscatter Models for Self-Affine Fractal Surfaces. *Remote Sens.* **2024**, *16*, 890. [CrossRef]

Short Biography of Author

Dr. Vassilis Karamanavis is a Projects Officer at the German Space Agency at DLR. Previously, he was a Staff Scientist and Project Manager at Fraunhofer FHR, specializing in space domain awareness with radar. His expertise includes satellite tracking, inverse synthetic aperture radar (ISAR) imaging, characterization of resident space objects, and radar-based analysis for space safety. He has also conducted research in radio astronomy at the Max Planck Institute for Radio Astronomy, focusing on very-long-baseline interferometry (VLBI) and multi-wavelength studies of cosmic radio sources. He holds a Ph.D. in Natural Sciences from the University of Cologne, Germany and a degree in Physics from the Aristotle University of Thessaloniki, Greece.

remote sensing

MDPI

Article

Space Domain Awareness Observations Using the Buckland Park VHF Radar

David A. Holdsworth [1,2,*], Andrew J. Spargo [2], Iain M. Reid [2,3] and Christian L. Adami [3]

1 Defence Science & Technology Group, Edinburgh, SA 5111, Australia
2 School of Physical Sciences, University of Adelaide, Adelaide, SA 5005, Australia; andrew.spargo@defence.gov.au (A.J.S.); ireid@atrad.com.au (I.M.R.)
3 ATRAD Pty. Ltd., Underdale, SA 5032, Australia
* Correspondence: david.holdsworth@defence.gov.au

Abstract: There is increasing interest in space domain awareness worldwide, motivating investigation of the use of non-traditional sensors for space surveillance. One such class of sensor is VHF wind profiling radars, which have a low cost relative to other radars typically applied to this task. These radars are ubiquitous throughout the world and may potentially offer complementary space surveillance capabilities to the Space Surveillance Network. This paper updates an initial investigation on the use of Buckland Park VHF wind profiling radars for observing resident space objects in low Earth orbit to further investigate the space surveillance capabilities of the sensor class. The radar was operated during the Australian Defence "SpaceFest" 2019 activity, incorporating new beam scheduling and signal processing functionality that extend upon the capabilities described in the initial investigation. The beam scheduling capability used two-line element propagations to determine the appropriate beam direction to use to observe transiting satellites. The signal processing capabilities used a technique based on the Keystone transform to correct for range migration, allowing the development of new signal processing modes that allow the coherent integration time to be increased to improve the SNR of the observed targets, thereby increasing the detection rate. The results reveal that 5874 objects were detected over 10 days, with 2202 unique objects detected, representing a three-fold increase in detection rate over previous single-beam direction observations. The maximum detection height was 2975.4 km, indicating a capability to detect objects in medium Earth orbit. A minimum detectable RCS at 1000 km of -10.97 dBm^2 (0.09 m^2) was observed. The effects of Faraday rotation resulting from the use of linearly polarised antennae are demonstrated. The radar's utility for providing total electron content (TEC) measurements is investigated using a high-range resolution mode and high-precision ephemeris data. The short-term Fourier transform is applied to demonstrate the radar's ability to investigate satellite rotation characteristics and monitor ionospheric plasma waves and instabilities.

Keywords: space domain awareness; space situational awareness; radar; signal processing

1. Introduction

There is increasing interest in space domain awareness (SDA) worldwide due to the increasing number of resident space objects (RSOs) [1]. This number has expanded rapidly in recent years due to the advent of mega-constellations such as the SpaceX Starlink constellation [2]. The increasing number of objects in orbit has led to renewed concerns over the so-called Kessler effect [3], where the density of objects in low Earth orbit (LEO) is sufficiently high that collisions between objects could cause a cascade where each collision generates space debris that increases the likelihood of further collisions. The risk of collision was brought into sharp focus in 2019 when the European Space Agency (ESA) performed a "collision avoidance maneuver" to protect its Aeolus Earth observation satellite from colliding with a Starlink satellite [4].

The increasing interest in SDA in Australia is demonstrated by the installation of a C-band radar at Exmouth, Western Australia, and the recent installation of an optical space telescope at a neighbouring site that has recently achieved Initial Operating Capability [5,6]. Additionally, LeoLabs has recently installed the Western Australian Space Radar, featuring two S-band phased array radar systems [7]. The use of non-traditional space sensors has also been investigated, such as event-based sensors [8], a DSTG experimental high-frequency (HF) line-of-sight (LOS) radar system [9,10], the Murchison Widefield radio astronomy array (MWA) for passive radar observations of satellite reflections from FM transmitters of opportunity [11,12], and the Buckland Park stratosphere–troposphere (BPST) radar, a civilian wind profiling radar operating in the lower VHF band [13].

The key factors to consider when discussing radar SDA capabilities are the detection sensitivity, field of view, and tracking capability, i.e., beam steering to follow satellite trajectories. As the BPST radar was not designed for SDA operations, it has a number of limitations compared to the purpose-built state-of-the-art (and significantly more expensive) radars of the Space Surveillance Network (SSN). A common first-order detection sensitivity figure of merit (FoM) is the signal-to-noise ratio (SNR) of a target with 0 dBm^2 RCS at a 1000 km range from both the radar transmitter and receiver. In this regard, the BPST radar is between 35 and 50 dB less sensitive than the SSN radars [13]. This is mostly a consequence of the SSN radars mostly operating in the UHF range in order to obtain high sensitivity against small objects (<10 cm) [14]. The BPST radar, operating at 55 MHz, has reduced sensitivity for small targets. For instance, Figure 1 of [13] shows that the RCS of a 20 cm diameter sphere is reduced by approximately 40 dB for a radar operating at 55 MHz (e.g., BPST radar: Rayleigh scatter) compared to 1.32 GHz (e.g., TRADEX L-band: Mie scatter). The field of view can be regarded as wide (WFoV) or narrow (NFoV) field of view. WFoV radars typically employ interferometric or beamforming techniques and are capable of detecting multiple objects simultaneously. These radars are useful for new object detection and are difficult to design in a low-cost form factor. NFoV radars are typically high-precision systems employing tracking capability and are cued by either a WFoV system or propagated state vectors. The BPST radar is classified as NFoV but lacks the range resolution and tracking capability of many of the other NFoV systems. A further limitation of the BPST radar that occurs due to the operating frequency of 55 MHz is the effects of signal propagation through the ionosphere (group retardation), which can result in positive range biases of between 0.5 and 1.5 km [13].

Despite these limitations, the BPST radar has attracted interest from within the Australian SDA community as a sensor that can complement (rather than compete with) the SSN. The radar has been shown to produce daily detection count rates ranging from 150 to 200, with a maximum detection height observed of 2491 km [13]. The radar's utility for object catalogue maintenance was demonstrated by its ability to determine propagation state vector errors and through observations of the Chinese space station Tiangong-1 in the last months of its return to Earth. The results also suggest that the measurements may be able to provide useful ionospheric parameters, such as total electron content (TEC) measurements, provided high-precision ephemeris data are available for the detected objects.

This paper follows up the earlier [13] by introducing new capabilities that have improved the detection rates three-fold. The radar's utility for providing total electron content (TEC) measurements is investigated using a high-range resolution mode and high-precision ephemeris data. Further, the short-term Fourier transform is applied to demonstrate a new radar capability to detect and measure RSO rotation characteristics and monitor ionospheric plasma waves and instabilities. Section 2 describes the Buckland Park stratosphere–troposphere (BPST) radar and the initial signal processing techniques and observations. Section 3 describes the new capability introduced into the radar, including revised operating parameters, beam scheduling, signal processing, and ionospheric correction, while Section 4 summarises the results. Section 5 presents satellite-derived ionospheric group retardation measurements, investigating the utility of the radar for providing total electron content (TEC) measurements. Section 6 presents micro-Doppler results, illustrating the utility of the radar for RSO characterisation and monitoring of ionospheric plasma waves and instabilities. Section 7 presents a discussion of the results. The paper closes with the conclusions.

2. Buckland Park VHF Radar RSO Observations

2.1. The Buckland Park VHF Radar

The Buckland Park stratosphere–troposphere (BPST) VHF radar (34°37′36.03″S, 138°28′3.91″E) is located 35 km north of Adelaide. The radar specifications are shown in Table 1. The radar is a pulsed monostatic system operating at 55 MHz, with a peak transmit power of 40 kW. The radar maximum duty cycle is 10%. The radar is able to change frequency on a pulse-to-pulse basis within ±50 kHz of the operating frequency. The radar was designed and manufactured by ATRAD and is operated by the University of Adelaide/ATRAD. The radar is designed primarily for measurement of tropospheric and stratospheric winds (0.5–20 km) [15], although similar radars have been used to observe polar mesosphere summer/winter echoes (PMSE and PMWE) [16,17] at high latitudes. Identical systems (albeit with a peak transmit power of 80 kW) are deployed throughout Australia by the Bureau of Meteorology [15]. The radar's oscillators/clocks are GPS-disciplined, allowing transmissions to be received using an interferometric Yagi antenna array at Mylor (55 km southeast of Buckland Park) for bistatic meteor observations [18].

Table 1. BPST radar parameters (from [13]).

Parameter	Value(s)
Frequency, f (MHz)	55
Maximum Transmit Power (kW)	40 (12 4 kW modules, i.e., 48 kW at the transmitter)
Maximum Duty Cycle (%)	10
Pulse Types	Monopulse, Barker, and Complementary Codes
Receiver Filter Widths (kHz)	4, 8, 16, 32
Pulse-to-pulse Frequency Extent	±50 kHz
Maximum Pulse Repetition Frequency (kHz)	20
Number of Transmit Antennae	144
Number of Receive Antennae	144 (Main array), 5 (Meteor array)
Combined Tx/Rx Main Array Beam Width (°)	7
Pulse Widths (m)	100–4000
Pulse Widths (μs)	0.67–26.67
Number of Receivers	6 (1–5: Meteor array, 6: Main array)
Range Sampling Resolution, Δr (km)	0.05–2
Beam Directions (Azimuth, Zenith) (degrees)	(0, 0) "Vertical", (0, 15) "North", (90, 15) "East" (180, 15) "South", (270, 15) "West"

The BPST radar transmit array consists of a 12 × 12 array (hereafter referred to as the "main array") of gamma-matched linearly polarised Yagi antennae. The inter-antenna spacing is 0.644 λ, with an aperture size of 38.6 by 38.6 m. The array allows five beam directions to be used: vertical and 15° off zenith in the cardinal directions (north, east, south, west). Reception is performed using the main array and/or five interferometric Yagi antennae for meteor observations. The radar has six receivers, with the main array connected to one receiver (typically receiver 6) and the five meteor interferometric antennae typically connected to the remaining five receivers. The combined transmit/receive main array beam width is 7° (half-power full-width). The radar has an auxiliary antenna array, which can be used for boundary layer radar observations [15]. Figure 1 shows a schematic of the main and auxiliary antenna arrays.

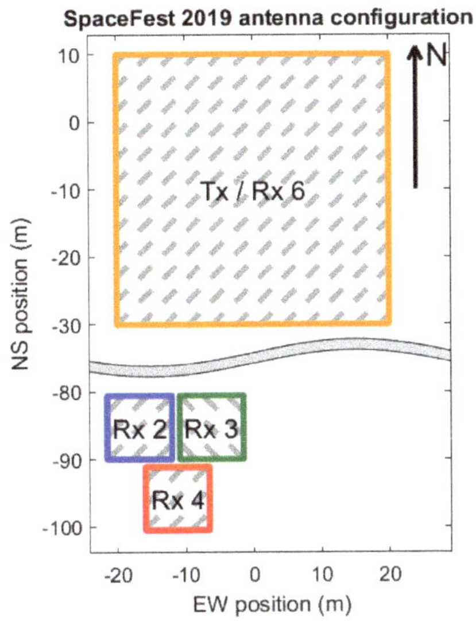

Figure 1. BPST radar main and auxiliary antenna arrays used for SpaceFest 2019.

The BPST radar records range–time (RT) data, $x(r, t)$. The radar typically dwells on a particular beam direction for 1–2 min, with 4 s of "dead" time to allow radar configuration for the next dwell. As the BPST radar is a coded pulsed monostatic radar that uses receiver blanking for the duration of the transmitted pulse, it has "blind ranges" where it is unable to observe RSOs with (ambiguous) ranges that are integer multiples of the radar's unambiguous range.

2.2. BPST RSO Observations

As the orbital velocity of a typical LEO object is around 8 kms^{-1}, two strategies were used in [13] to mitigate the effects of target motion:

1. Range migration is mitigated by dividing the RT data into overlapping coherent processing intervals (CPIs) of 4.096 s. Contiguous CPIs are spaced 0.512 s apart (i.e., an 8-times oversampling factor) to maintain acceptable temporal resolution;
2. Doppler broadening is mitigated by applying acceleration processing [19] to compensate for RSO motion throughout the CPI. This is achieved by processing each CPI using a bank of radial acceleration hypotheses covering the expected range of LEO RSO radial accelerations (0 to 250 ms^{-2}), yielding a three-dimensional range–time–radial acceleration (RTA) data-set $x(r, t, a)$.

The RTA data for each CPI are Doppler processed using a Hanning window, producing range–Doppler–radial acceleration (RDA) data $x(r, d, a)$. Peak processing is applied to find local maxima in each dimension, and interpolation is applied around local maxima to obtain sub-cell size peak RDA coordinates.

Two processing modes are described in [13]:

1. Detection mode (DM) uses radial acceleration processing to generate peaks over the entire RDA space. This allows matching with the RDA parameters obtained using either simplified generalised perturbation (SGP4) propagation of two-line elements (TLE) or Special Perturbations (SPs) data [20], both issued by SpaceTrack (https://www.space-track.org, accessed on 26 March 2024), in addition to the identification of objects that are not in the SpaceTrack catalogue;
2. Catalogue maintenance mode (CMM) uses radial acceleration processing using a subset of RDA space around each TLE/SP propagation: i.e., range ±5 km, Doppler ±20 Hz, radial acceleration ±0.05 kms^{-1}. This processing is consequently significantly

less computationally intensive than detection mode but only allows the identification of catalogued objects and may miss objects with erroneous propagation state vectors.

Once the peaks are produced, they are then associated with the propagation state vectors. If sufficient peaks are detected, an RSO detection is declared. The pseudo-code for DM and CMM is shown in Table 2.

Table 2. Pseudo-code for Detection Mode (DM) and Catalogue Maintenance Mode (CMM).

Detection Mode
Loop over overlap processing intervals
Data selection
Acceleration processing
Doppler processing
Peak detection and interpolation
Whitening
Peak selection
Loop over concurrent propagation state vectors
Associate peaks to propagation state vectors
Declare RSO detected if sufficient associated peaks
Catalogue Maintenance Mode
Loop over concurrent propagation state vectors
Data selection
Loop over overlap processing intervals
Acceleration processing
Doppler processing
Peak detection and interpolation
Whitening
Peak selection
Associate peaks to propagation state vectors
Declare RSO detected if sufficient associated peaks

The results presented in [13] reveal that 2410 RSOs were detected over 15 days in 2018, with 1392 unique objects detected. The daily detection count rates ranged from 150 to 200, and the maximum detection height observed was 2491 km. The radar's utility for object catalogue maintenance was demonstrated by its ability to determine propagation state vector errors and through observations of the Chinese space station Tiangong-1 in the last months of its return to Earth. For the latter, it was shown that for one particular transit, the two-line element was in error despite being issued only 37 min prior to the observation. The results also suggested that the measurements may be able to provide total electron content (TEC) measurements, provided high-precision ephemeris data (i.e., Special Perturbations [20]) data are available for the detected objects. Furthermore, rapid amplitude fluctuations were reported for some RSOs that were attributed to Faraday rotation effects due to the reception of ionospherically propagated signals using linearly polarised antennae [21].

3. New BPST Radar Capabilities Introduced for SpaceFest 2019

The observations described in this paper incorporated several new features introduced for evaluation during SpaceFest 2019. These features are discussed in the following subsections.

3.1. Revised Operating Parameters

The operating parameters used for SpaceFest 2019 are shown in Table 3. Two sets of parameters were used: "standard" and "high range resolution". The standard parameters differ from those used in [13], which mostly used a pulse repetition frequency (PRF) of 500 Hz. The change in parameters, including the use of a PRF of 2000 Hz, was motivated by radar modelling [22], which suggested that the revised parameters would yield an SNR increase of approximately 1.2 dB in addition to a small improvement in effective range resolution. The high-range resolution parameters were introduced to investigate the use of satellite observations to perform total electron content (TEC) measurements, as discussed in Section 5. Coherent averaging was used to reduce data rates. The rationale for the use of two sets of PRFs is discussed in Section 3.2.

Table 3. BPST radar parameters.

Parameter	Standard Mode	High Resolution
Pulse repetition frequency (Hz)	1953, 2000	9950, 10,000
Number of coherent averages	2	6
Dwell length (s)	56	56
Range ambiguity (km)	76.8, 75	15.08, 15
Radial velocity resolution [1] (ms^{-1})	1.36	1.36
Pulse type	13-bit Barker	13-bit Barker
Pulse width (km)	0.5	0.1
Pulse width (μs)	3.33	0.67
Receiver filter width (kHz)	128	640
Minimum range (km)	7.5	2.5
Maximum range (km)	69	12.5
Range sampling resolution (km)	0.25	0.1

[1] 2 s coherent processing interval.

3.2. Beam and PRF Scheduling

Unlike the SSN radars, the BPST radar does not possess the tracking capability of many of the other NFoV systems. The observation campaigns described in [13] were conducted using a single beam direction and single PRF. For SpaceFest 2019, "beam and PRF scheduling" was introduced to enable the radar to capture any RSOs that were predicted to transit any of the five radar beams. This was achieved using SGP4 propagations of TLEs to schedule the appropriate beam to use at any instant, as well as the appropriate PRF from a pair of candidate PRFs (as indicated in Table 3) to ensure that the RSO did not fall within radar blind ranges.

The beam scheduling routine calculates the expected (uncalibrated) power return for each RSO i and beam direction j using

$$P_{ij}(t) = \sigma_i R_i^{-4}(t)\exp\left(-\frac{\operatorname{acos}\left(\mathbf{a}_i(t)\cdot\mathbf{b}_j\right)^2}{\theta_b^2}\right), \quad (1)$$

where σ_i, $R_i(t)$, and $\mathbf{a}_i(t)$ are the radar cross-section (RCS), range, and direction cosine of the RSO i at time t; $\mathbf{b}_j$ is the direction cosine of the beam direction j; θ_b is the beam width; and $(\cdot)$ represents the inner product. This equation assumes a Gaussian beam pattern and incorporates RCS and range terms according to the radar equation. As there are no known published estimates of RSO RCS values in the lower VHF band, the microwave median radar RCS values of [23] have been used to approximate the corresponding values for 55 MHz using the technique described in [13]. This technique assumes each RSO can be represented by a sphere and calculates the corresponding spherical RCS at 55 MHz. The maximum value of $P_{ij}(t)$ for every minute is determined to obtain the beam direction to use for that minute. The appropriate PRF is determined by finding the PRF that minimises the duration that the (ambiguous) range of RSO yielding the maximum value of $P_{ij}(t)$ is within the radar blind ranges.

3.3. Quick Catalogue Maintenance Mode (QCMM) Processing

Keystone processing (KP) provides the ability to compensate for target range migration [24] and is typically applied prior to or in combination with pulse compression. For known satellites, TLE (or SP) propagations can be used to perform the KP. However, KP is not applicable to the BPST radar as the radar outputs range–time data $x(r, t)$ and does not allow access to the pre-matched filtered data. However, a "Keystone-like" processing (KLP) approach can be applied in two steps using the TLE (or SP) propagation range $R(t)$, as illustrated in Figure 2:

1. Range migration correction: Use the propagated range to convert the range–time data $x(r,\ t)$ to "range difference" (r') time data, $x_{r'}(r',\ t')$. This is achieved by interpolating $x(r,\ t)$ across the range dimension or through a circular shift of the range–time data through the range dimension by $\lceil (R(t)/\Delta r) \rceil$ samples, where $\lceil * \rceil$ represents the rounding-up operator and Δr is the range resolution. The former option is applied in this paper, with the interpolation achieved using a simple Fourier method.

$$x_{r'}(r',\ t) = \mathcal{F}(\mathcal{F}^{-1}(x(r,\ t)\exp\left(\frac{2\pi i R(t)}{R_a}\right))) \quad (2)$$

where $\mathcal{F}$ represents the fast Fourier transform and R_a is the range ambiguity.

2. Phase correction: The change in phase resulting from the change in the propagated range is corrected using the following:

$$x_k(r',\ t') = x_{r'}(r',\ t)\exp\left(\frac{4\pi i R(t)}{\lambda}\right) \quad (3)$$

The positive range difference illustrated in Figure 2 is typical of BPST observations and is due to signal propagation through the ionosphere (group retardation). This is discussed further in Section 3.5.

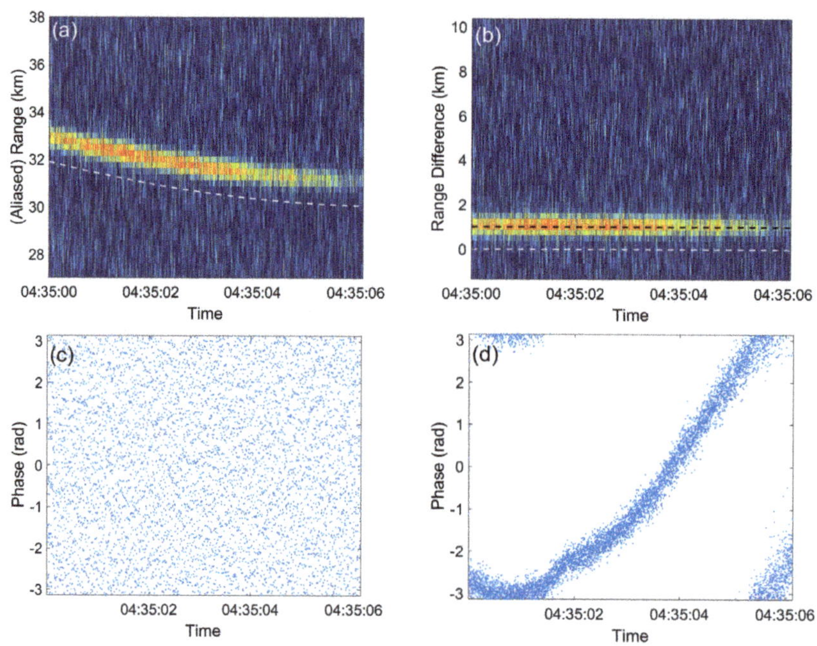

Figure 2. Example of application of "Keystone-like" processing. (**a**) Range–time power $|x(r,\ t)|^2$ for a transit of the rocket body SJ 16 02 (NORAD ID 41634) at 0:35 on 24 March 2019. The dashed white line indicates the predicted range obtained using SP propagation data. (**b**) Range difference time power $|x_{r'}(r',\ t)|^2$ after application of the first step of Keystone-like processing. The dashed white line indicates zero range, while the dashed black line indicates the range difference of maximum power. (**c**) Phase of the range difference time data $\angle x_{r'}(r',\ t)$ at the range difference indicated by the dashed black line in (**b**). (**d**) Phase of phase-corrected range difference time data $\angle x_k(r',\ t')$ obtained after applying the final step of Keystone-like processing.

Having applied the KLP, the range difference time data can then be divided into overlapping CPIs and Doppler processed using an appropriate window to yield "range-difference Doppler-difference" data, $x_k(r', d')$, which are then peak-processed in the manner described in [13]. Contiguous peaks with similar range and Doppler differences close to zero are then associated. If sufficient peaks are detected, an RSO detection is declared. The pseudo-code for QCMM processing is shown in Table 4.

Table 4. Pseudo-code for Quick Catalogue Maintenance Mode (QCMM).

```
Loop over Concurrent Propagation State Vectors
  Apply Keystone-Like Processing
  Loop over Overlap Processing Intervals
      Data selection
      Acceleration processing
      Doppler processing
      Peak detection and interpolation
      Whitening
      Peak selection
  Associate peaks to zero range and Doppler difference
  Declare RSO detected if sufficient associated peaks
```

The KLP obviates the need for acceleration processing and is consequently computationally less intensive than either DM or CMM. Like CMM, QCMM only allows the identification of catalogued objects and may miss objects with inaccurate propagation state vectors. As the KLP compensates for range migration, it allows the application of spectrograms to provide RSO micro-Doppler characterisation, as described in Section 6. As discussed in Section 2, the high orbital velocities of LEO RSOs motivate the use of overlapping CPIs for DM and CMM to mitigate range migration. Since the KLP removes range migration effects, QCMM can be applied using longer CPIs than DM and CMM, thereby increasing the coherent gain of the detections compared to DM and CMM and, consequently, the number of RSO detections.

An example of QCMM processing results for SL-8 R-B (NORAD ID 15032) on 22 March 2019 is shown in the top plots of Figure 3. These plots reveal a positive range difference (as also illustrated in Figure 1) and a negative radial velocity slope due to ionosphere signal propagation. These effects are discussed further in Section 3.5.

3.4. Matched Duration Quick Catalogue Maintenance Mode (MDQCMM) Processing

As previously discussed, KLP allows the CPI length to be increased without coherent loss due to range migration. This has allowed the development of a further signal processing strategy, matched duration quick catalogue maintenance mode (MDQCMM), that abandons the use of overlapping CPIs. MDQCMM is applied using an array of start and end times, which define candidate CPIs. Peak detection is applied for each candidate, with the CPI yielding the maximum SNR defining the transit duration. MDQCMM consequently yields significant SNR gains compared to QCMM and is therefore expected to enhance the detectability of low SNR RSOs and, hence, increase the detection rate. Furthermore, as MDQCMM may result in longer CPIs than the fixed-length CPIs used for DM, CMM, and QCMM, it will also reduce the radial velocity uncertainty. Like CMM and QCMM, MDQCMM only allows the identification of known objects and may not detect RSOs with erroneous TLE state vectors. Pseudo-code for MDQCMM processing is shown in Table 5.

Figure 4 provides two MDQCMM processing examples. For THEOS (NORAD ID 33396), the SNR versus start/end time plot shows a single broad peak, and the range–Doppler plot shows a single distinct peak. However, for CZ 2C R-B (NORAD ID 43031), the SNR versus start/end time plot shows two distinct maxima, and the range–Doppler plot shows a "defocussed" single distinct peak. This occurs because there is a Faraday fade during the transit, and the coherent gain is reduced when the fade is included in the CPI. For reference, the SNRs and detection durations obtained for THEOS and CZ 2C R-B using QCMM and

MDQCMM processing are shown in Table 6. While MDQCMM generates a single peak, the QCMM processing generates peaks every ≈0.25 s, and the median QCMM SNR is therefore used for this comparison. This comparison reveals that MDQCMM processing yields a larger SNR for both RSOs and a larger detection duration for THEOS. However, MDQCMM processing yields a smaller detection duration for CZ 2C R-B due to the aforementioned Faraday fade. These results emphasise that Faraday rotation can significantly compromise optimal BPST radar RSO detection.

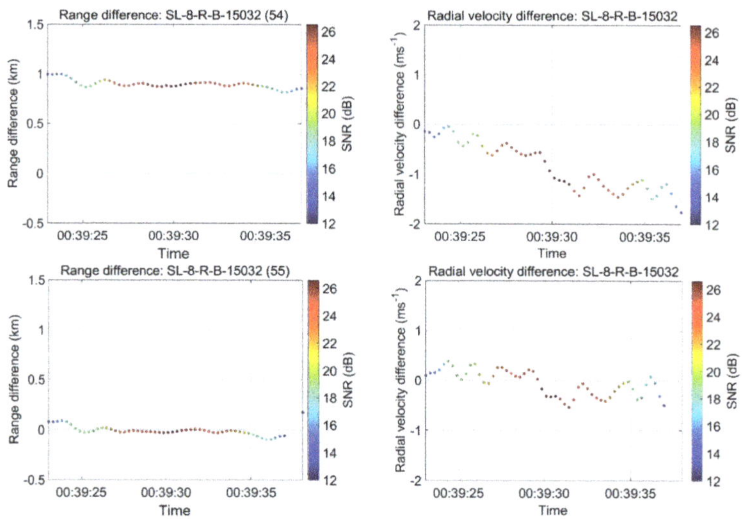

Figure 3. Example of peak range difference (**left**) and radial velocity difference (**right**) obtained using quick catalogue maintenance (QCMM) mode processing for SL-8 R-B (NORAD ID 15032) on 22 March 2019, both without (**top**) and with (**bottom**) ionospheric correction applied.

Table 5. Pseudo-code for Matched Duration Quick Catalogue Maintenance Mode (MDQCMM).

```
Loop over Concurrent Propagation State Vectors
  Apply Keystone-Like Processing
  Loop over Overlap Processing Interval Start Times
      Loop over Overlap Processing Interval End Times
        Data selection
        Acceleration processing
        Doppler processing
        Peak detection and interpolation
        Whitening
        Peak selection
        Associate peaks to zero range and Doppler difference
        Declare RSO detected if sufficient associated peaks
  If RSO detected, select start and end times with yielding maximum SNR
```

Table 6. SNRs and detection durations obtained using QCMM and MDQCMM processing for the THEOS (NORAD ID 33396) and CZ 2C R-B (NORAD ID 43031).

RSO	QCMM SNR (dB)	MDQCMM SNR (dB)	QCMM Duration (s)	MDQCMM Duration (s)
THEOS	23.32	29.90	12.84	16.77
CZ 2C R-B	26.19	30.96	16.90	10.24

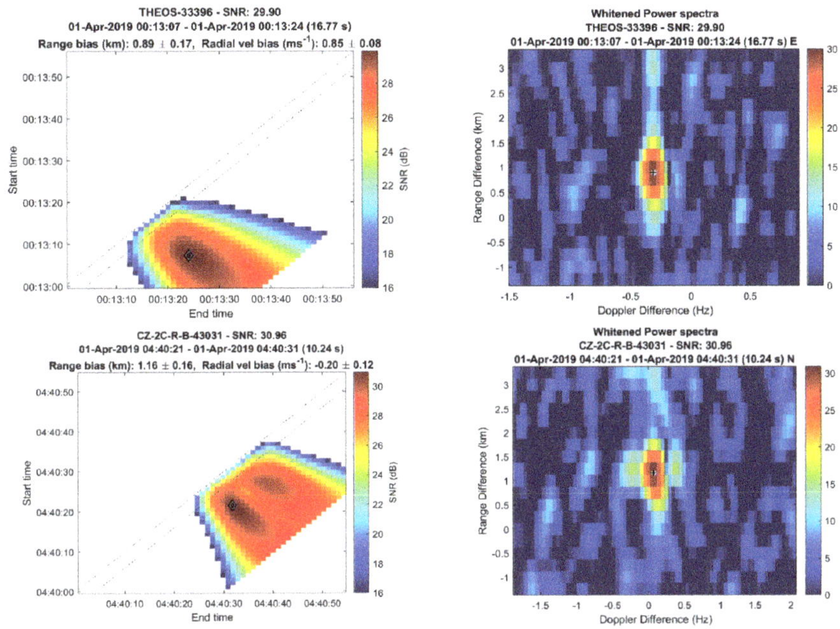

Figure 4. Example of application of matched duration quick catalogue maintenance mode (MDQCMM) for the THEOS satellite (NORAD ID 33396) at 00:13 (**top**) and for CZ 2C R-B (NORAD ID 43031) at 4:40 (**bottom**), both on 1 April 2019. The **left** plots show the SNR variation obtained over a grid of start and end times, with the diamond indicating the start/end time producing maximum SNR. The **right** plots show the range–Doppler display for the start/end time producing maximum SNR, with the plus sign indicating the interpolated peak position in the range difference/Doppler difference dimensions.

For reference, the computation times of the signal processing modes described in this paper for a typical 1 min dwell using MATLAB 2023b on a Dell Precision 7560 laptop (Adelaide, Australia) are shown in Table 7.

Table 7. Computation times of different signal processing modes for a typical 1 min dwell, using a coherent integration time (CIT) of 2 s (2048 samples), a CIT step of 0.25 s (256 samples), and 100 acceleration hypotheses (DM and CMM only).

Processing Type	DM	CMM	QCMM	MDQCMM
Processing time (s)	1204.8	39.9	5.7	77.6

3.5. Ionospheric Correction

Radars operating in the medium frequency to lower VHF bands suffer from group retardation and phase advance due to signal propagation through the ionosphere [20], where the measured range and phase exceed the actual values. The group R_g and phase R_p path ranges can be expressed as (from [21])

$$R_g = \frac{c}{2}\int_0^R \frac{dr}{U(r)}, \tag{4}$$

$$R_p = \frac{c}{2}\int_0^R \frac{dr}{v(r)}, \tag{5}$$

where c is the speed of light (ms^{-1}), $U(r)$ is the group velocity (ms^{-1}), and $v(r)$ is the phase velocity (ms^{-1}). In the absence of a magnetic field, the group and phase velocity can be expressed in terms of the radar operation frequency f(Hz) and the plasma frequency $f_p(r)$ (Hz) (from [21]):

$$U(r) = c\sqrt{1 - \frac{f_p^2(r)}{f^2}}, \tag{6}$$

$$v(r) = \frac{c}{\sqrt{1 - \frac{f_p^2(r)}{f^2}}}. \tag{7}$$

The plasma frequency is given by (from [20])

$$f_p(r) = \sqrt{\frac{N(r)e^2}{4\pi m\epsilon_0}}, \tag{8}$$

where $N(r)$ is the electron density (m^{-3}), e is the electron charge (Coulombs), m is the electron mass (g), and ϵ_0 is the permittivity of free space ($m^{-3}g^{-1}s^4A^2$).

Figure 5 illustrates the impact of ionospheric propagation using the SP propagations for SL-8 R-B (NORAD ID 15032) on 22 March 2019. These results were generated using the electron density profile obtained using the International Reference Ionosphere (IRI) 2016 model [25]. The top plots show the actual range and radial velocity and those obtained incorporating ionospheric propagation, hereafter the "effective" range and radial velocity. Although the effective range clearly exceeds the actual range (i.e., group retardation), the effective radial velocity appears to overlay the actual radial velocity. The bottom plots illustrate the range and radial velocity differences (effective minus actual). The range difference varies between 1.12 and 1.3 km, with the minimum range difference obtained when the RSO passes through the point of closest approach, where the ionospheric path length is minimised. Although the effective radial velocity appears to overlay the actual radial velocity, there is actually a small radial velocity difference that varies between ± 8 ms^{-1}, which is a small fraction (0.013) of the radial velocity variation. The radial velocity difference passes through zero as the RSO passes through the point of closest approach. The radial velocity difference exhibits a negative slope and shows excellent agreement with the radial velocity difference observed in the top right plot of Figure 3, indicating that the negative phase slope observed in that plot is the result of ionospheric propagation.

Compensation for the effects of group retardation and phase advance can be performed by replacing the propagation range $R(t)$ by $R_g(t)$ in (1) and by $R_p(t)$ in (2). The QCMM processing results for SL-8 R-B incorporating this correction using the International Reference Ionosphere (IRI) 2016 model are shown in the bottom plots of Figure 3. These plots reveal that the positive range difference and negative radial velocity slope observed without ionospheric correction have been rectified. Ionospheric correction also results in benefits for MDQCMM processing as the phase advance correction reduces the Doppler spread of the peak, thereby increasing the SNR and improving peak detectability.

3.6. Use of Auxiliary Antennae Arrays

As described in Section 2, the BPST main array uses only one of the six available receivers. It was, therefore, decided to reconfigure the radar to connect three of the additional receivers typically used for meteor observations to three sets of co-located 3 × 3 Yagi arrays installed for boundary layer radar (BLR) observations [15]. Unless otherwise indicated, all results presented in this paper use the main array only.

In order to investigate the aforementioned Faraday rotation effects in more detail, the polarisation of one group of nine antennae was rotated to be perpendicular to that of the main array and the other two groups of antennae. Figure 6 shows typical RSO observations obtained using the four antenna arrays. Where the three identically polarised arrays (indicated by yellow, blue, and red lines) show a null at around 4:39:42, the perpendicularly polarised antenna shows a peak. This is the behaviour expected for Faraday rotation, whereby the received signal amplitude maximises (minimises) when the plane of polarisation of the incident signal is identical (perpendicular) to the antenna polarisation [21]. The fact that the observations shown in Figure 6 are typical confirms that the rapid amplitude fading reported in [13] is most likely attributable to Faraday rotation.

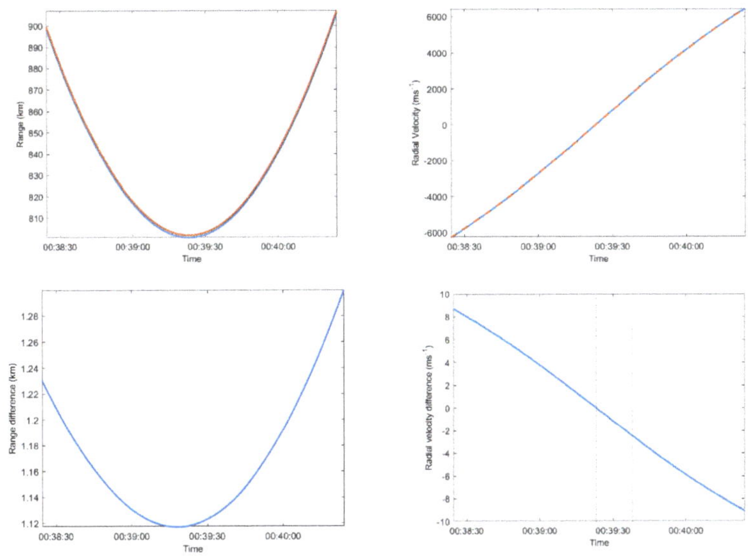

Figure 5. Illustration of the effects of ionospheric propagation on the SP propagation state vectors of SL-8 R-B (NORAD ID 15032) on 22 March 2019. (**Top left**) Actual (blue) and estimated (red) range resulting from ionospheric propagation. (**Top right**) Actual (blue) and estimated (dashed red) radial velocity resulting from ionospheric propagation. (**Bottom left**) range difference (estimate minus actual). (**Bottom right**) radial velocity difference (estimate minus actual). The dashed lines indicate the start and end times of the detections shown in Figure 4.

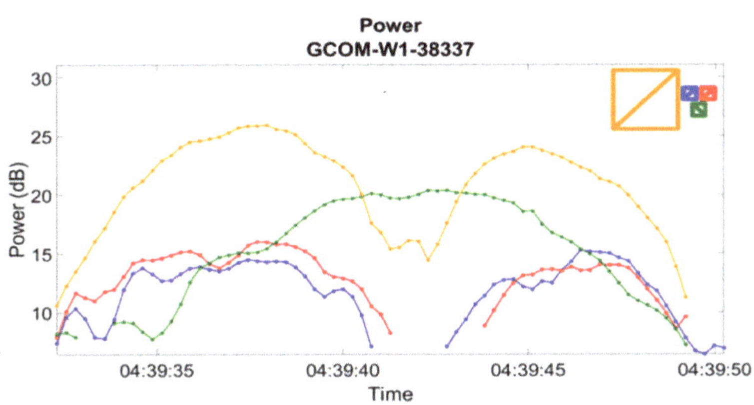

Figure 6. Illustration of the effects of Faraday rotation on the peak power estimates received on the BPST main array (yellow) and the three BLR arrays (blue, red, and green) for the GCOM-W1 satellite (NORAD ID 38337). The three identically polarised arrays (yellow, blue, and red) show a 10 dB null at around 4:39:42, while the array with perpendicular polarisation shows a peak.

4. Standard Mode Results

4.1. Special Perturbation (SP) Data Comparisons

The BPST radar was operated using standard mode parameters for ten days, including the five days of SpaceFest 2019, 22 to 26 March, 27 to 29 March, and 1 and 2 April. All processing was applied using a CIT length of 2048 samples ($\approx$2 s) with a 256 sample stepsize ($\approx$0.25 s), resulting in an oversampling factor of eight. MDQCMM was applied using CIT start/end time increments of 1024 samples ($\approx$1 s). A peak SNR threshold of 12 dB was applied. A minimum detection duration threshold of 5 s was required to constitute RSO detection, allowing discrimination of RSOs from long-lived meteor echoes, such as overdense or nonspecular echoes [13]. Unless otherwise indicated, the results presented in this section incorporate ionospheric correction, as described in Section 3.5.

The detection statistics for the ten days of standard mode observations are shown in Table 8. The results indicate average DM, QCMM, and MDQCMM daily count rates of 360, 450, and 590, respectively. The MDQCMM count rates represent a three-fold increase compared to the count rates reported in [13], indicating the success of the revised observation strategy and the signal processing improvements. The application of ionospheric correction increases the average QCMM and MDQCMM count rates by approximately 8 and 15 RSOs per day. A total of 2202 unique RSOs were detected using MDQCMM. Note that the variation in the daily detection rates is loosely correlated with the age of the TLEs used to produce the beam schedule.

Table 8. Daily count rates for DM, QCMM, and MDQCMM analyses. IC indicates that the ionospheric correction described in Section 3.5 has been applied.

Date	DM	QCMM	QCMM/IC	MDQCMM	MDQCMM/IC
22 March 2019	372	460	457	580	589
23 March 2019	302	406	399	493	520
24 March 2019	372	473	472	604	624
25 March 2019	383	484	484	615	636
26 March 2019	389	490	496	612	632
27 March 2019	358	447	441	583	595
28 March 2019	290	365	366	483	495
29 March 2019	316	397	395	497	501
1 April 2019	416	507	504	634	646
2 April 2019	400	481	579	627	636
Total	**3598**	**4510**	**4593**	**5728**	**5874**

Bold format indicates total number over all days.

In order to understand the relative daily count rates obtained using the DM, QCMM, and MDQCMM processing, Figure 7 shows histograms of the SNR and detection duration differences for common RSOs detected using QCMM and CMM and MDQCMM and QCMM. While MDQCMM generates a single peak, the CMM and QCMM processing generates peaks every $\approx$0.25 s. The median SNR is therefore used for the CMM and QCMM comparisons. For the QCMM and CMM comparison, the SNRs are roughly comparable, but QCMM yields a mean duration increase of 0.91 s. This results in a greater number of RSOs exceeding the 5 s detection threshold. For the MDQCMM and QCMM comparison, MDQCMM yields a mean SNR improvement of 4.81 dB, and the mean duration increase is 1.53 s. Note that the detection duration difference histogram exhibits a significant number of peaks with duration differences less than zero. This is due to the impact of Faraday fading on the MDQCMM processing, as discussed in Section 3.4.

The height, detection duration and RCS distribution of the MDQCMM results are shown in Figure 8. The lowest detection height was 341.2 km (SZ 11 MODULE, NORAD ID 41868). A total of 23 medium Earth objects (MEOs, 2000 km to 35,786 km) were detected, with a largest detection height of 2975.4 km (MIDAS 5 STRONGBACK 271), respectively. The histogram shows distinct peaks at 800, 1000, and 1200 and a broader peak at 1450 km. The 800 km peak is predominantly due to satellites with heights of 792 $\pm$ 2 km, comprising mainly Iridium (e.g., IRIDIUM-117, NORAD ID 42808) and Orbcomm-FM (e.g., OBRCOMM-FM-16, 25417) satellites. It is worth noting that the use of multiple PRFs avoids histogram "gaps" observed in [13], where pulse blanking resulted in gaps spaced at multiples of the range ambiguity. Detection durations between the minimum detection threshold of 5 s up to 50 s are observed. The larger durations are observed for objects with larger heights. As the beam area increases with height, higher objects consequently take longer to transit the BPST beams. The RCS values have been calculated using the technique described in [13], as described in Section 3.2. Figure 8 indicates that the MDQCMM processing is able to detect RSOs with RCS values as low as -14 dBm^2 (i.e., 0.04 m^2). Note that a common first-order detection sensitivity figure of merit of a radar is the minimum detectable RCS at a 1000 km range from both the radar transmitter and receiver. The minimum detectable RCS at 1000 km obtained using MDCMM processing is -10.97 dBm^2 (0.09 m^2).

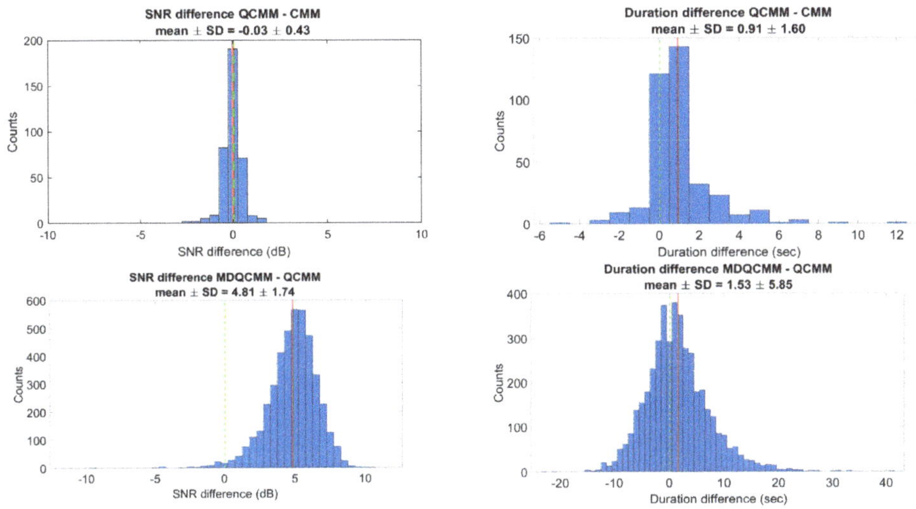

Figure 7. Detection duration (**left**) and RCS (**right**) histograms obtained using quick catalogue maintenance mode (QCMM, blue) and matched duration quick catalogue maintenance mode (MDQCMM) processing for standard mode observations. The red lines indicate the mean of the difference distribution, and the green dashed line indicates zero difference.

As discussed in Section 2, the BPST radar main array uses a single receiver and is unable to measure the angular position of the detected RSOs. However, the propagation state vectors can be used to infer the angular position, assuming the state vectors are accurate, which is generally the case for the SP perturbations. Figure 9 shows the angular positions expressed in terms of cartesian zenith angles and latitude/longitude obtained using QCMM for the ten days of observations. The cartesian zenith angles are colour coded by "uncalibrated power", which is the measured power (in squared A/D units) corrected for RCS and the fourth power of the RSO range in accordance with the radar equation. The uncalibrated power attempts to remove range and RCS effects from the power measurements, leaving only the effects associated with the radar antenna pattern. The colour coding clearly traces out the antenna pattern of the five BPST radar beams, confirming that the beam steering is functioning as intended.

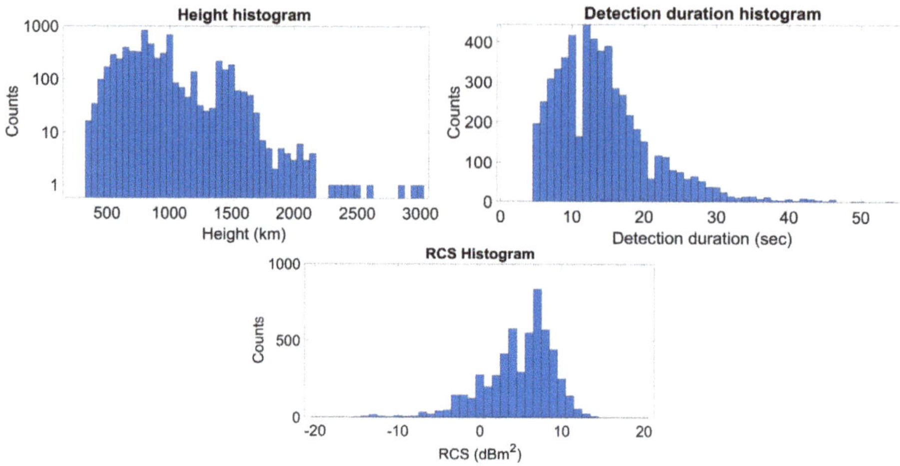

Figure 8. Histogram of RSO heights (**top left**), duration (**top right**), and radar cross-section (**bottom**) obtained using matched duration quick catalogue maintenance mode (MDQCMM) processing for standard mode observations.

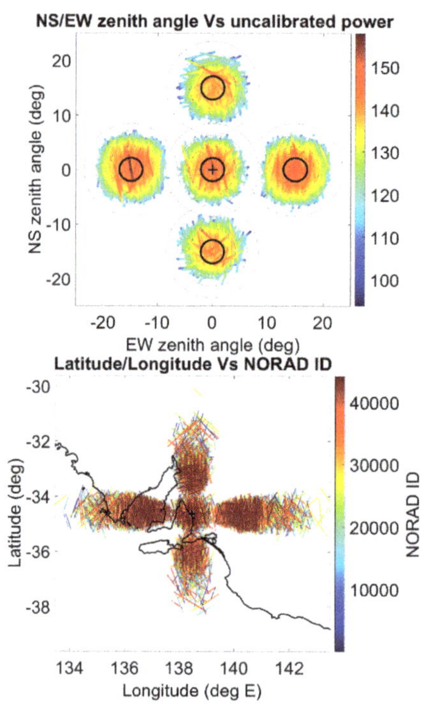

Figure 9. Illustration of angular parameters of RSOs detected by the BPST radar from 22 to 29 March and 1 and 2 April using quick catalogue maintenance mode (QCMM). (**Top**) Cartesian zenith angle components colour coded by "uncalibrated power" (measured power corrected for the satellite range and RCS). The thick black circles indicates the theoretical half-power beam width of the radar, and the dashed circles indicate multiples or two, three and four times the beam width. (**Bottom**) Latitude and longitude colour coded by NORAD ID.

In order to re-emphasise the advantages of using SP data in comparison to TLE/SGP4 data as highlighted in [13], Figure 10 shows a comparison of the radial velocity differences using SP perturbations and to TLE/SGP4 obtained using QCMM processing. The mean and standard deviation obtained using the SP estimates are substantially smaller than the TLE/SGP4 estimates. The SP estimates indicate that the errors in the BPST radial velocity estimates are of the order of 1 ms^{-1}. The accuracy of the BPST radial velocity estimates is investigated further in the following section.

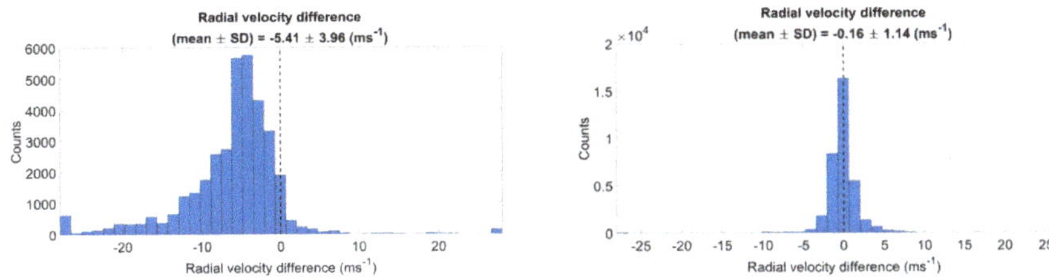

Figure 10. Histograms of QCMM radial velocity differences obtained using TLE/SGP4 (**left**) and SP propagations (**right**) for the ten days of standard mode measurements. The means and standard deviations of the distributions are indicated in the plot titles.

4.2. *International Laser Ranging Service Data Comparisons*

The International Laser Ranging Service (ILRS) was established in September 1998 to support programs in geodetic, geophysical, and lunar research activities and to provide the International Earth Rotation Service with products important to the maintenance of an accurate International Terrestrial Reference Frame [26]. The ILRS currently includes more than 40 satellite laser ranging stations and routinely tracks about 20 retroreflector-equipped satellites, which are used to provide precise ephemerides that are made available at https://ilrs.gsfc.nasa.gov (accessed on 26 March 2024).

Table 9 shows a comparison of the radial velocity differences obtained using TLE/SGP4 data, SP data, and ILRS ephemerides, the latter both without and with ionospheric correction, as described in Section 3.5. Although the mean (i.e., bias) of ILRS radial velocity differences are larger than the SP estimates, the standard deviation obtained using the ILRS ephemerides is substantially smaller than both the SP and TLE/SGP4 estimates. These results suggest that the radial velocity accuracy of the BPST radar is better than 1 ms^{-1}.

Table 9. Mean and standard deviation of QCMM Doppler differences.

Propagation Type	Mean Doppler Difference	Standard Deviation of Doppler Difference
TLE/SGP4	−5.41	3.96
SP	−0.01	1.13
SP with ionospheric correction	−0.02	0.86
ILRS	−0.19	0.37
ILRS with ionospheric correction	−0.05	0.24

TerraSAR-X (NORAD ID 33605) and TanDEM-X (NORAD ID 31698) are a pair of retroreflector-equipped satellites that contribute to the ILRS data-set. The satellites fly in a closely controlled formation with a typical separation of between 250 and 500 m. Figure 11 illustrates the range and radial velocity difference peaks obtained for QCMM processing using the TanDEM-X ILRS ephemerides on 1 April 2019 and clearly shows two sets of peaks. Given the use of TanDEM-X ephemerides and assuming the ionospheric correction is correct, the range and radial velocity difference peaks for TanDEM-X should be close to zero, as indicated by the solid grey line. The range and radial velocity difference peaks should correspond to the difference between the TanDEM-X and TerraSAR-X ephemerides, as indicated by the dashed line. Thus, despite the small separation between the two satellites, both satellites have been detected and are easily distinguishable, with Figure 11 indicating that the two sets of peaks correspond to TanDEM-X (matching the solid line) and TerraSAR-X (matching the dashed line).

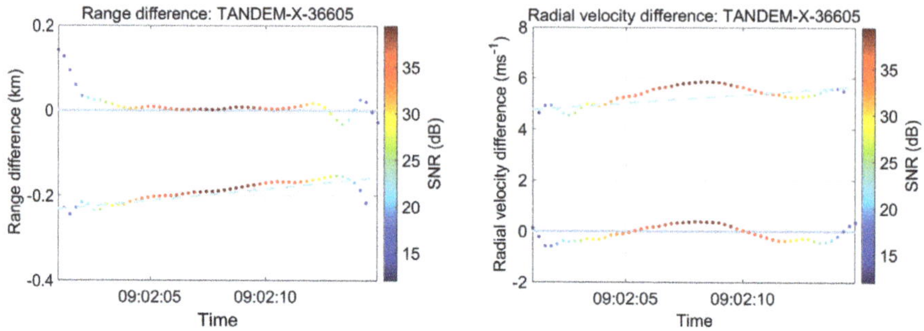

Figure 11. Range (**left**) and radial velocity difference (**right**) peaks obtained using quick catalogue maintenance mode (QCMM) processing using the ILRS ephemerides for TanDEM-X (NORAD ID 33605) on 1 April 2019. The solid grey lines show zero range and radial velocity difference for TanDEM-X, while the grey dashed lines show the corresponding zero range and Doppler differences for TerraSAR-X (NORAD ID 31698).

5. High-Range Resolution Mode Results

In [13], it was suggested that the BPST radar may be able to accurately measure total electron content (TEC) using measurements of the group retardation of LEO RSOs, i.e., the range difference between the estimated and predicted RSO ranges. Further, it was also suggested that the diversity of RSO heights may allow the BPST radar to estimate electron density profiles of the top side of the ionospheric F2 layer.

TEC measurements are typically made using ground-based dual frequency reception of GNSS satellite signal transmissions [27]. The TEC is defined as the number of free electrons in a column of a unit cross-sectional area along the ray path from the transmitter to the receiver end. The TEC measurement that characterises the ionosphere and plasmasphere for a particular location is obtained when the satellite is exactly overhead or above the location and is known as the vertical TEC (VTEC). Slant TEC is defined as the electron column density along an (oblique) line of sight. In the conversion of slant TEC to vertical TEC, it is often assumed that the ionosphere and plasmasphere are horizontally stratified and spatially uniform. Further, the ionosphere is typically simplified to be a thin layer at a fixed altitude of 350–400 km above the Earth's surface [28]. GNSS TEC measurements are also subject to biases due to differential code biases and cycle slips.

The measurement of the range difference of RSOs allows the TEC to be estimated using (from [21])

$$\text{TEC} \approx \frac{r' f^2}{40.3 \times 10^{16}}, \tag{9}$$

where r' is the estimated range difference, f is the radar operating frequency, and TEC is measured in TEC units (TECUs). This equation assumes $f \gg f_H$, where f_H is the gyro-frequency. This condition is easily satisfied for the BPST radar, which operates at f = 55 MHz at a location where $f_H \approx 1.5$ MHz. The RSO range difference measurement avoids the aforementioned assumptions and biases inherent in GNSS TEC measurements. However, accurate measurements are reliant on both accurate range difference measurements and precise propagation state vectors.

A high-range resolution experiment was performed on 30 and 31 March 2019 using the high-resolution mode parameters shown in Table 3 to investigate the ability of the BPST radar to accurately measure TEC using measurements of the group retardation of LEO RSOs. A pulse width of 100 m was used to improve the range estimation accuracy compared to the standard mode experiment. The high-resolution mode parameters reduced the effective SNR of the radar measurements by approximately 3.5 dB compared to the standard mode experiment and consequently reduced the daily detection rates. In order to further improve the accurate group retardation measurements, Special Perturbations (SPs) propagations were used to provide accurate estimates of the RSO positions.

The MDQCMM peak range difference measurements obtained without ionospheric correction for 31 March 2019 are shown in Figure 12. Although the number of RSO observations is sparse, it is clear that the range difference (or group retardation) measurements follow the expected behaviour. The group retardation is larger during the day when solar radiation is available to produce ionisation. The maximum group retardation at around 0600 (1530 LT) is in agreement with IRI predictions. The group retardation increases monotonically with height, as most clearly observed around 0 UT, where it increases from 0.6 at 500 km to 0.85 at 1100 km. Finally, we note the presence of a pre-dawn group retardation minimum, as expected due to the pre-dawn foF2 minimum typically observed in the Australian region [29].

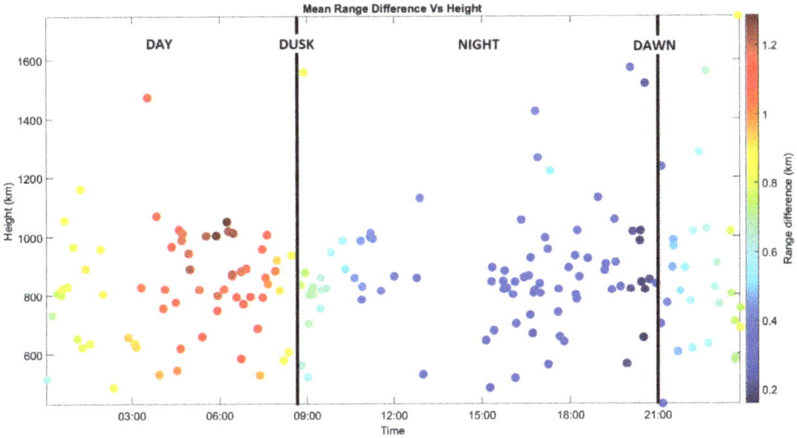

Figure 12. Range difference (i.e., ionospheric group retardation) as a function of height and time obtained using matched duration quick catalogue maintenance mode (MDQCMM) processing on 31 March 2019.

6. Spectrogram Analysis

In the situation where the TLE/SGP4 or SP propagations are sufficiently accurate, the application of QCMM and MDQCMM results in the RSOs being confined to a single range difference cell. In this case, a short-time Fourier transform (STFT) or spectrogram can be applied to observe the evolution of the Doppler characteristics of the RSO. In most cases, the spectrogram exhibits a single peak close to zero Doppler, as indicated for COSMOS-1844-1793 (NORAD ID 17973) in Figure 13. However, in some circumstances, the peak is displaced from 0 Hz. Such displacements are attributed to inaccuracies in the TLE/SG4 or SP propagations used in the analysis. In some circumstances, the spectrogram exhibits either periodic or random variation. As discussed in the following, we interpret the periodic variation as being due to RSO rotation and the random variation as being due to ionospheric effects.

The periodic variations observed in spectrograms obtained using the BPST radar have previously been used to investigate the micro-Doppler signatures of three non-operational satellites: RHESSI (NORAD ID 27370), TOPEX/Poseidon (NORAD ID 22076), and Telkom 3 (NORAD ID:38744) [30,31]. These papers also use computational electromagnetic modelling of simple computer-aided design (CAD) models to provide information on the satellite orientation and spin axis. Further, ref. [30] illustrates that the micro-Doppler periodicity of Telkom 3 decreased from 12 to 6.5 s between 2017 and 2020, indicating an increase in rotation rate over time.

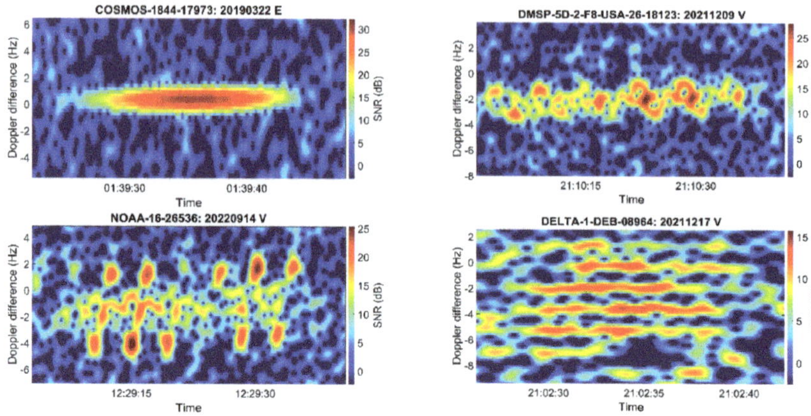

Figure 13. Example of spectrograms indicating non-rotating and rotating RSOs. (**Top left**) COSMOS-1844-1793 (NORAD ID 17973). (**Top right**) DMSP 5D 2 F8 USA (18123). (**Bottom left**) NOAA-16 (26536). (**Bottom right**) DELTA 1 DEB (8964). The top left plot represents an example of a non-rotating RSO, while the remaining plots indicate examples of rotating RSOs.

Figure 13 illustrates two further examples of BPST radar spectrograms exhibiting micro-Doppler periodicity consistent with RSO rotation. The spectrogram for DMSP 5D 2 F8 USA (NORAD ID 18123) exhibits a periodicity of approximately 6 s, best illustrated in the segment between 21:10:20 and 21:10:29. The spectrogram for NOAA-16 (26536) exhibits a periodicity of approximately 4 s. In this example, Faraday fading around 12:29:23 appears to slightly disrupt the periodicity. The spectrogram for DELTA 1 DEB (8964) does not exhibit micro-Doppler periodicity; rather, it indicates symmetric sidebands spaced at approximately 1.6 Hz intervals. This is consistent with a 1.6 Hz amplitude modulation of the radar cross-section (RCS), suggesting that the RSO is rotating. Although this spectrogram cannot be used to distinguish micro-Doppler features, it provides information on the rotation rate of the RSO. Determination of the actual rotation rate requires additional information on the shape of the RSO. For instance, if the RSO has a cylindrical shape, the 1.6 Hz amplitude modulation would correspond to a rotation rate of 0.8 Hz.

Figure 14 illustrates two examples of BPST radar spectrograms exhibiting random variations obtained from observations of two Starlink satellites. Since numerous transits of the same satellites have been observed without such random variations, we attribute the variations to ionospheric phenomena occurring during the observations. The random variations obtained using the BPST and DSTG HF LOS radars have been investigated [32,33] and attributed to two sources. The first source is ionospheric plasma instabilities in the 0–0.46 Hz frequency range. The second source was identified as Electromagnetic Ion Cyclotron (EMIC) plasma waves in the 0.1–0.5 Hz frequency range, propagating parallel to the Earth's surface in a waveguide centred at an altitude of 250 km. EMIC waves can last several hours, while plasma irregularities last only 10 s [33]. These phenomena produce electron density perturbations along the signal path between the radars and the RSO, which manifest as variations in the measured Doppler. These variations result in a small increase in the uncertainty of the Doppler measurements [32].

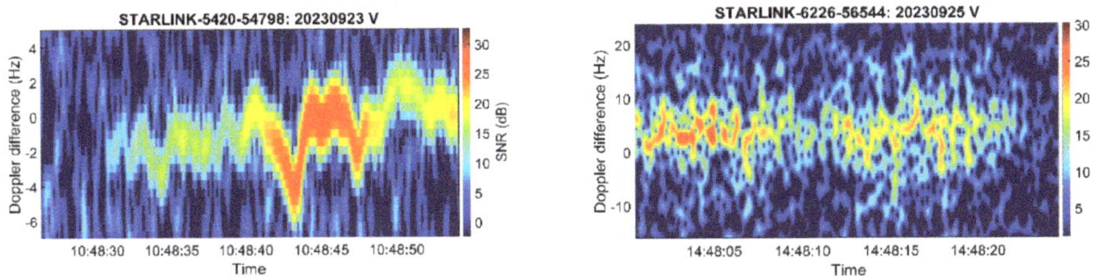

Figure 14. Example of spectrograms for two Starlink satellites illustrating random Doppler variations. (**Left**) Starlink 5420 (NORAD ID 54798), (**right**) Starlink 6226 (56544).

7. Discussion

This current paper introduces new experimental parameters, beam scheduling, and signal processing capabilities that increased the daily count rates observed by the radar from between 150 and 200 RSOs to between 480 and 630 RSOs, representing a three-fold increase. As the introduction of the new experimental parameters and beam scheduling were made simultaneously, it is difficult to isolate the contribution of these two new components. However, Table 8 illustrates the contribution of the signal processing advances, illustrating that QCMM and MDQCMM analyses increase the count rates by approximately 1.3 and 1.6 compared to DM. The results indicate that a minimum detectable RCS at 1000 km of -10.97 dBm^2 (0.09 m^2) was observed. The results also indicate that the radial velocity error measured by the BPST radar is less than 1 ms^{-1}. This is a remarkable result, given that the orbital velocity of a typical LEO object is around 8 kms^{-1}. The results further indicate that the TanDEM-X and TerraSAR-X satellites, which are typically separated by between 250 and 500 m, were both detectable and distinguishable. Despite the increased count rates provided by the new QCMM and MDQCMM analyses, they are unable to detect uncatalogued RSOs or RSOs with inaccurate propagation state vectors. This suggests that there is a role for both DM (detection of uncatalogued RSOs) and MDQCMM (catalogue maintenance) for future BPST RSO observations.

This paper also demonstrates a method to compensate for ionospheric group retardation and phase advance. The success of this compensation is dependent on the accuracy of the electron density profile used to perform the compensation. Ideally, the electron density profile could be provided by a co-located vertical incidence ionospheric sounder (VIS) [34] or a real-time ionospheric model (RTIM) [35,36]. However, for locations such as Buckland Park, where neither a VIS or RTIM is available, a climatological model such as the International Reference Ionosphere (IRI) must be used. However, we note that comparisons between the IRI and ionospheric sounder electron densities have highlighted limitations with the IRI model [37].

An auxiliary antenna array, including three sets of nine antennae, including one set of antennae with dipoles aligned perpendicularly to the main array, was used to confirm that signal fading observed by the BPST radar is most likely the result of Faraday rotation effects. We note that Faraday rotation could potentially be eliminated through the transmission of circularly polarised signals, representing a potential avenue for further research.

The earlier study [13] suggests that BPST RSO ionospheric group retardation measurements may provide a means to perform total electron content (TEC) measurements. However, the results indicated that a further increase in range resolution would be required to accurately measure TEC. Observations using a significantly increased range resolution were performed prior to the conclusion of SpaceFest 2019. However, the increased range resolution was achieved at the expense of a sensitivity reduction of 3.5 dB. Although the number of RSO observations was sparse, the group retardation measurements clearly followed the expected diurnal behaviour. The results also suggest that electron density profile determination will require either a significantly higher number of RSOs than were in orbit at the time of these measurements or the use of a significantly higher-power aperture product than that currently used by the BPST radar. Regarding the former, we note that there are currently a significantly larger number of RSOs in LEO orbits than at the time of these measurements, including approximately 5500 Starlink satellites. However, as will be discussed in a future paper, the frequent manoeuvres made by Starlink satellites to maintain constellation result in the propagation estimates becoming "stale", such that they do not accurately predict the state vectors at the observation time. Regarding the use of a significantly higher transmit power by the BPST radar, we note that there are a number of mesospheric–stratospheric–tropospheric (MST) radars operating at similar frequencies to the BPST radar that use significantly higher transmit power. The middle and upper atmosphere (MU) radar is located at the Shigaraki Observatory, Kyoto University, Japan [38], and operates at 46.5 MHz, with a peak power of 1 MW (25 times that of the BPST radar) and a beam width of 3.6°. The MAARSY mesosphere–stratosphere–troposphere (MST) is located in Andenes, Norway [39], and operates at 53.5 MHz, with a peak power of 800 kW (20 times that of the BPST radar) and a beam width of 3.6°. The MU and MAARSY radars may provide sufficient power aperture product to allow the detection of sufficient RSOs to allow the estimation of electron density profiles. We also note that the analysis described in this paper has been successfully applied to a small subset of data from the MAARSY radar.

The application of QCMM and MDQCMM results in the RSOs being confined to a single range difference cell. This allows a short-time Fourier transform (STFT), or spectrogram, to be applied to observe the evolution of the Doppler characteristics of the RSO. In some circumstances, the spectrogram exhibits either periodic variations attributed to RSO rotation or random variation due to ionospheric plasma instabilities and Electromagnetic Ion Cyclotron (EMIC) plasma waves [32,33]. The periodic variations provide micro-Doppler information that can be combined with simulated signatures from computational electromagnetic modelling of simple computer-aided design (CAD) models to provide further interpretation of the RSOs, such as orientation and spin axis. We believe the EMIC wave observations represent the first time that radar has been used to observe such waves, and we also believe that, during summer, radar may provide a more sensitive measure of the plasma waves than ground and space-borne magnetometers normally used to detect these waves.

The QCMM analysis is computationally non-intensive and can be applied in real time using a PC with modest computational capabilities. This has been exploited by introducing a second dedicated RSO analysis PC to the BPST radar to perform real-time QCMM analysis. This has allowed the QCMM analysis to be applied to data collected for the Doppler beam steering (DBS) analysis of tropospheric and stratospheric winds for which the radar was originally designed [15]. Despite the data acquisition parameters being non-optimal for RSO observations, daily count rates of up to 100 RSOs have been obtained. These count rates could potentially be increased by using the beam scheduling strategy described in this paper, subject to meeting the operational requirements of the DBS analysis, which requires each of the five available beam directions to be used once every five minutes. Noting that the Bureau of Meteorology (BoM) operates four ST radars similar to the BPST throughout Australia, as well as nine smaller, lower-powered boundary layer radars, the BPST QCMM analysis could potentially be applied to data collected by the BoM network to improve Australia's SDA measurement capabilities.

A limitation of the BPST radar configuration used in this paper is that the main array is connected to a single receiver, and the radar is consequently unable to estimate the RSO angle of arrival (AOA). For the subsequent SpaceFest 2020 campaign, the BPST radar main array was separated into a central 6×6 array connected to one receiver, surrounded by eight 3×3 arrays, each connected to a separate receiver, thereby providing an RSO AOA estimation capability. Further, a remote receiving array of three 3×3 antenna arrays was deployed 35 km south of the BPST radar to allow bistatic measurements. These results will be described in a subsequent paper.

8. Conclusions

This paper follows up on an initial investigation of the use of a VHF wind profiling radar for observing RSOs. The radar was operated during the Australian Defence "Space-Fest" 2019 activity, incorporating new beam scheduling and signal processing capabilities. The beam scheduling capability used two-line element propagations to determine the appropriate beam direction to use to observe the transiting RSOs. The signal processing capabilities used a Keystone transform-like approach to correct for range migration, allowing the coherent integration time to be increased to improve the SNR of the observed targets, thereby increasing the detection rate.

The results reveal that 5874 objects were detected over 10 days, with 2202 unique objects detected, representing a three-fold increase in the detection rate over previous observations. A minimum detectable RCS at 1000 km of -10.97 dB m^2 (0.09 m^2) was observed. The effects of Faraday rotation resulting from the use of linearly polarised antennae was demonstrated. The radar's utility for providing total electron content (TEC) measurements was illustrated using a high-range resolution mode and high-precision ephemeris data. The short-term Fourier transform was applied to demonstrate the radar's ability to determine RSO rotation characteristics and short-period ionospheric perturbations.

Author Contributions: Conceptualisation, D.A.H.; methodology, D.A.H., A.J.S. and C.L.A.; software, D.A.H., A.J.S. and C.L.A.; validation, D.A.H.; formal analysis, D.A.H.; investigation, D.A.H.; resources, I.M.R.; data curation, D.A.H. and A.J.S.; writing—original draft preparation, D.A.H.; writing—review and editing, D.A.H., A.J.S., C.L.A. and I.M.R.; visualisation, D.A.H.; supervision, D.A.H.; project administration, I.M.R.; funding acquisition, I.M.R. All authors have read and agreed to the published version of the manuscript.

Funding: Funding for the BP ST radar was provided by the University of Adelaide, ATRAD Pty Ltd., the Australian Research Council grants DP0878144 and DP1096901 and by the Adelaide University ARC Small Grants Scheme. A.J.S. was supported by an Australian Government Research Training Program scholarship.

Data Availability Statement: Data supporting reported results can be found at https://doi.org/10.25909/25044581 (accessed on 26 March 2024).

Acknowledgments: The authors thank Brendan Hennessy for his insightful comments. The involvement of I.M.R. and C.L.A. was supported by ATRAD Pty Ltd. Data. The TLE and SP data used in this paper were obtained from the SpaceTrack web site https://www.space-track.org (accessed on 26 March 2024) The ILRS data were obtained from https://ilrs.gsfc.nasa.gov (accessed on 26 March 2024).

Conflicts of Interest: Authors I.M.R. and C.L.A. are employed by ATRAD Pty. Ltd. The remaining authors declare that the research was conducted in the absence of any commercial or financial relationships that could be construed as a potential conflict of interest.

References

1. Boley, A.C.; Byers, M. Satellite mega-constellations create risks in Low Earth Orbit, the atmosphere and on Earth. *Sci. Rep.* **2021**, *11*, 10642. [CrossRef] [PubMed]
2. Mann, A. Starlink: SpaceX's Satellite Internet Project. 2021. Available online: https://www.space.com/spacex-starlink-satellites.html (accessed on 26 March 2024).
3. Patel, N.V. Averting space doom. *IEEE Spectr.* **2015**, *52*, 16–17. Available online: https://ieeexplore.ieee.org/abstract/document/7024495 (accessed on 26 March 2024).
4. ESA. Spacecraft Dodges Large Constellation. Available online: https://www.esa.int/Safety_Security/ESA_spacecraft_dodges_large_constellation (accessed on 26 March 2024).
5. Smith, C.H.; Greene, B.; Bold, M.; Drury, R. Development of a new SDA facility at Learmonth Australia. In Proceedings of the Advanced Maui Optical and Space Surveillance Technologies Conference, Maui, HI, USA, 11–14 September 2018.
6. Space Surveillance Telescope Declared Operational—Australian Defence Magazine. Available online: https://www.australiandefence.com.au/news/space-surveillance-telescope-declared-operational (accessed on 26 March 2024).
7. LeoLabs Commits to WA for Next Space Radar. October 2021. Available online: https://www.wa.gov.au/government/announcements/leolabs-commits-wa-next-space-radar (accessed on 26 March 2024).

8. Cohen, G.; Afshar, S.; van Schaik, A.; Wabnitz, A.; Bessell, T.; Rutten, M.; Morreale, B. Event-based sensing for space situational awareness. In Proceedings of the Advanced Maui Optical and Space Surveillance Technologies Conference (AMOS), Maui, HI, USA, 19–22 September 2017.
9. Frazer, G.J.; Meehan, D.H.; Warne, G.M. Decametric measurements of the ISS using an experimental HF line-of-sight radar. In Proceedings of the 2013 International Conference on Radar, Adelaide, SA, Australia, 9–12 September 2013.
10. Frazer, G.J.; Rutten, M.; Cheung, B.; Cervera, M.A. Orbit determination using a decametric line-of-site radar. In Proceedings of the Advanced Maui Optical and Space Surveillance Technologies Conference (AMOS), Maui, HI, USA, 9–12 September 2014.
11. Palmer, J.E.; Hennessy, B.; Rutten, M.; Merrett, D.; Tingay, S.; Kaplan, D.; Tremblay, S.; Ord, S.M.; Morgan, J.; Wayth, R.B. Surveillance of Space using passive radar and the Murchison Widefield Array. In Proceedings of the IEEE Radar Conference, Seattle, WA, USA, 8–12 May 2017; pp. 1715–1720.
12. Hennessy, B.; Rutten, M.; Young, R.; Tingay, S.; Summers, A.; Gustainis, D.; Crosse, B.; Sokolowski, M. Establishing the Capabilities of the Murchison Widefield Array as a Passive Radar for the Surveillance of Space. *Remote Sens.* **2022**, *14*, 2571. [CrossRef]
13. Holdsworth, D.A.; Spargo, A.J.; Reid, I.M.; Adami, C. Low Earth Orbit Object Observations Using the Buckland Park VHF Radar. *Radio Sci.* **2020**, *55*, 2019RS006873. [CrossRef]
14. Hall, T.D.; Duff, G.F.; Maciel, L.J. The Space Mission at Kwajalein. *Linc. Lab. J.* **2012**, *19*, 48–63.
15. Dolman, B.K.; Reid, I.M.; Tingwell, C. Stratospheric tropospheric wind profiling radars in the Australian network. *Earth Planets Space* **2018**, *70*, 170. [CrossRef]
16. Morris, R.J.; Murphy, D.J.; Vincent, R.A.; Holdsworth, D.A.; Klekociuk, A.R.; Reid, I.M. Characteristics of the wind, temperature and PMSE field above Davis, Antarctica. *J. Atmos. Sol.-Terr. Phys.* **2006**, *68*, 418–435. [CrossRef]
17. Morris, R.J.; Klekociuk, A.R.; Holdsworth, D.A. First observations of Southern Hemisphere polar mesosphere winter echoes including conjugate occurrences at~69 S latitude. *Geophys. Res. Lett.* **2011**. [CrossRef]
18. Spargo, A.J.; Reid, I.M.; MacKinnon, A.D. Multistatic meteor radar observations of gravity-wave–tidal interaction over southern Australia. *Atmos. Meas. Tech.* **2019**, *12*, 4791–4812. [CrossRef]
19. McGeogh, J.; Jensen, J. *Acceleration and Velocity Processing of HF Radar Signals*; Technical Report; Naval Research Laboratory: Washington, DC, USA, 1967.
20. Bird, D. Sharing Space Situational Awareness Data. In Proceedings of the Advanced Maui Optical and Space Surveillance Technologies Conference, Maui, HI, USA, 14–17 September 2010.
21. Davies, K. *Ionospheric Radio*; IET: London, UK, 1990.
22. Holdsworth, D.A.; Reid, I.M. Preliminary Count Rate Maximisation of the Buckland Park Interferometric Meteor Radar. In Proceedings of the Workshop on the Applications of Radio Science, Hobart, TAS, Australia, 18–20 February 2004.
23. McCant, M. Median RCS Values. 2017. Available online: https://www.prismnet.com/~mmccants/catalogs/index.html (accessed on 21 May 2021).
24. Perry, R.P.; DiPietro, R.C.; Fante, R.L. Coherent integration with range migration using keystone formatting. In Proceedings of the 2007 IEEE Radar Conference, Waltham, MA, USA, 17–20 April 2007; IEEE: Piscataway, NJ, USA, 2007; pp. 863–868.
25. Bilitza, D.; Altadill, D.; Truhlik, V.; Shubin, V.; Galkin, I.; Reinisch, B.; Huang, X. International Reference Ionosphere 2016: From ionospheric climate to real-time weather predictions. *Space Weather.* **2017**, *15*, 418–429. [CrossRef]
26. Pearlman, M.R.; Degnan, J.J.; Bosworth, J.M. The international laser ranging service. *Adv. Space Res.* **2002**, *30*, 135–143. [CrossRef]
27. Klobuchar, J.A. Ionospheric effects on GPS. In *Chapter 12 of Global Positioning System: Theory and Applications*; American Institute of Aeronautics and Astronautics: Reston, VA, USA, 1996; Volume 1, p. 30.
28. Seemala, G.K. Estimation of ionospheric total electron content (TEC) from GNSS observations. In *Atmospheric Remote Sensing*; Elsevier: Amsterdam, The Netherlands, 2023; pp. 63–84.
29. Wilkinson, P.J. Predictability of ionospheric variations for quiet and disturbed conditions. *J. Atmos. Terr. Phys.* **1995**, *57*, 1469–1481. [CrossRef]
30. Heading, E.; Nguyen, S.T.; Holdsworth, D.; Field, D.; Reid, I. Analysis of RF Signatures for Space Domain Awareness using VHF radar. In Proceedings of the 2022 IEEE Radar Conference (RadarConf22), New York City, NY, USA, 21–25 March 2022; IEEE: Piscataway, NJ, USA, 2022; pp. 1–6.
31. Heading, E.; Nguyen, S.T.; Holdsworth, D.; Field, D.; Reid, I. Micro-Doppler Signature Analysis for Space Domain Awareness using VHF radar. *Remote Sens.* **2024**, *in press*.
32. Jonker, J.R.; Cervera, M.A.; Holdsworth, D.A.; Harris, T.J.; MacKinnon, A.D.; Neudegg, D.; Reid, I.M. Impact of Ionospheric Doppler Perturbations on Space Domain Awareness Observations. In Proceedings of the International Conference on Radar, Sydney, NSW, Australia, 6–10 November 2023.
33. Jonker, J.R.; Cervera, M.A.; Harris, T.J.; Holdsworth, D.A.; MacKinnon, A.D.; Neudegg, D.; Reid, I.M. Observational evidence of EMIC waves propagating in the mid-latitude ionospheric waveguide using ground-based HF and VHF radar. *Adv. Space Res.* **2024**.
34. Harris, T.J.; Quinn, A.D.; Pederick, L.H. The DST group ionospheric sounder replacement for JORN. *Radio Sci.* **2016**, *51*, 563–572. [CrossRef]
35. Barnes, R.I.; Gardiner-Garden, R.S.; Harris, T.J. Real time ionospheric models for the Australian Defence Force. In Proceedings of the Workshop for Applications of Radio Science (WARS), Leura, NSW, Australia, 20–22 February 2002.

36. Reinisch, B.; Galkin, I.; Huang, X.; Vesnin, A.; Bilitza, D. The Ionosphere Real-Time Assimilative Model, IRTAM-A Status Report. In Proceedings of the EGU General Assembly Conference Abstracts, Vienna, Austria, 27 April–2 May 2014; Volume 16.
37. Edwards, D.J.; Chambers, T.E.; Cervera, M.A. CLIMF2: A Climatological Model of the Ionospheric F2 Layer. Part 2: Validation and Comparison with IRI. *Adv. Space Res.* **2023**. [CrossRef]
38. Sato, T.; Kimura, I.; Kayama, H.; Furusawa, A. MU radar measurements of orbital debris. *J. Spacecr. Rocket.* **1991**, *28*, 677–682. [CrossRef]
39. Renkwitz, T.; Stober, G.; Latteck, R.; Singer, W.; Rapp, M. New experiments to validate the radiation pattern of the Middle Atmosphere Alomar Radar System (MAARSY). *Adv. Radio Sci.* **2013**, *11*, 283–289. [CrossRef]

remote sensing

MDPI

Article

The Future of Radar Space Observation in Europe—Major Upgrade of the Tracking and Imaging Radar (TIRA)

Jens Klare *, Florian Behner, Claudio Carloni, Delphine Cerutti-Maori, Lars Fuhrmann, Clemens Hoppenau, Vassilis Karamanavis, Marcel Laubach, Alexander Marek, Robert Perkuhn, Simon Reuter and Felix Rosebrock

Fraunhofer Institute for High Frequency Physics and Radar Techniques FHR, Fraunhoferstraße 20, 53343 Wachtberg, Germany; florian.behner@fhr.fraunhofer.de (F.B.); claudio.carloni@fhr.fraunhofer.de (C.C.); delphine.cerutti-maori@fhr.fraunhofer.de (D.C.-M.); lars.fuhrmann@fhr.fraunhofer.de (L.F.); clemens.hoppenau@fhr.fraunhofer.de (C.H.); vasileios.karamanavis@fhr.fraunhofer.de (V.K.); marcel.laubach@fhr.fraunhofer.de (M.L.); alexander.marek@fhr.fraunhofer.de (A.M.); robert.perkuhn@fhr.fraunhofer.de (R.P.); simon.reuter@fhr.fraunhofer.de (S.R.); felix.rosebrock@fhr.fraunhofer.de (F.R.)

* Correspondence: jens.klare@fhr.fraunhofer.de

Abstract: The use of near-Earth space has grown dramatically during the last decades, resulting in thousands of active and inactive satellites and a huge amount of space debris. To observe and monitor the near-Earth space environment, radar systems play a major role as they can be operated at any time and under any weather conditions. The Tracking and Imaging Radar (TIRA) is one of the largest space observation radars in the world. It consists of a 34 m Cassegrain antenna, a precise tracking radar, and a high-resolution imaging radar. Since the 1990s, TIRA contributes to the field of space domain awareness by tracking and imaging space objects and by monitoring the debris population. Due to new technologies, modern satellites become smaller, and satellite extensions become more compact. Thus, sensitive high-resolution space observation systems are needed to detect, track, and image these space objects. To fulfill these requirements, TIRA is undergoing a major upgrade. The current imaging radar in the Ku band will be replaced by a new radar with improved geometrical and radiometric resolution operating in the Ka band. Due to its wideband fully polarimetric capability, the new imaging radar will increase the analysis and characterization of space objects. In addition, the tracking radar in the L band is also being currently refurbished. Through its novel modular structure and open design, highly flexible radar modes and precise tracking concepts can be efficiently implemented for enhanced space domain awareness. The new TIRA system will mark the start of a new era for space observation with radar in Europe.

Keywords: radar; inverse synthetic aperture radar; space domain awareness; space reconnaissance; space surveillance and tracking; signal processing; resident space objects; space debris; Tracking and Imaging Radar; TIRA

1. Introduction to the Tracking and Imaging Radar TIRA

One of the world's largest radar domes (radomes), visible from afar, is situated in Wachtberg near Bonn, Germany. The radome has a diameter of 47.5 m and protects the radar dish of the TIRA system from all weather influences (Figure 1). TIRA is one of the most powerful radar systems for space observation in Europe. Its three main components are the fully movable Cassegrain antenna system with a diameter of 34 m, the tracking radar in the L band and the imaging radar in the Ku band (Figure 2). The 240 t antenna structure rotates with up to 24 deg/s in azimuth. This is necessary to track any artificial space object almost seamlessly, as tilting of the antenna above 90° in elevation is not possible. With this ability, the pass of a resident space object (RSO) can be observed without interruption close to its zenith. Note, that due to confidentiality, not all radar parameters may be listed in this paper.

Among the currently operating space radar systems, TIRA is recognized worldwide as a powerful weather-independent sensor for the observation of the near-Earth space environment [1,2]. For a comprehensive review of this field, see [3] and references therein. Recent advances are highlighted in the Special Issue at hand (*Radar for Space Observation: Systems, Methods and Applications*, a special issue of *Remote Sensing* https://www.mdpi.com/journal/remotesensing/special_issues/7C43H2J2V2, accessed on 29 October 2024) . The development and construction of TIRA started by the end of the 1960s. As an experimental system, TIRA was modified and upgraded continuously over the decades. Its focus changed to space observation in the 1980s, followed by enhancing capabilities, e.g., to measure the space debris population in the 1990s or to increase the bandwidth of the imaging radar in the early 2000s. In addition, individual experiments like measurements on airplanes or jamming were conducted on the side. The latest upgrades were the implementation of a new antenna drive and control system in 2012 and the replacement of the tracking radar transmitter one year later. Finally, after 46 years of operation, the radome was renewed in 2014. For this, the new radome was constructed inside the old one up to a certain height, followed by taking off the cap and replacing it with the new cap, which was pre-assembled on the ground. At the end, the lower part of the old radome was dismantled, revealing the new radome step by step. The new radome consists of statistically distributed triangles with side lengths between 2.1 m and 4.5 m which ensures that interferences are minimized. A membrane between the aluminum rods allows the radar waves to pass through nearly undisturbed.

In its current configuration, TIRA is able to accurately determine RSO orbits while also generating detailed radar images of them [4–7]. In the following paragraphs, the present state of the system and its capabilities are introduced. Later, the plans for the complete overhaul of TIRA's two radars are discussed in detail. This transformative system upgrade will be a game changer in space observation with radar.

Figure 1. Illustration of the Tracking and Imaging Radar, TIRA. The system's 34 m fully movable parabolic antenna and radar systems are protected by a 47.5 m radome. The surrounding 3-story building houses offices and laboratories associated with FHR's research activities in SSA as well as technical infrastructure necessary for the operation of the TIRA system.

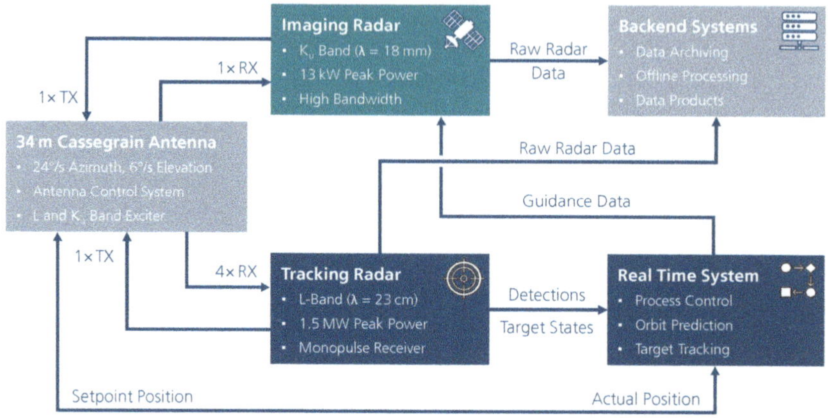

Figure 2. Block diagram illustrating the current system architecture of TIRA. Components of the tracking radar are shown in dark blue, the imaging radar is show in turquoise, and common system components have a light blue background.

1.1. Current Configuration of the Tracking Radar

The tracking radar operates in the L band in a frequency band around 1.3 GHz. Radar pulses are sent to space objects and received after reflection of the electromagnetic waves by the large radar antenna. Range, range rate, azimuth angle, elevation angle, and time are calculated from the radar echoes by using a monopulse system in order to estimate the orbits of the space objects with high precision. The tracking of space objects starts when they cross the horizon and stops just before they disappear below the horizon. For this, the tracking radar uses a monopulse system which consists of four horns for transmit and receive. The ratio of the sum channel to the respective difference channels is a measure of the deviation of the object direction from the antenna axis. This allows the antenna position to be continuously corrected so that space objects are always close to the center of the main lobe and can be precisely tracked. TIRA's mechanical alignment accuracy corresponds to 3 m at a distance of 1000 km. Since the true object positions differ from the measured positions due to the refraction of the radar waves in the Earth's atmosphere, the data are corrected after the measurement using recorded environmental parameters (e.g., temperature, atmospheric pressure, and humidity) with a suitable atmospheric model. The main lobe has a width of almost half a degree (about the diameter of the full moon), which corresponds to a lobe diameter of 8.6 km at a distance of 1000 km. Thanks to the high transmission power in combination with the large antenna area to achieve high sensitivity, even objects with a diameter of just a few centimeters can be detected, tracked, and measured at a distance of 1000 km.

For a standard RSO observation, the TIRA system uses a two-line-element (TLE) dataset to initially acquire and track a selected space object. The knowledge of the space object position can be more or less accurate, depending on different factors. As the orbit of a space object is affected by diverse forces, e.g., solar radiation pressure (SRP), atmospheric drag, and third-body attraction forces from the Moon and the Sun, the age of the TLE dataset plays a fundamental role. The largest deviation between the actual orbit and the propagated one using the TLE dataset is in the along-track direction. It can already be about 1 s (i.e., about 7 km to 8 km along-track distance in low Earth orbit (LEO)) for a one/two-day old TLE. The tracking method, which is currently implemented in the L-band radar, estimates the time offset (along-track) between the actual orbit of the space object and the one derived from the TLE dataset in real time. This time offset is then compensated continuously for each pulse to ensure that the object remains in the main lobe of the antenna beam. Without this correction, the object could be missed as the antenna lobe of the L-band radar at a range of 1000 km is 8.6 km.

1.2. Current Configuration of the Imaging Radar

The TIRA system also has a broadband, coherent imaging radar system which works in the Ku band with a center frequency of 16.7 GHz. It is always used when not only very precise orbit data are required, but also the imaging of a space object is needed.

This is the case, for instance, during a contingency support when a satellite is not behaving nominally, and a visual analysis is needed. Indeed, two-dimensional radar images can be used to detect mechanical damage to the satellite that may have been caused by a collision with space debris. If one processes 2D radar images one after the other over the whole flyover, one can even generate a radar video which shows the rotational motion of the satellite and eventually different sides of the object. From the radar image series, one can calculate the attitude motion of a satellite and assess a potential tumbling motion. Due to the higher frequency compared to the L-band tracking radar, the antenna lobe is significantly narrower with 0.031°, which corresponds to a lobe diameter of 540 m at a distance of 1000 km. Since only the L-band radar has a tracking capability, it is needed to steer the imaging radar. Due to the significantly narrower antenna lobe in the Ku band, the tracking radar must track space objects extremely precisely so that the Ku-band antenna lobe stays on the object.

2. Radar Techniques for Space Observation and Products

2.1. Generation of High-Quality Observation Vectors for Orbit Determination

The tracking radar of the TIRA system measures for each transmitted pulse the range of a space object as well as its range rate, azimuth angle, and elevation angle. The estimation accuracy of each measured parameter depends on its specific resolution and on the signal-to-noise ratio (SNR). It is well known that averaging over several measurements improves the estimation accuracy of a random variable. Indeed, the estimation accuracy is improved by a factor $\sqrt{N}$ when averaging over N realizations. The latter assumes that the process of the random variable is wide-sense stationary, i.e., the variance is finite, the expected value is time-invariant, and the autocovariance is translationally invariant.

However, for LEO objects, the different parameters strongly change over time due to the fast motion of these objects relative to the radar. Averaging over a large number of pulses is counterproductive as it deteriorates the estimates due to the underlying dynamics. This effect is examined in the framework of orbit determination [8] where a processing time of about 1 s is derived as a good compromise between improvement in the estimation accuracy and process stationarity.

As introduced in [6], a coherent processing over N subsequent pulses can be applied to increase the estimation accuracy of the range rate and to derive a new parameter, the range-rate rate (Figure 3c). The latter is determined independently of the aforementioned observables. This parameter is particularly interesting for initial orbit determination (IOD) [6]. Currently, the tracking radar uses a correlator quartet to determine the range and range rate of the object. This method works similarly to a monopulse system by comparing the outputs of the matched filter at different delays and Doppler frequencies. The major advantage of this method is its rapidity as it avoids a grid search to estimate the Doppler frequency. The price to pay is that the estimated values are slightly biased (Figure 3b, red dots). The bias is much more important than the center of the correlator quartet being shifted from the true object position in the range/Doppler plane. This solution was chosen in the past as the best tradeoff for a real-time implementation. It will be replaced by the realization of a matched filter bank in the future development stage of TIRA to obtain unbiased estimates and better control the residuals. The range rate bias can be corrected by the coherent integration filter [6] (Figure 3b, black dots). The refined range rate and the range-rate rate can be used to improve the range estimate (Figure 3a, black dots).

If the object was also observed with the imaging radar, a more accurate range could be determined from the imaging radar data. Indeed, the used bandwidth of the imaging radar is much larger compared to the one of the tracking radar, yielding a much finer range resolution. Since the imaging radar transmits chirp waveforms, a coupling between the range and the Doppler exists. This ambiguity has to be first solved in order to estimate accurately the range of the object [6] (Figure 3a, blue dots).

A high-quality observation vector at a given epoch can be derived, which regroups the new range estimate with the refined range rate, the range rate, and the angular parameters. The benefit and longevity of these highly accurate measurements for orbit determination and cataloging are currently under evaluation.

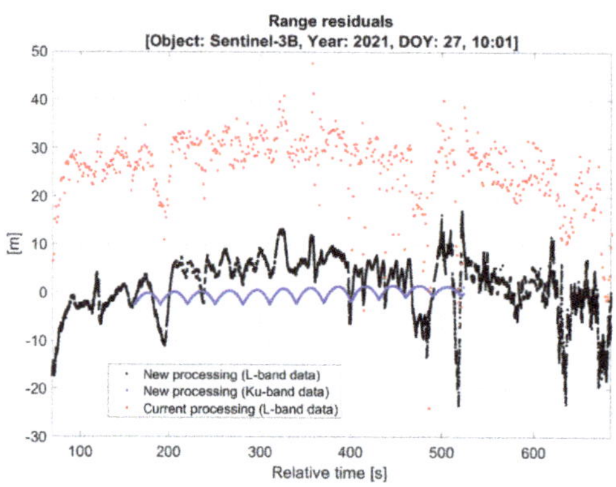

(**a**) Range residuals.

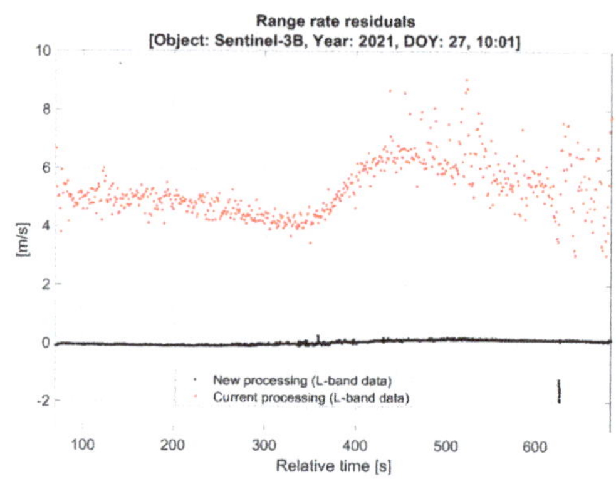

(**b**) Range rate residuals.

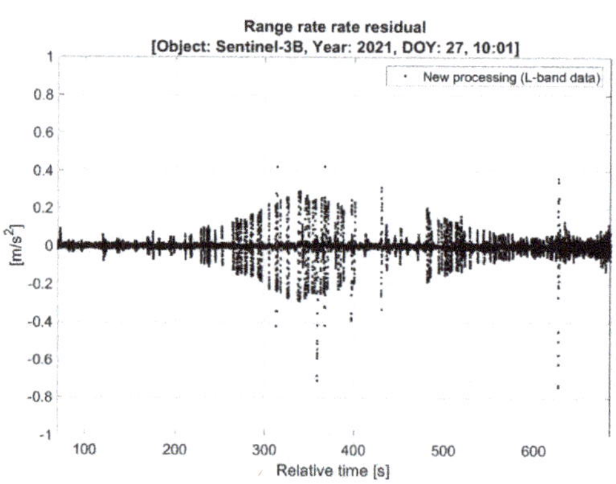

(**c**) Range-rate rate residuals.

Figure 3. Sentinel−3B (27 January 2021) [6].

2.2. *Separation of Multiple Objects*

In certain situations, space objects are orbiting very closely to each other. This can happen, for example, during the release of satellites, during robotics missions, or during a spying attack. In this case, it is important to detect and monitor the maneuver from the ground and to estimate the distance between the objects. This can be achieved using the imaging radar. However, for the current imaging radar, the distance between the objects should be less than 100 m in range and 500 m in cross range at a range of 1000 km. This strong constraint restricts the use of the imaging radar only to close proximity maneuvers. It is worth mentioning that if the range/Doppler histories of the objects differ from each other, the objects cannot be all focused simultaneously. Indeed, different processing filters are required to image the objects. Depending on the processing, one of the objects will be sharply imaged, while all the others will be blurred.

It is possible to discriminate objects with the tracking radar without the use of radar imaging. Even if space objects are located jointly in the antenna beam and in the same range/Doppler resolution cell after matched filtering, coherent processing can be applied to the multi-channel data acquired by the monopulse system in order to discriminate the objects in the 4D space range rate/range-rate rate (the rate of change of the range rate)/azimuth angle/elevation angle. The separation in angle can be achieved, e.g., by using the monopulse dechirping method [6], which is an extension of the method presented in [9].

This multi-channel coherent processing can be applied to discriminate objects in different orbital regimes. The example of an Astroscale exercise [10] during which an encounter maneuver was performed between two LEO objects, ELSA-D (larger object), and ELSA-D CLIENT (smaller object, see Figure 4), is shown in Figure 5. Figure 5 plots the measured radar cross section (RCS). During a limited time, both objects were located in the same radar beam. This corresponds to the region indicated by the orange box in Figure 5, which presents an oscillating pattern revealing the presence of several close objects. Applying coherent processing enables us to identify both objects (e.g., at time B and C), whereas only one object can be detected in the green regions (e.g., at time A and D). The right picture shows the relative location of the two objects in the antenna beam after monopulse dechirping processing. It has to be noted here that the TIRA system uses an amplitude monopulse to estimate the angular position. Contrary to a phase processing, it is not possible to correct for the object motion within the antenna beam during the integration time. The fast object motion in LEO, the required beam tracking, and the variation in the relative position of the object within the antenna beam cause an amplitude modulation in the received signals affecting the estimation of the angular position. Nevertheless, the relative position of the objects in the antenna beam can be coarsely followed over time.

The same type of processing can be applied to the geosynchronous equatorial orbit (GEO) orbital regime. One major difference is that the single pulse SNR is so low that space objects cannot be directly detected. Integrating over several pulses is mandatory for target detection. GEO objects orbit at the same angular velocity as the Earth. They thus appear to be stationary for an observer on the Earth and their location in the antenna beam is fixed. Figure 6 shows the results of an observation of the NILESAT cluster. All four satellites could be successfully detected, and their parameters estimated.

Figure 4. ELSA-D and ELSA-D CLIENT [10].

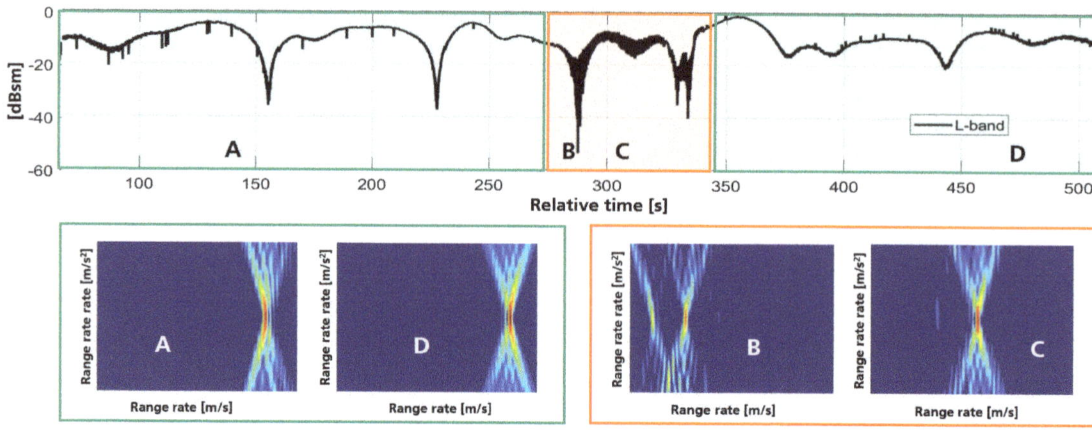

(a) Measured RCS and matched filter output.

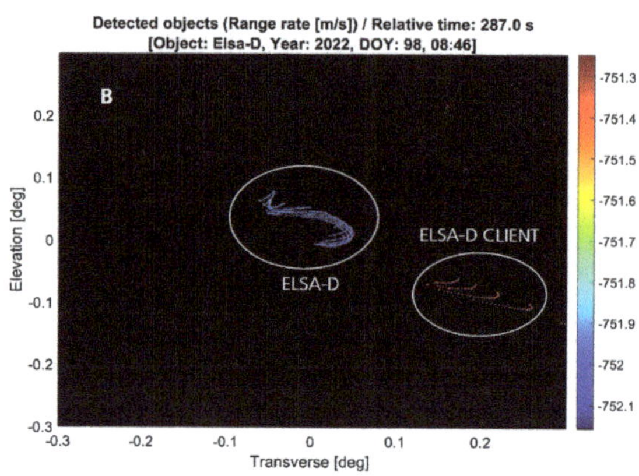

(b) Position in the antenna beam.

Figure 5. Astroscale exercise (8 April 2022) [11].

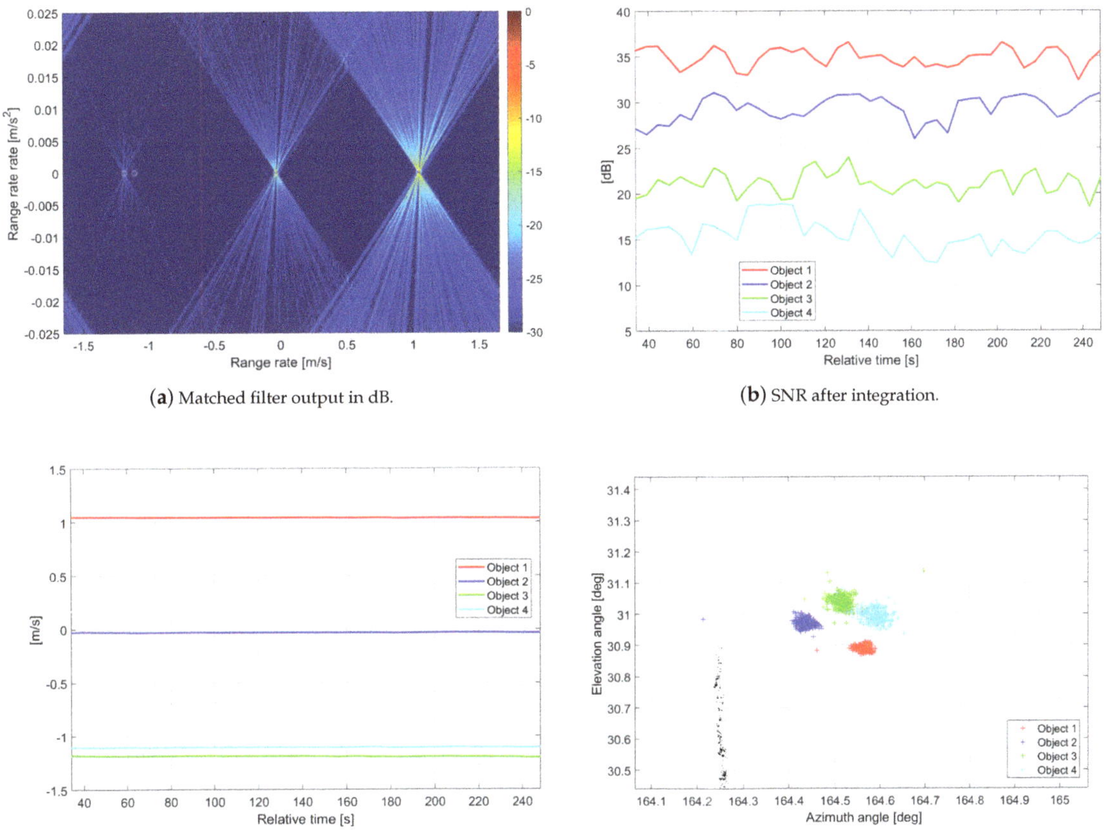

(**a**) Matched filter output in dB.

(**b**) SNR after integration.

(**c**) Range rate.

(**d**) Position in the antenna beam.

Figure 6. NILESAT cluster (16 March 2023).

2.3. Radar Imaging of Space Objects

Different focusing techniques can be used to image space objects. The choice of the method depends on the bandwidth of the transmitted radar waveform, on the attitude motion of the object (i.e., slow or fast rotating), and on the spanned k-space. Conditioned by these parameters, either a fully coherent, a partially coherent, or an incoherent processing is possible. The basic prerequisite of all these reconstruction techniques is similar, as they exploit the variation in the aspect angle between the space object and the radar during the data acquisition to image the object. This variation leads to a change in the Doppler frequency over time. For all these methods, the spatial resolutions achieved in both image dimensions are independent of the object range. This is a main advantage compared to optical systems.

The major challenge of radar imaging for space observation is the accurate estimation of the translational and rotational motion of the object. Indeed, the knowledge of the object kinematic is a prerequisite to form sharp images of the rotating space object. This dynamic information has to be estimated directly from the radar data as space objects are uncooperative targets, meaning that accurate enough information about their attitude motion and positions (requiring cm/mm-precise ephemerides depending on the center frequency used) is not available. During the processing time required to compute a radar image, the rotation vector should be constant. Otherwise, the imaging plane varies, and the radar images are distorted.

2.3.1. Imaging with the Ku-Band Wideband Radar

Inverse synthetic aperture radar (ISAR) is a fully coherent processing method, which is the most commonly used technique to produce high-resolution images of rotating objects [12]. It is particularly adapted to intermediate tumblers. A fine resolution in range is achieved with a large signal bandwidth. In order to separate backscatterers in cross range, a spectral analysis is performed on the range profiles acquired over consecutive pulses. The cross-range scaling of the ISAR image is set by the effective angular velocity of the space object. The integration time is usually chosen to obtain similar resolutions in range and cross range.

To illustrate the efficiency of the ISAR technique, Figure 7 shows the last ISAR image series of the European Space Agency (ESA)'s satellite AEOLUS before its re-entry. The time between two consecutive radar images is about 6 s. The effect of the rotational motion of the object is particularly visible in the image sequence. The different parts of the satellite (e.g., main body, solar panels) can be clearly identified. The reflections in the center part (see, e.g., the marked region with an ellipse in Figure 7f) are not caused by a system artifact but by the multiple reflections of the radar waves in the instrument payload Aladin (range extended returns).

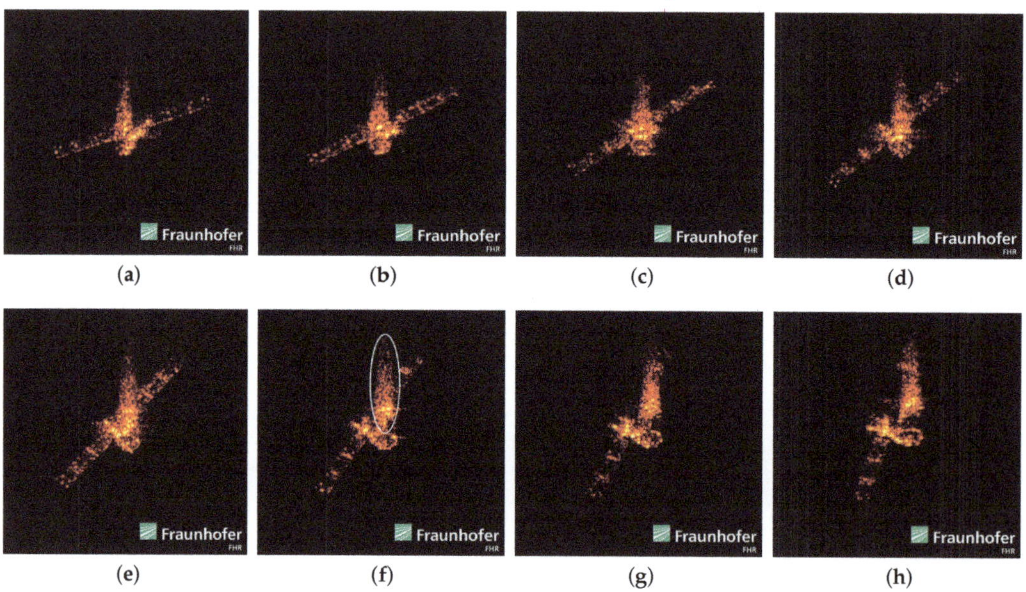

Figure 7. (**a–h**) Series of ISAR images of AEOLUS obtained during its last pass observed by TIRA on 28 July 2023 [13].

Due to the limited pulse repetition frequency (PRF), the raw data of fast rotating objects might be under-sampled. This violation of the Nyquist condition affects the signal phase over slow time (aliasing effect). As a result, the radar image processed with the ISAR principle is the superposition of the true image of the object (unambiguous signal part) with its ghost images (ambiguous signal parts). The distance between the unambiguous and ambiguous images depends on the rotational motion of the object. For high angular velocities, all the images are overlapping, and the object cannot be imaged properly. To illustrate this effect, Figure 8 shows an ISAR image of Jason-2. The apparent angular velocity was estimated to be about 9 deg/s. The additional solar panels at the edge of the image can be clearly identified as ghost images.

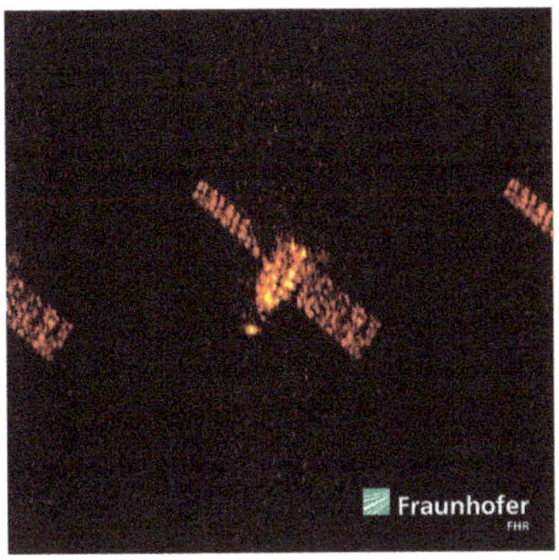

Figure 8. Jason-2 (13 October 2022).

The radar data of fast tumblers can be processed using the same tomographic approach that is used for medical imaging [14]. The tomography technique exploits the amplitude values of the range profiles over time. The achievable resolution depends on the transmitted signal bandwidth and on the covered aspect angle. The Russian satellite COSMOS-1408 is taken as example in the following. This satellite was hit by a Russian anti-satellite (ASAT) weapon in November 2021 [15]. The largest debris piece of the generated fragmentation cloud was observed with the TIRA system two weeks after the event. From the range profiles presented in Figure 9 [15], an apparent angular velocity of about 90 deg/s could be extracted. Because of the strong under-sampling, meaningful ISAR images could not be computed. Figure 9 shows a radar image generated with the tomographic approach. The figure reveals that the solar panels, the gravitational boom, and the signals intelligence antennas were destroyed by the ASAT weapon. Only a solar panel attachment is still visible in Figure 9 when comparing it with Figure 10. Radar images computed with tomography differ from the ones computed with ISAR. The longer integration time raises the requirement on the range profile alignment and on the stability of the rotational vector. Temporal variation in the angular velocity causes a smearing of the radar images and should be accounted for. This is also true for ISAR; however, it is less critical due to the shorter processing interval.

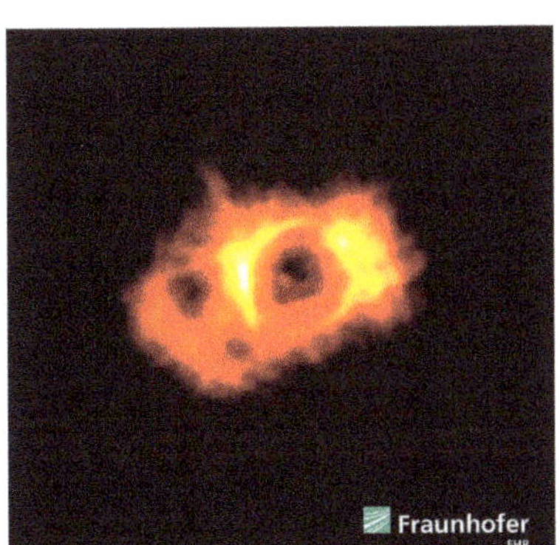

Figure 9. Cosmos-1408 (1 December 2021) computed with an angular velocity of 92.3 deg/s.

Figure 10. Tselina-D model [16].

2.3.2. Imaging with the L-Band Narrowband Radar

Recent work has demonstrated the possibility of imaging with the narrowband tracking radar of the TIRA system [17]. Again, the angular velocity of the object is a key parameter for imaging, as it determines the integration time and the sampling positions. Different techniques can be applied.

The first method is a partially coherent processing. It exploits the micro-Doppler signature of the space object using a tomographic approach. This Doppler tomography technique was presented in [17] and applied to the monostatic and bistatic radar data of an Atlas-5 Centaur rocket body orbiting in the medium Earth orbit (MEO) region. The spatial resolution is a function of the PRF, the angular velocity, and the used radar wavelength. Figure 11a shows a Doppler tomography image [17] of the rocket body including a zoom of its upper part. The shape of the object is clearly recognizable. The radar waves were not backscattered by the center part of the rocket body towards the radar, explaining the dark area in the center of the radar image, except for a small region corresponding to the reflection off of a cable duct.

The data can be also fully coherently processed similar to the ISAR reconstruction using a larger aperture angle [17]. The minimum achievable resolution is given by a quarter of the used wavelength. This processing technique can also be applied to the data of the imaging radar. Figure 11b presents the reconstructed radar image using a back-projection algorithm [17]. The resolution is finer compared to the Doppler tomographic reconstruction at the cost of high side lobes. Fine structures of the object are visible in the zoom area, in particular the structured pattern of the adapter ring (see Figure 12). To better analyze the object, a composite image merging the results of both reconstruction methods can be generated using different thresholds for imaging, as presented in Figure 11c.

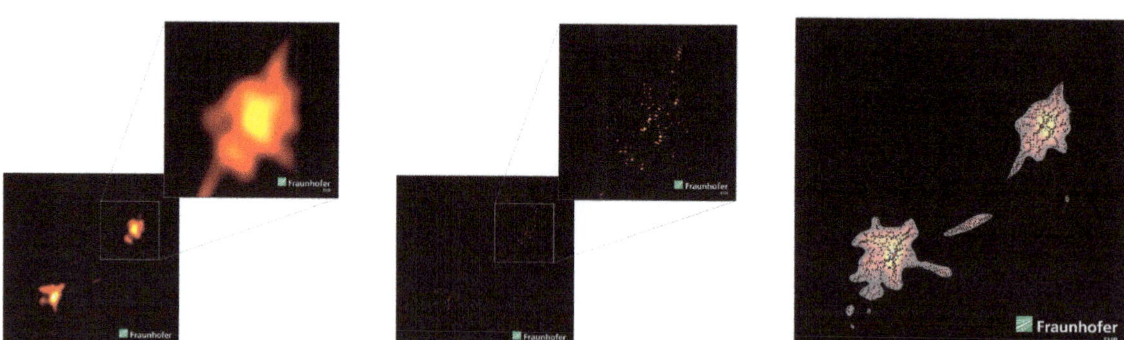

(**a**) Doppler tomography. (**b**) Back-projection. (**c**) Composite image.

Figure 11. Atlas-5 Centaur rocket body (16 March 2023) [17].

Figure 12. Photo of an Atlas-5 Centaur rocket body [18].

3. Challenges for a Future Imaging Radar for Space Observation

3.1. Future Situation in Earth Orbit

The use of near-Earth space has grown drastically since the first artificial satellite Sputnik was launched in 1957. Sixty years later, the annual number of launched objects is raising exponentially (see Figure 13). As of 2024, more than 19,000 satellites have been placed into Earth orbit [19]. In addition, there is a constantly increasing amount of space debris. It is estimated that around 4500 objects orbiting Earth are larger than 10 cm, around 1,100,000 objects between 1 cm and 10 cm, and around 130 million objects between 1 mm and 1 cm, as of August 2024 [19]. The amount of space debris is constantly rising due to debris breaking apart, explosions of spent rocket stages and collisions of debris between each other or from debris with satellites. When two space objects collide, thousands of new pieces of debris can be created. This is what happened, for instance in 2009, when the defunct Russian military satellite Cosmos-2251 collided with the communications satellite Iridium-33. Many thousands of new pieces of space debris were detected. This large amount of space debris around Earth is a significant threat to all active satellites as any collision can damage or completely destroy a satellite generating even more debris.

In recent years, so-called mega constellations have also been launched. In each of these constellations, thousands of small satellites are placed into orbit. The growing object density increases the overall collision probability, as any of these satellites can collide with each other or with space debris, creating a new cloud of debris.

Finally, every active satellite becomes space debris at the end of its life if not actively de-orbited, or even earlier in the event of a failure. In this case the defunct satellite stays as debris until it re-enters in the Earth's atmosphere after spending many years in orbit or until the satellite is taken out of orbit using special de-orbiting missions [7].

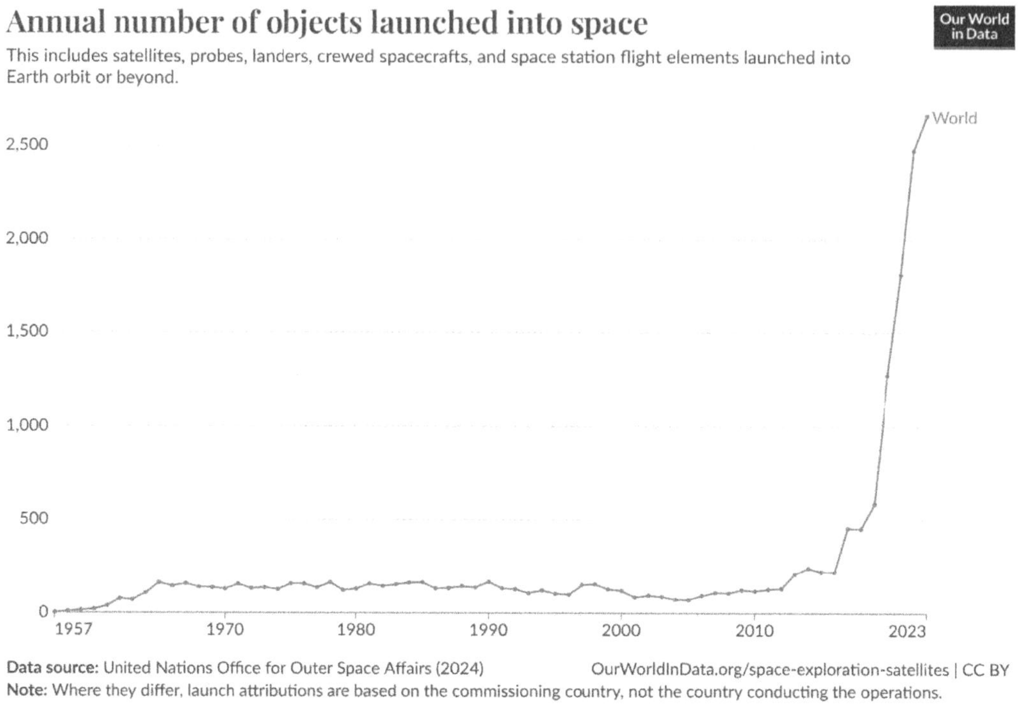

Figure 13. United Nations Office for Outer Space Affairs' (2024), with major processing by Our World in Data, "Annual number of objects launched into space—NOOSA" [dataset]. United Nations Office for Outer Space Affairs' "Online Index of Objects Launched into Outer Space" [original data] [20].

3.2. The Motivation for Ground-Based High-Resolution Fully Polarimetric Space Imaging Systems

One driver of change in the near-Earth orbit is the rapid progress in the development of new technologies. In the past decades, large, complex and very expensive monolithic individual satellites were built. Currently, the trend is towards the deployment of small and cost-effective systems. The trend towards constellations of small, less expensive, and simple satellites replacing large satellites allows for much shorter development and deployment times. Also, a better global coverage can be achieved with a constellation of small satellites. This development is also supported by standardized satellite buses and falling costs for satellite launches by private companies such as SpaceX. Inevitably, the amount of space debris will increase considerably in the near future, as small satellites do not have a long operational lifetime and are usually not equipped with an active or passive de-orbiting system. This trend is accelerated by the deployment of mega-constellations. For instance, Starlink satellites for global internet access might reach 12,000 objects in 2025. In addition, the reliability and thus the operational safety of these small satellites are also expected to decrease. This raises the risk of collisions with other space objects.

The trend towards smaller satellites poses a great challenge to ground-based reconnaissance systems. Moreover, the mountings and instruments are also becoming more compact for large satellites, so that their ground-based reconnaissance is also becoming increasingly difficult. This problem is heightened by the fact that some satellites have to be operated at significantly higher orbits. One reason is that LEO is already extremely populated at altitudes between 500 km and 800 km. Another reason is that it is often not possible to raise the orbit of small satellites when the orbit height decreases because of friction in the atmosphere.

Due to the future development of artificial satellites in near-Earth space described above, the reconnaissance capability of ground-based sensors has to be improved to meet the requirements of future reconnaissance systems. Future ground-based space reconnaissance systems must operate with a significantly higher spatial and radiometric resolution in order to be able to image small satellites at a far range together with their instruments and small attachments in order to perform technical analyses.

Additional information about space objects can be gained by using polarimetric ISAR. Especially in the case of a fully polarimetric space radar, the full electromagnetic vector field is available as opposed to a single polarized ISAR image. Fully polarimetric radar imaging takes advantage of illuminating objects with both horizontally (H) and vertically (V) oriented electric field vectors and also record both polarized returns. This acquires all four components of the scattering matrix and allows us to investigate the structure of space objects due to the interaction with the polarized electromagnetic waves [21].

Indeed, depending on the transmitted and received polarization directions, structures (e.g., bars) of an object might be clearly visible in the radar image or suppressed. This allows us to separate different areas or components of a space object in radar images in order to investigate them independently of each other in more detail. Moreover, very bright and prominent scatterers which outshine faint nearby structures could be suppressed to make adjacent faint scatterers visible.

A well-known consequence for ISAR imaging is the layover effect. This means that spatially separated resolution cells of a 3D space object can superimpose in the processed 2D radar image. As a result, the interpretation of 2D radar images of a space object might be difficult, since scatterers which physically and spatially do not belong to each other appear at the same image position. Polarized radar images might help to solve this problem by separating these superimposed pixels in different polarization directions.

Another prominent effect in imaging space objects is the occurrence of multipath reflections, e.g., from the satellite body itself or from mountings like optics of cameras. For instance, single or double reflections appear in different polarization regimes.

Figure 14 shows exemplarily a space object which was observed in a bistatic polarized configuration by the Office national d'études et de recherches aérospatiales (ONERA) and Fraunhofer Institute for High Frequency Physics and Radar Techniques (FHR). One sees very clearly that different regions are prominently visible in both polarizations. This helps to investigate and to interpret the radar image much better than with just a single polarized ISAR image.

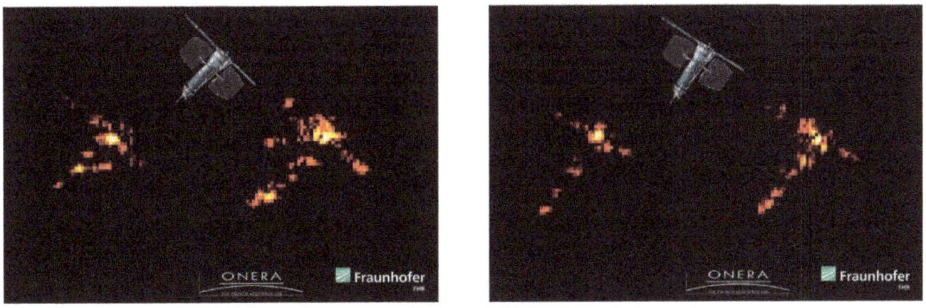

(**a**) Horizontal polarization (**b**) Vertical polarization

Figure 14. Bistatic polarized ISAR images of a space object.

4. New TIRA System Design

With the currently ongoing upgrades, the TIRA system is facing a major overhaul concerning both radar instruments and the complete process control infrastructure. This extensive endeavor offers the unique opportunity to design a new overall system architecture. This architecture, conceptualized to be modular and open, will enable continuous incremental upgrades of TIRA in the future. During the current phase, one key is the utilization of commercial off-the-shelf (COTS) components adapted to and interacting via robust interfaces. Figure 15 shows the prospective system architecture after completion of the major upgrades to the radar subsystems. These upgrades will enhance TIRA's capabilities as a research instrument and meet the increased requirements emerging in the context of space situational awareness (SSA).

The future system is composed of three main subsystems: the fully movable parabolic antenna, the high-power L band radar, and the broadband imaging radar in the Ka band. A key point in the new architecture is to enable the Ka band imaging radar to track space objects in addition to the tracking capabilities of the primary tracking radar in the L band. The overall process control will be implemented as a distributed real-time system with emphasis on composability, scalability, and dependability. Real-time control of subsystems is localized as needed to achieve fast response times and autonomous subsystem functionality along parameters is defined and continuously updated by higher systems. Both radars and the antenna infrastructure are controlled by a central real-time processing system.

This system handles orbit prediction and target tracking and automates standard measurement scenarios, while extensive customization allows users to conduct specialized experiments. Data storage and offline processing of the radar data are handled on common back-end systems. These also aggregate and record monitoring data sent by the various subsystems to maintain an overall picture of past and current system status and health.

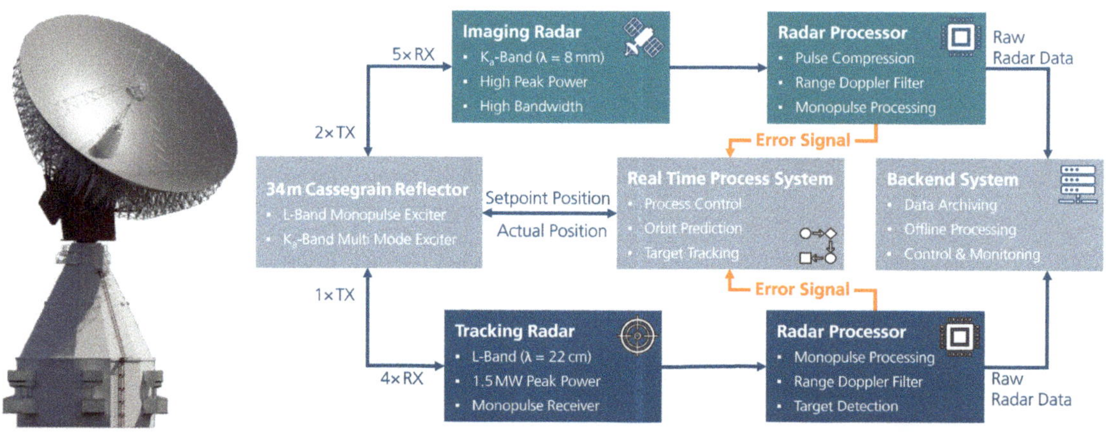

Figure 15. Block diagram illustrating the future TIRA system architecture for enhanced space observation capabilities as a research instrument and an operational radar system.

4.1. Model-Based System Engineering

To cope with the complexity of the future TIRA system during design, realization, and also operation, it was decided to apply methods from the field of systems engineering, particularly model-based system engineering (MBSE). The specific method and tooling used was the ARChitecture Analysis and Design Integrated Approach (ARCADIA) [22] along with the Capella open-source modeling software.

Figure 16 summarizes the principle of the ARCADIA method. The development process starts with the operational analysis evaluating the needs of the users focusing on the system. The view is then reversed in the functional analysis providing what the system has to accomplish for these users. Following these analyses, the architectural design is first considered in the logical architecture layer, which refines the system's functions and allocates these functions to local components. Finally, the development cycle leads to the physical architecture and the respective component specifications.

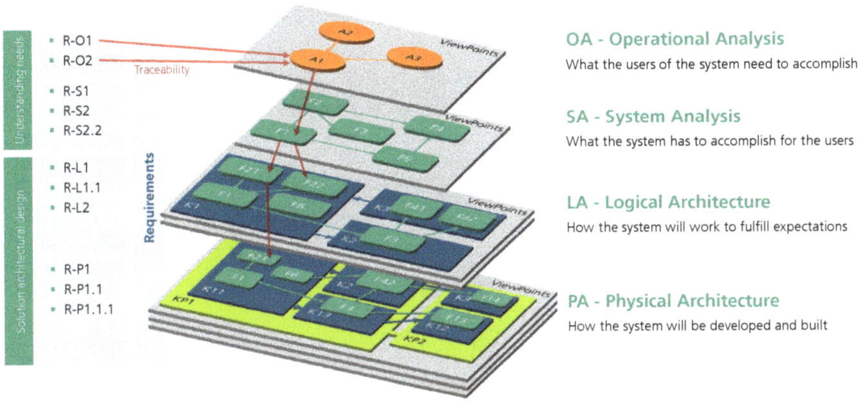

Figure 16. Overview of the ARCADIA method for MBSE.

In addition to this structured development, MBSE also facilitates documentation of the requirements, here denoted as R-*, system design, and the system itself. These requirements are traceable from operational analysis to the physical architecture, thus consequences of design changes can be easily assessed.

This approach, if consequently applied, will lead to a well-defined system architecture with consistent interfaces to obtain an extensible research system adaptable to new requirements.

4.2. System Monitoring

Monitoring system health and status permanently is important to maintain high availability. It further gives crucial information in case of repair and maintenance. This includes system health parameters such as device temperature, voltages and currents, application parameters like encoder values of the rotating axes, set and error values of the antenna controller and meteorological measurements like air temperature/humidity and liquid water content to estimate the signal propagation parameters like the zenith tropospheric delay.

Every module reports environmental and system parameters via a network to a time series database (InfluxDB, open-source, MIT license, https://www.influxdata.com/ accessed on 29 October 2024) . This solution enables us to analyze a high-rate time series of measurements with high performance and presents the result on a configurable dashboard system (Grafana, open-source, AGPLv3 license, https://grafana.com/, accessed on 29 October 2024) in real time. Collecting a high-rate time series requires high storage capacity and slows down the querying and processing of such kinds of data. Time series databases are designed to overcome this drawback by rule-based aggregation of the high-rate data using statistical downsampling methods and fast compression algorithms to reduce data size and query time.

Alongside MBSE, we also aim to develop a comprehensive digital twin of our radar system. The basic concepts are shown in Figure 17 by the example of the antenna's positional control system. The digital twin can be based on physical models with parameters estimated from monitoring data or even black-box models. New positional controller designs, for example, can be tested and optimized on the digital twin, before they are implemented on the real system. Having a complete system model, new methods of space observation can be developed with the digital twin before applying them on the real system.

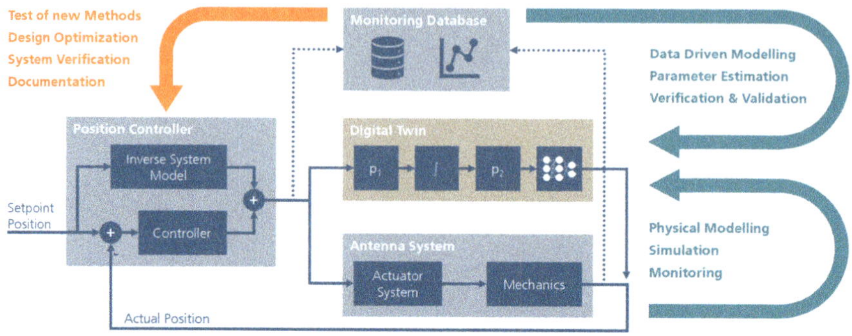

Figure 17. Illustration of the digital twin and model-based design concepts using TIRA's high-precision position control as an example.

To not be tied to a specific modeling tool, the open Functional Mock-up Interface (FMI) standard is used to integrate simulation models into the system as illustrated in Figure 18. Such models can be exported from both open-source and commercial modeling tools.

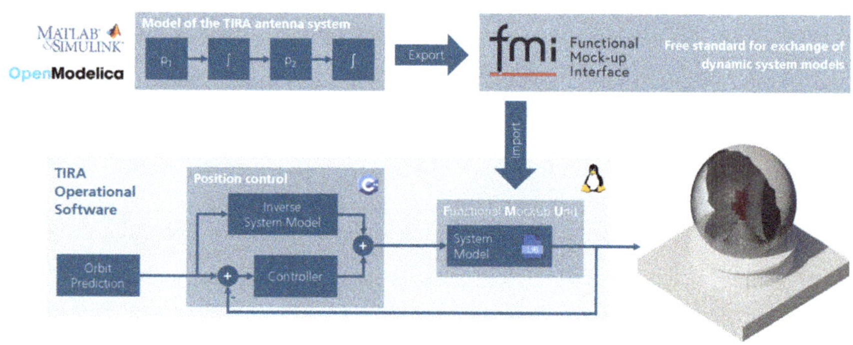

Figure 18. Principle integration of system models into the operational software of TIRA utilizing the open FMI standard.

4.3. Software Architecture

With TIRA being an experimental system, it is expected that parts of the monitoring and control software will be modified frequently. It must be possible to adapt or replace individual software components without affecting the entire system. In particular, changes in one component must not impair the proper functioning of other components. This requires well-defined interfaces and good isolation between the software components. These requirements lead to a microservice-based architecture, which also integrates well with the distributed nature of the various control, monitoring, and signal processing components.

The communication between the services uses the ZeroMQ messaging library and the Protocol Buffers' data serialization format. ZeroMQ offers a high degree of flexibility in the design of the communication architecture, while Protocol Buffers implements an efficient and compact format for structured data. Both technologies are well established and have previously been used in the context of large steerable antennas and telescopes [23,24]. To ensure that the service interfaces are well defined, they are specified using a combination of the Protocol Buffers interface description language and a custom, YAML-based format.

4.4. Access to the Global Reference Frame and Time Scale

The radar measurements of the TIRA instruments are acquired in their native coordinates given by signal delay and Doppler frequency or (pseudo)-range and range rate, respectively, and the monopulse direction of arrival (DOA) estimates of the targets echoes. Besides these measurements, the azimuth and elevation of the mechanical pointing as axis positions of the reflector are logged by the antenna positioning system.

On the other hand, the targets' motion, usually that of space objects, have to be modeled in a global inertial reference system like the International Celestial Reference System (ICRS). To obtain the measurements in the respective ICRS realization, the International Celestial Reference Frame (ICRF), we have to transform the acquired data with the help of auxiliary measurements and through intermediate systems.

TIRA maintains a local time scale operating a cesium beam primary clock to which timestamps and frequencies of the system are tied to give coherent measurements. This local time is attached to coordinated universal time (UTC) via common-view GNSS time transfer or precise point positioning (PPP). As the local clock is clocking a time-transfer global navigation satellite system (GNSS) receiver in the same setup as that maintained by metrology institutions, we can directly tie the local time scale to the UTC (PTB) which is realized in Braunschweig by jointly processing the GNSS raw data of both stations.

Measurements of the radar are given in their respective feed's fixed coordinate frames. The measured propagation time is represented as pseudo-range and referenced to the intersection of the antenna rotational axes. Doppler frequencies are interpreted as range rates. To obtain the geometric range of the target, multiple effects of signal propagation, for example, from special relativity, troposphere, and ionosphere, have to be taken into account. The estimation of the respective models is supported by a meteorological measurement as well as observations of the GNSS permanent station. The monopulse system is calibrated and aligned along with measurements of the one-way and two-way antenna patterns using a terrestrial remote station at a distance of 23 km. Further calibration and validation are established through the in-orbit measurement of suitable objects with known precise ephemerides.

The derived target position in the feed's coordinate system is then transformed to the antenna coordinates with knowledge of the reflectors' pointing vector. In order to access this vector, the mechanical boresight reference of the antenna is given by an optical mirror which was used to align the parabolic reflector panels during initial setup. The orientation of that mirror and thus the pointing of the antenna is measured by encoders on the axes. Measurements on the elevation angles are referenced to the local tangent plane by autocollimation with an automatic level towards the reference mirror; the azimuth encoders are referenced by transferring the normal of the reference mirror to a theodolite on the north monument outside the radome structure via collimation, which then measures the horizontal angle towards a marker on the GNSS station. As this baseline is known by astronomical optical measurements, a direct relation between the azimuth reading and true north is established.

As indicated in Figure 19, three vectors are involved to find the relation between the reflector system and the global terrestrial system. The true orientation of the azimuth axis, the local vertical as normal to the geoid, which is accessed by leveling, and the normal to the ellipsoid as reference in the global system. The deviation between the axis and the local vertical is determined by measurement, while the so-called deflection of the vertical between the reference surface normals are evaluated from the geoid model.

Although the axes of the antenna were aligned to the local vertical or plumb line during construction, the azimuth axis tilted during the first decades, probably due to settlement.

Figure 20 shows the history [25] of the measured deviation of the azimuth axis from the true vertical. The symbol δ denotes the magnitude of the angular deviation in arcseconds, δ_N and δ_E give its components in the north and east direction, respectively. Until recently, the measurement was performed using an automatic level in combination with a parallel plate micrometer observing the vertical movement of a point fixed to the turning azimuth platform over a full revolution. The amplitude and phase of that sinusoidal quantity along the azimuth angle then gave the magnitude and direction of the axis deviation.

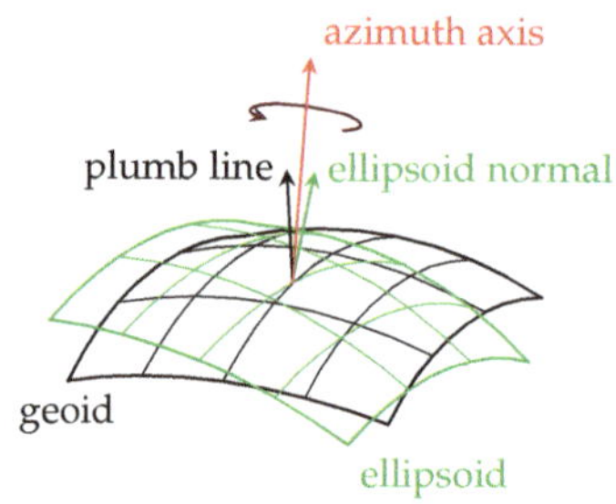

Figure 19. Orientation of the system's azimuth axis in a global reference frame via local measurements against the true vertical.

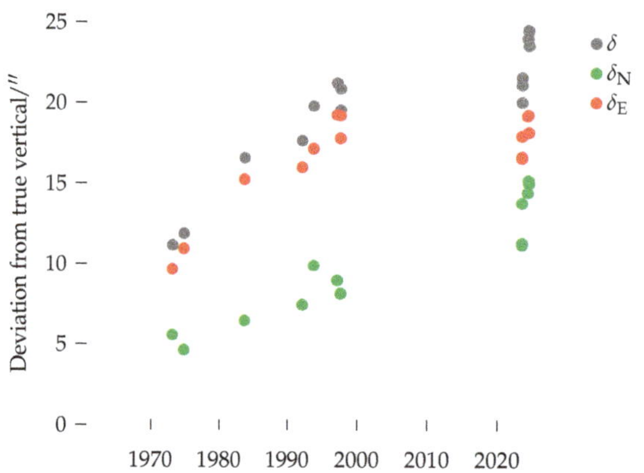

Figure 20. History of the measured deviation of the antenna's azimuth axis from the true vertical.

Today, this deviation is regularly monitored using a high-performance accelerometer with a noise density of $2\,\mu\text{m}\,\text{s}^{-2}\,\text{Hz}^{-0.5}$. Mounted near the center of the azimuth rotation, it directly observes the precession of the vector of gravity, yielding measurements with arcsecond accuracy.

The deflection of the vertical is estimated as a gradient of the German Combined Quasigeoid (GCG2016), yielding $\xi = 5.35''$ in the north–south direction and $\eta = -1.13''$ in the west–east direction, respectively. These values are in good agreement with evaluations by ICGEM [26] using the global EIGEN-6C4 model to degree 2190 giving $\xi = 5.04''$ and $\eta = -1.42''$.

As the target positions are now available in local tangent plane coordinates on the reference ellipsoid of the International Terrestrial Reference Frame (ITRF), they can be directly converted to Cartesian coordinates. The origin of the local tangent plane system is maintained by GNSS-PPP and monitored by GNSS–real-time kinematics (RTK) to the nearest SAPOS station. To estimate satellite positions from measurements in the local system, tidal effects at the ground have to be taken into account to compute the instantaneous station position. Utilizing the Earth rotation parameters (ERPs), these positions [27] may be transformed finally to the inertial ICRF.

5. Tracking Radar

The combination of the dynamic antenna and highly sensitive tracking radar in the L band enables unique capabilities for detection, discrimination, and orbit determination of space objects from the low Earth orbit (LEO) to geosynchronous orbit (GSO) regime. A new signal processing system based on the software-defined radio principle is being developed for the tracking radar to utilize the full potential of the system in meeting future requirements originating from scientific questions relating to space situational awareness.

The resulting system architecture is illustrated in Figure 21. By placing the analog-to-digital boundary as far as possible towards the radio frequency (RF) front end of the radar, all further signal processing steps can be defined in software. In this flexible approach, a modular software architecture enables these processing steps to be replaceable and swappable in the future to expand the implementation for new applications and research.

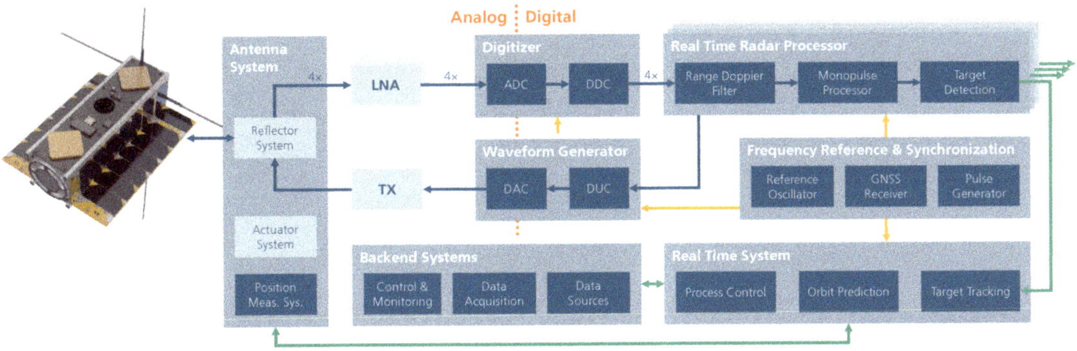

Figure 21. Block diagram of the new system architecture for the TIRA tracking radar implementing the software-defined radio principle (Depicted satellite: Trisat, CC BY-SA 4.0 https://creativecommons.org/licenses/by-sa/4.0, accessed on 29 October 2024, via Wikimedia Commons).

5.1. Early Digitization

One key aspect in the design of the new tracking radar is the early conversion of the RF signal into a discrete signal vector for further processing. This *software-defined radar* approach allows us to flexibly test novel signal processing algorithms in the real-time environment of radar target tracking. For this purpose, the RF signal is sampled directly in the bandpass domain.

To reduce costs and development time, COTS components are procured and customized to our needs. Two synchronized two-channel *ADQ32* digitizer systems from Teledyne SPDevices are used to sample the four channels of the monopulse system in the second Nyquist zone. This system converts the signal to baseband by digital down-conversion (DDC). The following signal processing steps are performed on a graphical processing unit (GPU) to meet real-time requirements for the closed-loop tracking of targets.

As the signal is sampled directly after the receiver protection and the low-noise signal amplification module, the demanding environment of the module location on the moving part behind the apex of the antenna paraboloid (red platform in Figure 1) requires modifications to the cooling concept of the COTS signal conversion module, which was built for a regular workstation computer in a controlled lab environment. This module is integrated to an edge computing device developed by DELL, which itself is housed in a custom-designed 19-inch rack shelter.

5.2. Real-Time Radar Processor

The recording system allows continuous recording of the received signal. The current processing concept features a target acquisition channel searching in a wide range-Doppler band for semi-blind target recognition by Doppler variant pulse compression, as well as several narrowband tracking channels on the sum channel of the monopulse system. This enables multiple target detection and tracking.

For each detected target, a DOA-estimation using the monopulse ratio after evaluating the prompt sample of the monopulse receiving channels can be calculated to estimate range, Doppler, azimuth and elevation angles as observations for further processing in a tracking filter.

6. Imaging Radar

To meet the challenges of the future space situation with more and smaller satellites, it is desirable to improve the system's capabilities to image these satellites. Therefore, the imaging radar's resolution in the range and cross-range direction will be enhanced by increasing the bandwidth and center frequency, respectively. Due to the higher frequency compared to the current imaging system, a higher atmospheric attenuation is expected.

Figure 22 shows the attenuating effect of different atmospheric phenomena, like precipitation, fog, or clouds, and air of various humidity and temperature values across the ITU-assigned IEEE radar bands. The models used are according to ITU recommendations [28–30].

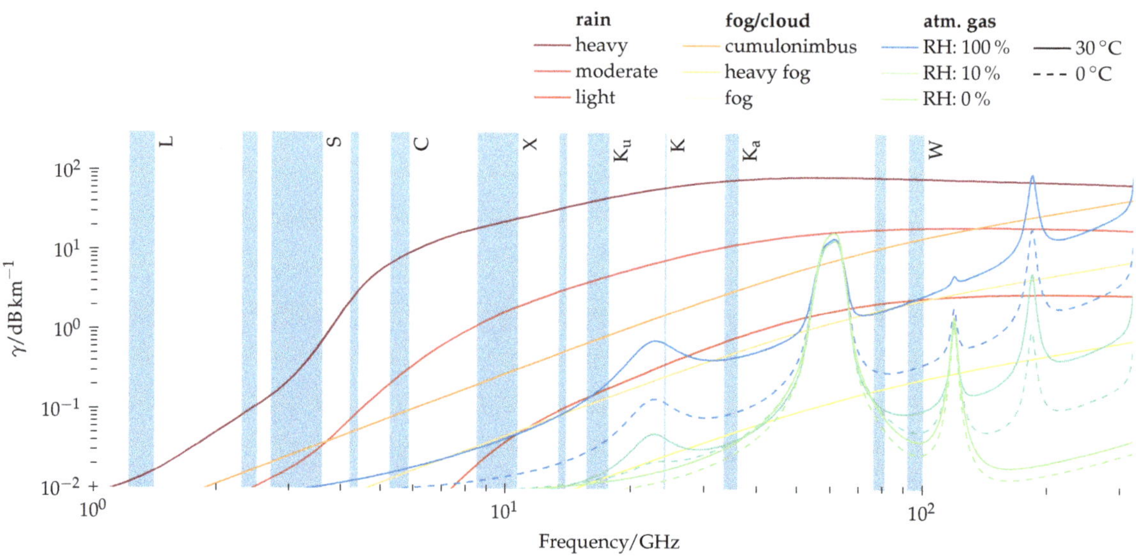

Figure 22. Modeled specific attenuation for precipitation, liquid water, and atmospheric gases with different relative humidity and temperature values along the ITU-assigned IEEE radar bands.

In addition to these challenging conditions, the antenna beam width is reduced by the smaller wavelength-to-diameter ratio of the antenna, which makes it necessary to have a tracking capability for the imaging radar in order to ensure optimum illumination of the target.

These prerequisites lead directly to the following system requirements. To maximize the received response energy, long pulses with a high pulse rate must be transmitted at high transmission power. In contrast to the previous system, a multi-channel receiver must be developed such that the necessary target tracking capability required for antenna guidance by means of monopulse tracking and the two orthogonal polarizations can be measured at the same time while imaging.

Since this sensor is an experimental system, a flexible, extensible signal processing system is implemented, capable of utilizing arbitrary radar waveforms.

6.1. Challenges of Tracking for the New Imaging Radar

The new imaging radar, like its predecessor, requires guidance from the tracking radar. To ensure high-quality imaging, the object must be within the 3 dB beam width of the antenna for a sufficient observation interval. This requirement concerns the angle α between the antenna direction and the line of sight to the satellite, illustrated in Figure 23. Instead, the angle ϑ is the antenna 3 dB beam width, which, for the new imaging radar, is smaller than the current imaging radar due to the higher frequency band. The condition $\alpha \leq \vartheta/2$ must be satisfied to guarantee the object is within the antenna beam.

Figure 24 displays the beam's pointing accuracy for a standard observation of TIRA. Two thresholds for angle α are depicted: the green one for the current imaging radar and the red one for the new imaging radar. Compared to the current imaging radar, the new system, due to its narrower beam width, requires improved tracking capabilities. Therefore, a new tracking filter has to be developed and implemented in the TIRA system.

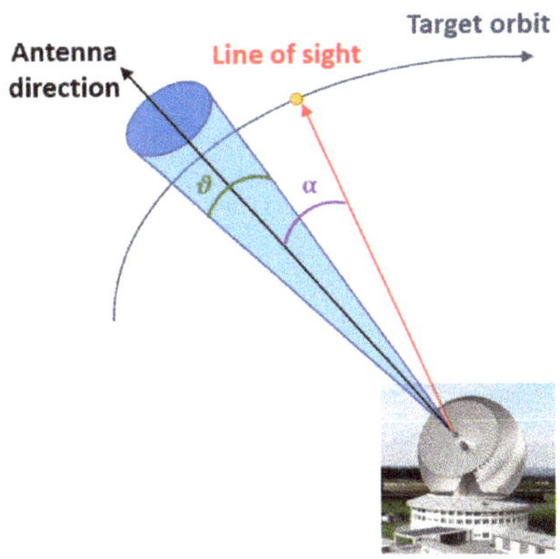

Figure 23. Geometry of the angles ϑ and α. ϑ is the 3 dB beam width of the antenna. α is the angle between the antenna pointing direction and the line of sight to the satellite with the origin at the radar.

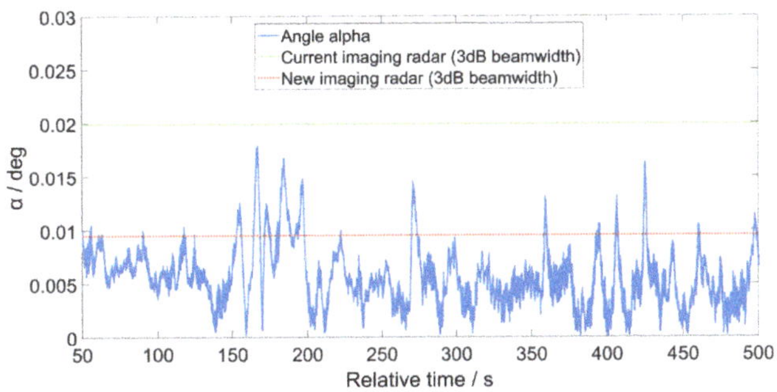

Figure 24. Beam's pointing accuracy of the TIRA system during a standard tracking observation.

As previously mentioned, TIRA measures the range, range rate, azimuth, and elevation of a space object in real time. A sensitivity analysis was performed to understand the impact of the accuracy of each radar observable on the tracking performance. Another investigation was conducted by including an additional parameter, the range-rate rate, as input to the tracking filter. This parameter can be estimated from the phase independently of the other observables [6]. An extended Kalman filter (EKF) was used for the simulations. Results showed that the tracking performance could be strongly improved by increasing the angular accuracy of the radar measurements. Instead, the other parameters (range, range rate, range-rate rate) only increased slightly the tracking performance [31].

A possibility to obtain angular measurements independent of the L-band radar is to take advantage of the imaging radar of the TIRA system directly during tracking. To achieve this, the imaging radar needs to be equipped with a monopulse system. The benefit of this is to provide more accurate angular data than those of the tracking radar, thanks to the higher frequency band. Therefore, it is possible to enhance the tracking performance by combining the information of the two radars of TIRA. A common technique is the multi-sensor data fusion that consists in combining the output of several sensors to describe a single probabilistic state [32].

A first investigation of the multi-sensor data fusion between the tracking and the imaging radar of the TIRA system was performed in [31]. Again, an EKF was used for the simulations. Results confirmed that the multi-sensor data fusion could be a solution to achieve a more accurate tracking system by ensuring the target object within the 3 dB beam width of the antenna of the future imaging radar.

Currently, different scenarios are being investigated based on the idea of a two-step tracking filter. It consists of two tracking filters running in parallel: an EKF and a batch least-squares filter. While the EKF tracks the object in real-time, the batch least-squares filter works in near real time by collecting data over a certain period of time. Every time the batch filter provides an output, it is injected in the EKF with the purpose of improving the tracking performance. Some first results are shown in [33]. Lastly, the implementation of an antenna steering function is needed to regulate the antenna pointing direction over time.

6.2. Challenges in High-Bandwidth Pulsed Radars

For ground-based imaging and tracking radars for space observation, the combination of stability, high transmit power, and bandwidth is essential to achieve long-range detection, a high resolution, and accurate velocity measurements.

In radar, the range resolution is directly related to the signal bandwidth, and a higher bandwidth therefore improves the spatial resolution in radar imaging while a high-power transmitter provides the SNR necessary for both accurate tracking and imaging.

Since the relative bandwidth available from high-power microwave tube amplifiers is in the order of 5 to 15 %, it is clear that an enhancement of the range resolution requires a shift of the TIRA system imaging radar to a higher frequency band.

6.3. Real-Time Processing

Due to the narrow antenna lobe in the Ka band, the imaging radar is planned to also have a tracking capability to ensure optimum target illumination. Therefore, real-time processing of the acquired signals is required. Since arbitrary waveforms should be used, pulse compression using the deramping technique is not feasible. In order for any pulse compression filter to be applied, the received signal must be sampled in its entire bandwidth. Maximizing the pulse energy by increasing the pulse length to the maximum limit of the amplifier gives us a large number of samples per pulse which have to be Fourier transformed using a fast Fourier transform (FFT) for fast filtering. The calculation of one thousand FFTs per second with a size of about one million points is not reasonable to implement in a field-programmable gate array (FPGA). On the other hand, GPUs are very flexible and capable of handling these calculations on large datasets. The selected GPU-based solution enables a flexible data processing chain that implements pulse compression and tracking filters and even real-time ISAR imaging algorithms.

6.4. Ka-Band Transmitter

The transmission of microwave power, particularly in the higher frequency bands, is constrained by the availability of sufficiently powerful microwave amplifiers. To achieve a greater transmitted power than that permitted by commercially available high-power amplifiers, the output power of two electron vacuum tubes will be combined.

In addition to the RF high-power amplifier, an appropriate antenna front end is required that can handle and transmit the high power and bandwidth to the antenna. At the same time, it implements a duplexer that separates the sensitive receiver channel from the high-power transmission channel. Several radar systems have demonstrated that quasi-optical antenna front ends can handle the highest RF power and minimize the microwave losses compared to traditional waveguide-based systems [34–36].

In essence, a quasi-optical front end is a quasi-optical transmission line combined with a quasi-optical duplexer. The quasi-optical transmission line is a configuration of focusing mirrors that periodically refocus a diverging Gaussian microwave beam [37]. The quasi-optical duplexer uses the polarization of the Gaussian beam to separate the TX and RX signals. Key component of the duplexer is a quasi-optical Faraday rotator, which is a nonreciprocal polarization rotation device that can be used together with polarizing grids to construct the duplexer [38,39].

For the TIRA system upgrade, a quasi-optical antenna front end suitable for fully polarimetric radar operation (see Section 3.2) with a broad bandwidth and high transmission power will be implemented. A block diagram of the antenna front-end concept is shown in Figure 25.

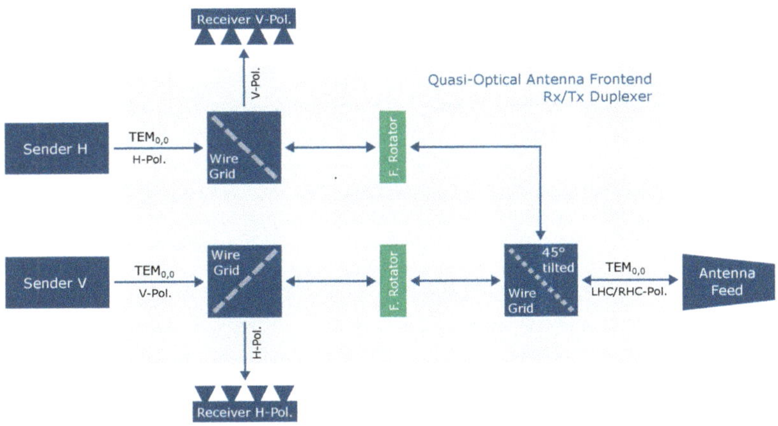

Figure 25. Principle block diagram of the high-power polarimetric quasi-optical monopulse front end.

The radar signal is individually amplified by the two high-power vacuum tube amplifiers and launched by two horn antennas as fundamental Gaussian beams ($TEM_{0,0}$ mode) into the quasi-optical antenna front end which guides the RF to the antenna. In the transmission channel, the linearly polarized Gaussian beams are transmitted through wire grids following the horn antennas. The subsequent Faraday rotators are designed to rotate the polarization plane by 45°. In combination with the polarization-splitting wire grids, the Faraday rotators function as an efficient quasi-optical duplexer (see [34,36,39] for details). At the third wire grid, the power signals from both high-power amplifiers are combined.

The existing antenna geometry prevents direct quasi-optical coupling into the antenna as implemented, e.g., in [34]. Therefore, the power-combined signal is coupled by two adjustable mirrors into an oversized corrugated circular waveguide, which is used as the antenna feed due to limited space. However, oversized corrugated circular waveguides permit low-loss signal transmission similar to a quasi-optical beam line and support comparable high power levels [40].

The presented quasi-optical duplexer approach, which utilizes Faraday rotators and polarization-splitting wire grids, enables the simultaneous transmission and reception of RF signals with arbitrary polarization. The polarization of the transmitted signal is determined by the relative phase between the two high-power signals when they are combined at the third wire grid. This feature allows for the polarization of the transmit signal to be freely selected and switched from pulse to pulse or even within a pulse. The ability to freely select polarization with inter-pulse switching is a unique feature of the new TIRA system that will enable fully polarimetric measurements in the future (see Section 3.2).

As previously mentioned (see Section 6.3), target tracking for antenna guidance is required simultaneously to the imaging for the new Ka-band system. Therefore, in addition to its imaging and polarimetric radar capabilities, the quasi-optical antenna front end is designed to support simultaneous multimode monopulse tracking [34,41]. For this, the quasi-optical components are designed to support the propagation of the relevant higher-order Gaussian modes in the antenna front end, and a special multimode RX antenna similar to [42] is under development.

A further advantage of the quasi-optical duplexer are the low reflections in the transmit path and therefore a high isolation between the transmit and receive channels of $> 40\,\text{dB}$. This reduces the demands on the power-limiters installed upstream of the receiver electronics, which are only required to absorb a small amount of power during transmission. Consequently, limiters with low transmission losses can be used, resulting in an increased SNR.

7. Conclusions

The space observation radar TIRA is undergoing a transformative upgrade in order to fulfill upcoming requirements for space domain awareness, space safety, and space sustainability. For this, the current imaging radar working in the Ku band will be replaced by a new imaging radar which will operate in the Ka band. With this new frequency band, a higher bandwidth will enable a higher spatial resolution for the radar images. The new Ka-band amplifiers will also provide significantly increased transmit power compared to the current Ku-band system. Together with the higher center frequency, a gain in sensitivity will be reached. Please note that some radar parameters are classified and cannot be listed in this paper. Another benefit of the new imaging radar will be the availability of fully polarimetric radar images. This will allow a much more detailed analysis of space objects. In addition, the complete processing control infrastructure of the current tracking radar in the L band is facing a major overhaul, which will enable continuous incremental upgrades in the future through a modular and open design. Tracking will be carried out in a new combined mode fusing the data acquired by the tracking and the imaging radars. Also, new signal processing techniques are under development to improve the characterization of space objects by improving the imaging quality and exploiting the polarimetric information.

Author Contributions: Writing—original draft, review and editing, J.K., F.B., C.C., D.C.-M., L.F., C.H., V.K., M.L., A.M., R.P., S.R. and F.R. All authors have read and agreed to the published version of the manuscript.

Funding: The projects detailed in this article are funded by the German Federal Ministry of Defence (BMVg).

Data Availability Statement: The original contributions presented in the study are included in the article, further inquiries can be directed to the corresponding author.

Conflicts of Interest: The authors declare no conflicts of interest.

Abbreviations

The following abbreviations are used in this manuscript:

ADC	Analog-to-digital conversion
ARCADIA	ARChitecture Analysis and Design Integrated Approach
ASAT	Anti-satellite
COTS	Commercial off-the-shelf
DAC	Digital-to-analog conversion
DDC	Digital down-conversion
DOA	Direction of arrival
DUC	Digital up-conversion
ERPs	Earth rotation parameters
ESA	European Space Agency
FFT	Fast Fourier transform

FHR	Fraunhofer Institute for High Frequency Physics and Radar Techniques
FMI	Functional Mock-up Interface
FPGA	Field-programmable gate array
GEO	Geosynchronous equatorial orbit
GNSS	Global navigation satellite system
GPU	Graphical processing unit
ICRF	International Celestial Reference Frame
ICRS	International Celestial Reference System
IOD	Initial orbit determination
ISAR	Inverse synthetic aperture radar
ITRF	International Terrestrial Reference Frame
LEO	Low Earth orbit
MBSE	Model-based system engineering
MEO	Medium Earth orbit
ONERA	Office national d'études et de recherches aérospatiales
PPP	Precise point positioning
PRF	Pulse repetition frequency
RCS	Radar cross section
RF	Radio frequency
RSO	Resident space object
RTK	Real-time kinematics
SAPOS	Satellite positioning service of the German surveying authorities
SNR	Signal-to-noise ratio
SRP	Solar radiation pressure
SSA	Space situational awareness
TIRA	Tracking and Imaging Radar
TLE	Two-line-element
UTC	Coordinated universal time

References

1. Mehrholz, D. Radar techniques for the characterization of meter—Sized objects in space. *Adv. Space Res.* **2001**, *28*, 1259–1268. [CrossRef]
2. Mehrholz, D.; Leushacke, F.W.; Jehn, R.; Klinkrad, H.; Landgraf, M. Detecting, tracking and imaging space debris. *ESA Bull.* **2002**, *109*, 128–134.
3. Muntoni, G.; Montisci, G.; Pisanu, T.; Andronico, P.; Valente, G. Crowded Space: A Review on Radar Measurements for Space Debris Monitoring and Tracking. *Appl. Sci.* **2021**, *11*, 1364. [CrossRef]
4. Sommer, S.; Rosebrock, J.; Cerutti-Maori, D.; Leushacke, L. Temporal analysis of ENVISAT's rotational motion. In Proceedings of the 7th European Conference on Space Debris 2017, Darmstadt, Germany, 18–21 April 2017; Volume 7.
5. Sommer, S.; Karamanavis, V.; Schlichthaber, F.; Patzelt, T.; Rosebrock, J.; Cerutti-Maori, D.; Leushacke, L. Analysis of the attitude motion and cross sectional area of TIANGONG-1 during its uncontrolled re-entry. In Proceedings of the 1st NEO and Debris Detection Conference, Darmstadt, Germany, 22–24 January 2019; Volume 1, p. 469.
6. Cerutti-Maori, D.; Rosebrock, J.; Carloni, C.; Budoni, M.; Maouloud, I.; Klare, J. A novel High-Precision Observation Mode for the Tracking and Imaging Radar TIRA–Principle and Performance Evaluation. In Proceedings of the 8th European Conference on Space Debris, ESA/ESOC, Darmstadt, Germany, 20–23 April 2021.
7. Karamanavis, V.; Dirks, H.; Fuhrmann, L.; Schlichthaber, F.; Egli, N.; Patzelt, T.; Klare, J. Characterization of deorbiting satellites and space debris with radar. *Adv. Space Res.* **2023**, *72*, 3269–3281. [CrossRef]
8. Carloni, C.; Cerutti-Maori, D.; Budoni, M.; Rosebrock, J. Performance analysis of tracking filters according to input data processing. In Proceedings of the 8th European Conference on Space Debris, Darmstadt, Germany, 20–23 April 2021; Volume 8.
9. Leushacke, L.; Czeslik, C. High doppler resolution monopulse techniques for the radar observation of targets in the geosynchronous ring. In Proceedings of the Proc. DGON/ITG International Radar Symposium, Munich, Germany, 15–17 September 1998; Volume 1, pp. 457–464.
10. The End-of-Life Services by Astroscale Demonstrator (ELSA-d). Available online: https://astroscale.com/missions/elsa-d/ (accessed on 29 October 2024).
11. Cerutti-Maori, D.; Carloni, C.; Rosebrock, J.; Maouloud, I. Improving the Performance of the Space Observation Radar TIRA through Dedicated Signal Processing Techniques and Advanced Experimental Modes. In Proceedings of the Sensors and Electronics Technology Panel and Systems Concepts and Integration Panel Specialists Meeting SET-SCI-297: "Space Sensors and Space Situational Awareness", Interlaken, Switzerland, 10–11 October 2022. [CrossRef]
12. Chen, V.C.; Martorella, M. *Inverse Synthetic Aperture Radar Imaging: Principles, Algorithms and Applications*; Radar, Sonar and Navigation, Institution of Engineering and Technology; Wiley: Hoboken, NJ, USA, 2014.

13. Final Images Before Aeolus Demise. Available online: https://www.esa.int/Space_Safety/Space_Debris/Final_images_before_Aeolus_demise (accessed on 29 October 2024).
14. Kak, A.C.; Slaney, M. *Principles of Computerized Tomographic Imaging*; Society for Industrial and Applied Mathematics: Philadelphia, PA, USA, 2001. [CrossRef]
15. Cerutti-Maori, D.; Carloni, C.; Rosebrock, J.; Siminski, J. Observation of COSMOS-1408 Debris Cloud with the Tracking and Imaging Radar (TIRA) System. In Proceedings of the 2nd NEO and Debris Detection Conference, Darmstadt, Germany, 24–26 January 2023.
16. Available online: https://web.archive.org/web/20220225183304/https://www.yuzhnoye.com/en/company/history/electronic-surveillance-spacecraft.html (accessed on 17 January 2024).
17. Serrano, A.; Kobsa, A.; Uysal, F.; Cerutti-Maori, D.; Ghio, S.; Kintz, A.; Morrison, R.L. Jr.; Welch, S.; van Dorp, P.; Hogan, G.; et al. Long baseline bistatic radar imaging of tumbling space objects for enhancing space domain awareness. *IET Radar Sonar Navig.* **2024**, *18*, 598–619. [CrossRef]
18. Available online: https://commons.wikimedia.org/wiki/File:Landsat-9_Centaur_1_(cropped).jpg (accessed on 29 October 2024).
19. Available online: https://www.esa.int/Space_Safety/Space_Debris/Space_debris_by_the_numbers (accessed on 15 August 2024).
20. Available online: https://ourworldindata.org/grapher/yearly-number-of-objects-launched-into-outer-space (accessed on 21 August 2024).
21. Van Zyl, J.; Kim, Y. Mathematical representations of scatterers. In *Synthetic Aperture Radar Polarimetry*; Yuen, J.H., Ed.; Wiley: Hoboken, NY, USA, 2011; p. 29.
22. Voirin, J.L. (Ed.) *Model-Based System and Architecture Engineering with the Arcadia Method*; Implementation of model based system engineering set; ISTE Press: London, UK, 2018.
23. Creager, R.; Whitehead, M. Refactoring GBT software to support high data rate instruments using data streaming technology. *Softw. Cyberinfrastructure Astron. III* **2014**, *9152*, 91520X. [CrossRef]
24. Lyard, E.; Walter, R.; Kosack, K.; Jacquemier, J.; Oya, I.; Wegner, P.; Fuessling, M.; Wu, X.; Consortium, F. Modern middleware for the data acquisition of the Cherenkov Telescope Array. *arXiv* **2015**, arXiv:1508.06473.
25. Spata, M.; Ruf, B.; Mehrholz, D.; Steinheuer, B.; Buck, G. Die Bestimmung der Großradaranlage des FGAN-Forschungsinstituts in Werthhoven im Bezugssystem ETRS 89. *Vermess. Und Raumordn.* **1998**, *60*, 430–437.
26. Ince, E.S.; Barthelmes, F.; Reißland, S.; Elger, K.; Förste, C.; Flechtner, F.; Schuh, H. ICGEM—15 years of successful collection and distribution of global gravitational models, associated services, and future plans. *Earth Syst. Sci. Data* **2019**, *11*, 647–674. [CrossRef]
27. Petit, G.; Luzum, B. (Eds.) *IERS Conventions (2010)*; Number 36 in IERS technical note; Verl. des Bundesamtes für Kartographie und Geodäsie: Frankfurt am Main, Germany, 2010.
28. International Telecommunication Union. *ITU-R P.676: Attenuation by Atmospheric Gases and Related Effects*; International Telecommunication Union: Geneva, Switzerland, 2019.
29. International Telecommunication Union. *ITU-R P.840: Attenuation due to Clouds and Fog*; International Telecommunication Union: Geneva, Switzerland, 2019.
30. International Telecommunication Union. *ITU-R P.838: Specific Attenuation Model for Rain for Use in Prediction Methods*; International Telecommunication Union: Geneva, Switzerland, 2005.
31. Carloni, C.; Cerutti-Maori, D.; Budoni, M.; Rosebrock, J.; Maouloud, I.; Stoll, E. Improving the tracking performance of space observation radar systems. *Adv. Space Res.* **2024**, *73*, 6066–6078. [CrossRef]
32. DeMars, K.; McCabe, J.; Darling, J. Collaborative Multi-Sensor Tracking and Data Fusion. In Proceedings of the 5th AAS/AIAA Space Flight Mechanics Meeting, Williamsburg, VA, USA, 11–15 January 2015.
33. Carloni, C.; Cerutti-Maori, D.; Maouloud, I.O.; Stoll, E. A novel tracking approach based on data fusion for multi-frequency space observation radar systems. *Acta Astronaut.* 2024, *to be submitted*.
34. Fitzgerald, W.D. A 35-GHz beam waveguide system for the millimeter-wave radar. *Linc. Lab. J.* **1992**, *5*, 245–272.
35. Stambaugh, J.J.; Lee, R.K.; Cantrell, W.H. The 4 GHz bandwidth millimeter-wave radar. *Linc. Lab. J.* **2012**, *19*, 64–76.
36. Jamnejad, V.; Long, E.; Durden, S. Design of a quasi optical transmission line for Cloud and precipitation radar system of ACE mission. In Proceedings of the 2015 IEEE Aerospace Conference, Big Sky, MT, USA, 7–14 March 2015; pp. 1–6. [CrossRef]
37. Goldsmith, P.F. Gaussian Beams and Antenna Feed Systems. In *Quasioptical Systems: Gaussian Beam Quasioptical Propagation and Applications*; Wiley-IEEE Press: Hoboken, NJ, USA, 1998; pp. 125–156. [CrossRef]
38. Webb, M.R. A mm-wave four-port quasi-optical circulator. *Int. J. Infrared Millim. Waves* **1991**, *12*, 45–63. [CrossRef]
39. Hunter, R.I.; Robertson, D.A.; Goy, P.; Smith, G.M. Design of High-Performance Millimeter Wave and Sub-Millimeter Wave Quasi-Optical Isolators and Circulators. *IEEE Trans. Microw. Theory Tech.* **2007**, *55*, 890–898. [CrossRef]
40. Clarricoats, P.; Saha, P. Attenuation in corrugated circular wave-guide. *Electron. Lett.* **1970**, *6*, 370–372. [CrossRef]

41. Clarricoats, P.J.B.; Elliot, R.D. Multimode corrugated waveguide feed for monopulse radar. *IEE Proc. Microwaves Opt. Antennas* **1981**, *128*, 102. [CrossRef]
42. Dubrovka, F.F.; Piltyay, S.; Ovsyanik, Y.A.; Dubrovka, R. Eight-channel directional coupler of orthogonal H21 modes in circular waveguide for X-band quasi-monopulse antenna systems. *Radioelectron. Commun. Syst.* **2020**, *63*, 656–665. [CrossRef]

remote sensing

MDPI

Article

ATLAS: Latest Advancements and First Observations

João Pandeirada [1,2,*], Miguel Bergano [3], Paulo Marques [2,4], Bruno Coelho [3,5], Domingos Barbosa [6] and Mário Figueiredo [1,2]

1 Instituto Superior Técnico, Universidade de Lisboa, 1049-001 Lisboa, Portugal; mario.figueiredo@tecnico.ulisboa.pt
2 Instituto de Telecomunicações, 1049-001 Lisboa, Portugal; pmarques@isel.pt
3 ATLAR Innovation, 3320-229 Pampilhosa da Serra, Portugal; miguel.bergano@atlar.pt (M.B.); bruno.coelho@atlar.pt (B.C.)
4 Instituto Superior de Engenharia de Lisboa, 1959-007 Lisboa, Portugal
5 CFisUC, Departamento de Física, Universidade de Coimbra, 3004-516 Coimbra, Portugal
6 Instituto de Telecomunicações, Universidade de Aveiro, 3810-164 Aveiro, Portugal; dbarbosa@av.it.pt
* Correspondence: joao.pandeirada@tecnico.ulisboa.pt

Abstract: The increasing amount of space debris poses a significant threat to operational satellites and space-based services. This article updates the community on the current status of the development of ATLAS, a tracking radar that is part of the EUSST network and aims to detect space objects in low Earth orbits. This article focuses on the latest activities performed: calibration of the pointing system and initial observations of space objects. The calibration procedure consisted of cross-scanning the Solar disk and yielded great results, obtaining an offset of 5.3° in azimuth and 0.10° in elevation. The first observation campaign resulted in 33 range detections of the International Space Station (ISS) with a probability of false alarm of 10^{-9}. The observations were then used to readjust the radar equation to assess the real-world performance of the system.

Keywords: SST; radar; tracking; pulse; Doppler; ISS; satellite; LEO

1. Introduction

Since the beginning of the Space Age in the 1950s, humanity has launched an increasing amount of objects into the Earth's orbit [1]. A consequence of the unregulated launch of objects into orbit is the increasing amount of space debris—defunct, non-operational man-made objects and fragments that orbit the Earth. Many space-based assets are the backbone of an array of services essential for the progress and safety of economies and societies, such as communication, navigation, and Earth observation. Space debris poses a significant threat to these assets, as their presence can lead to collisions with active satellites.

To mitigate the risk of collision and ensure the long-term operation of current satellites, many organizations have created sensor networks comprising telescopes, radars, and lasers to monitor the orbits of space objects and warn satellite operators. This proactive approach is central to the field of Space Situational Awareness (SSA) and Space Surveillance and Tracking (SST) and is vital for safeguarding the interests of nations and upholding international agreements set forth by the United Nations for the peaceful use of outer space [2]. In this context, understanding and addressing the issue of space debris has become imperative for the safe and sustainable utilization of Earth's near-space environment.

In the European effort, there are mainly two initiatives: the ESA SSA Programme and European Space Surveillance and Tracking (EUSST). The ESA SSA Programme is a multifaceted initiative with many objectives, including monitoring space weather, tracking near-Earth objects, cataloging space debris, and providing essential services like satellite collision avoidance [3]. EUSST is a support framework designed to establish an independent monitoring network dedicated to detecting and tracking space debris, as well as issuing alerts when preventative maneuvers may be required [4]. The primary objective of EUSST is to create an SST capability at the European level, ensuring a suitable degree of European self-sufficiency. Portugal is a participant in the EUSST program, actively developing expertise and capabilities in both optical and radar sensors to contribute to this initiative [5,6].

This paper updates the community on the current status of the development of ATLAS (rAdio TeLescope pAmpilhosa Serra), a monostatic tracking radar that is part of the EUSST network. Section 1 briefly explains the problem of space debris. Section 2.1 introduces ATLAS, explains how it has evolved since its infancy in 2020, and outlines its current operational status. Section 2.2 provides a technical summary of the transmitting and receiving chains. Section 2.3 explores the task of tracking space objects with ground-based radar, the main constraints and problems that arise, and how to approach them to perform successful observations. Section 2.4 is dedicated to explaining the procedure used in ATLAS to calibrate the pointing system. Section 2.5 explains the signal model assumed and used to develop the signal processing chain. Section 2.6 is dedicated to the signal processing chain: echo processing and detection. Section 3 showcases the results obtained during the calibration and first space object observations. Section 4 is dedicated to discussing the results and drawing conclusions from them. Section 5 concludes this article and reveals the next steps in the development of ATLAS.

2. Materials and Methods

2.1. ATLAS: Overview and Current Status

ATLAS was designed to consolidate national resources in preparation for the establishment of an SST sensor network. In 2019, Instituto de Telecomunicações initiated a major retrofit of its 9-m Cassegrain Antenna located in Pampilhosa da Serra Space Observatory (PASO), Portugal, to attach ATLAS, a complete radar system operating in the C-band (5.56 GHz) (Figure 1). A comprehensive technical overview of the system was presented in [5], accompanied by simulation studies on its expected operational performance. At the time of publication [5] in 2021, the generation and processing of the baseband signals were being thoroughly tested in a laboratory. These experiments led to several studies on the design of waveforms and signal processing chains for the system [5] (Section 4) [7]. The studies culminated in the first real-life scenario tests, where the whole system (except the antenna and power amplifiers) was tested in a setup with a major reflector at a well-defined distance [8].

Figure 1. Antenna located in PASO, with the radar system inside the hub.

After completing the preliminary functional tests in the lab and the field, we proceeded to the deployment phase. The deployment consisted of connecting the output of the power amplifier to the antenna feed through a polarizer. The polarizer underwent a matching procedure to maximize the isolation between the input/output ports and minimize reflections [9]. As part of the developments in 2022, we also presented the ATLAS Cloud Service API (ACSA) in [9]. ACSA enables monitoring and controlling of the radar through a web-based API, allowing the system to be accessed by multiple clients remotely. PASO houses many other sensors and systems [6,10], allowing ATLAS to act as a complete SST ground station with the following features:

- Monitoring and prediction of weather and sky conditions [11];
- Generation of schedules for optimal observation of a given catalog of space objects;
- Data fusion of optical and radar observations [12];

2.2. ATLAS: Technical Summary

ATLAS is divided into a transmitting part and a receiving part. The transmitter, illustrated in a simplified depiction in Figure 2, is the subsystem responsible for generating and outputting the RF signal. Signal generation commences with a frequency synthesizer that generates the carrier frequency set at 5.56 GHz. This carrier signal is subsequently directed into a PIN modulator, which modulates the carrier in amplitude. The PIN modulator is controlled by a user-defined pulse shape. Following modulation, the signal proceeds to the power amplifier, where its peak power is adjustable within a range spanning from 155 W to 2 kW, facilitating versatile power configurations. For the observations shown in this article, the peak power was set to 1.6 kW. Table 1 summarizes all relevant parameters of the transmitter chain.

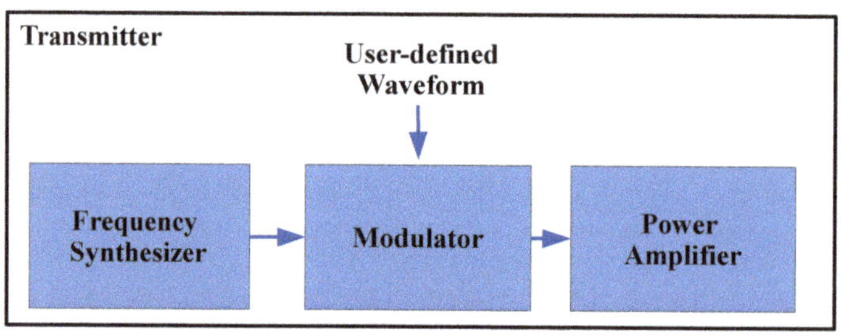

Figure 2. Functional diagram of the transmitter subsystem.

Table 1. ATLAS transmitter characteristics.

Operating Frequency	5. 56 GHz
Peak power	1.6 kW
Waveform	AM Chirp [5] (Section 4)
Pulse repetition frequency	33 Hz
Number of pulses	100
Emitter bandwidth	2 MHz

The receiver, depicted in Figure 3, includes an analog chain that is responsible for the demodulation and amplification of the received echoes. It starts with a low-noise amplifier (LNA), followed by a downconversion to a 400 MHz intermediate frequency (IF) and filtering (80 MHz of bandwidth). The local oscillator is provided by the same reference signal from the transmitter, guaranteeing a coherent conversion. The IQ detector converts the signal to a baseband and separates it into the I/Q components, which are then digitized at a maximum sample rate of 100 MS/s and stored in a computer for further processing. Table 2 summarizes all relevant parameters of the receiver chain.

Table 2. ATLAS receiver characteristics.

LNA noise figure (<15 °C) (N_f)	0.7 dB
IF filter BW (B)	80 MHz @ −3dB
Maximum sample rate	100 MS/s

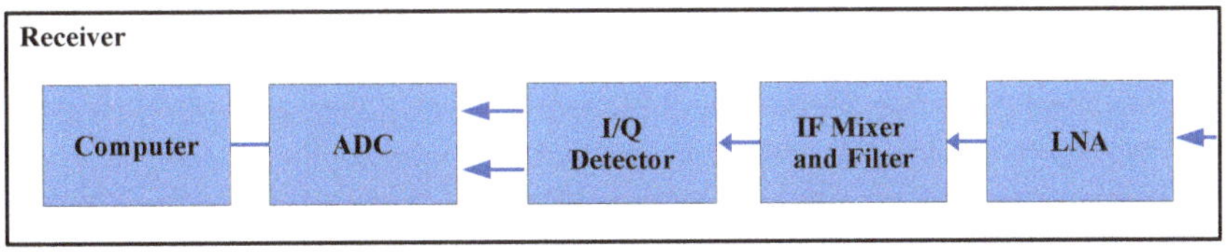

Figure 3. Functional diagram of the receiver subsystem.

2.3. Space Object Tracking in LEO: Error Sources and Observation Strategy

Observing space objects in low Earth orbit (LEO) with a monostatic tracking radar is no trivial task due to the dynamics between the radar and the space objects and because of the multiple sources of error that arise. Tracking radars feature narrow beamwidths to increase the resolution and signal-to-noise ratio (SNR). On the other hand, objects in LEO have very high velocities [13]. These two facts make the window of opportunity for observation very short. Errors in both the pointing system and the orbit prediction can easily lead to missing objects.

The most complete source of orbital information is the US Two-Line Element (TLE) catalog. Given the sheer number of objects present in this catalog and the speed at which the orbital information is updated, the majority of SST stations, including ours, use it to perform predictions of the objects to observe.

The correct method for propagating TLE orbital data is by using an SGP propagator, with the most popular one being the SGP4 [14]. Although using TLE+SGP4 is a fast, convenient, and sufficient approach for other use cases, our experience suggests that it is less than ideal for uncooperative tracking in LEO given the errors associated with the usage of mean elements and simplified propagation models. Accuracy depends on sensor quality, data quantity, and space conditions, none of which are disclosed by the US Space Surveillance Network. ESA published a notable article, where it was estimated that TLE orbit errors in LEO were around 0.102 km (radial), 0.471 km (along-track), and 0.126 km (out-of-plane) [14]. These errors increase as we propagate further from the epoch, but it is difficult to quantify how much. TLE data do not include any information related to maneuvers, making it very difficult to predict a satellite's position after one. The history of the TLE sets of a given object can be used to understand whether it has undergone maneuvers, which can be useful information when dealing with uncooperative satellites [15].

Inaccurate geo-referencing of the antenna homing system may lead to biases in the azimuth and elevation measurements. Once the homing system is correctly geo-referenced, a finer calibration can be performed using celestial sources like the Sun/Moon to further decrease the offset errors [16]. Periodic calibration routines involving celestial sources should be performed to ensure the pointing system remains accurate.

The selection of resident space objects (RSOs) to track during the calibration phase can also play an essential role in mitigating these sources of error. The International Space Station (ISS) serves as an ideal candidate for initial measurements due to its significantly large Radar Cross-Section (RCS), approximately 400 m^2, and its ability to orbit within the lower LEO region (below 500 kilometers in altitude), resulting in the reception of higher-energy echoes compared to other RSOs. Position and velocity state vectors for the ISS are provided by NASA and are publicly available [17,18]. These vectors are listed at four-minute intervals over the last 15 days, and during maneuvers, the vectors are reported at two-second intervals [18]. The ephemeris files also contain other important information, such as the mass of the ISS, drag area, drag coefficient, solar radiation area, and solar radiation coefficient, allowing orbit propagation with special perturbation techniques [19].

Other LEO operational satellites such as CRYOSAT-2 and Jason-3 are equipped with telemetry and retroreflectors, which provide precise ephemeris information [20]. This information can be used for highly accurate predictions of their passages, offering valuable data for both calibration procedures and evaluating the performance of ground-based sensors [21]. Passive calibration satellites, like STELLA and STARLETTE, were designed to have high reflectivity and spherical shapes, ensuring that their RCS remains independent of the viewing angle [22]. They play a critical role in fine-tuning RCS measurements obtained by the sensors, as they provide a reliable reference point for RCS values.

In light of the foregoing discussion, the following are some key ideas that every SST ground station should be aware of:

- Use the most recent TLE sets available to reduce propagation errors and unaccounted maneuvers;
- Perform regular calibration routines using celestial sources;
- Use well-behaved satellites that provide precise ephemeris information and/or constantly updated TLEs to refine the pointing system;

Finally, the observation strategy used can also help mitigate some of the aforementioned errors. We suggest implementing a semi-surveillance approach. This approach entails positioning the antenna at a predicted location in the sky for the target satellite and illuminating it both before and after the expected time of the object's passage within the antenna's beamwidth. By adopting this method, the likelihood of successfully capturing the target increases, and the resulting echoes yield valuable insights into how to rectify discrepancies in orbit predictions and offsets of the system. By assuming a symmetric illumination time before and after the expected time of passage, it is expected that the echoes will appear centered around the middle of the illumination time window. By assuming a range acquisition window centered on the object's expected range, the echoes are expected to appear centered in the range window. Deviations from the center of the range-pulse map can be used to compensate for pointing, timing, and orbit prediction errors.

2.4. Calibration Procedure

Calibration of both the radar performance and the pointing system requires well-characterized and stable sources of electromagnetic energy, preferably located in the far-field region of the antenna. The far field of an antenna is the distance from the source necessary to approximate the electromagnetic wavefronts as approximately planar [23] (Chapter 1). This condition is met if the distance from the source to the aperture is at least $\frac{2D^2}{\lambda}$. This approximation is typically used as a condition in the radar equation, so we need to ensure that targets are in this region. The far-field distance of ATLAS is approximately 3 km, so we consider all satellites in orbit and celestial bodies (including the Sun) to meet this criterion.

The use of EM energy from the Sun is a well-established method for checking antenna alignment and receiver sensitivity, and it is quite common for manufacturers to provide Sun calibration tools as part of their software packages. In [24], the authors presented a method for locating the Sun's coordinates and determining the pointing offsets. A method for determining the true elevation of a target using observations of the Sun was introduced in [25]. In [26], the authors presented an improved method for pointing calibration with the Sun using operational weather radar scans made at very low elevation angles. When calibrating radar using the Sun, the radar is, in fact, used as a radiometer. The transmitter is not used since we expect no echo from the Sun due to its distance from the Earth [27]. Because of this, only the receiving chain can be calibrated using the Sun, and to calibrate the whole system, other techniques have to be used, such as the ones explained in Section 2.3.

The Sun is one of the preferred candidates for calibration because of the following:

- It can be regularly observed from any point on Earth;
- Its position in the sky can be precisely calculated at any time and location [26,28];
- It radiates at all wavelengths, and the radiation is generally unpolarized [29];
- The solar flux incident on the Earth's surface varies between 100 and 300 solar flux units. The solar flux is measured constantly by many stations worldwide, and these measurements are public;

The diameter of the Sun in the visible wavelengths is approximately 695,700 km, as defined by the International Astronomical Union (IAU) [30], which results in an apparent arc of approximately 0.53°. At centimeter wavelengths, this value is different, and an effective RF diameter can be obtained [31]. For ATLAS's operating frequency, the Sun can be considered a homogeneous disc with a diameter of approximately 0.57°[27] (and the references therein).

It is common to use the horizontal coordinate system [32] when working on a radar ground station. This system uses the station's (observer's) local horizon as the plane to define two angles: azimuth and elevation. Azimuth is the angle around the horizon; it starts at 0° at true north and increases eastward. Elevation is the angle above the horizon starting at 0° at the horizon and increasing until 90° at the zenith, for visible objects. The horizontal coordinate system is always referenced to the station position, which is the origin, so to obtain the position of the object referenced to the Earth, the location of the station on the Earth is necessary. The equations for the elevation and azimuth calculations of the Sun are:

$$E_l = \sin^{-1}(\sin\theta_{lat}\sin\delta + \cos\theta_{lat}\cos\delta\cos T_0), T_0 = (t_s - 12) \times 15°, \quad (1)$$

$$A_z = \cos^{-1}\frac{\sin E_l \sin\theta_{lat} - \sin\delta}{\cos E_l \cos\theta_{lat}}, \quad (2)$$

where E_l and A_z are the elevation and azimuth of the Sun, respectively. δ is the declination of the Sun, θ_{lat} is the latitude of the ground station, and t_s is the hour angle of the Sun. The up-to-date declination and hour angle of the Sun are made public by many observatories. The elevation angle is affected by changes in the refractivity of the atmosphere with altitude. The authors of [28] provided an extensive explanation of the equations necessary for calculating the refractivity index and compensating for the apparent elevation obtained in (1). These equations usually require access to local meteorological parameters such as ambient temperature and pressure. The effect of refraction should be taken into account, especially when collecting data at low elevations (<10 deg [26]).

Measurement of solar noise is a relatively straightforward, accurate, reproducible, and convenient method for assessing receiver performance. It consists of measuring the ratio between the power received by the antenna when pointed at the Sun and when pointed at the cold sky [33]. This metric proves valuable in receiver optimization because a higher measured solar noise correlates with improved G/T, allowing for the detection of weaker signals.

The basic principle of pointing calibration with the Sun consists of scanning the azimuth/elevation positions around the predicted position of the Sun and measuring the received power [16]. When the antenna is pointing to the center of the solar disk, it will receive the maximum power. The difference between the sensor angle readings at the peak power and the predicted position of the Sun corresponds to the bias in the positioning system.

The cross-scanning routine consists of the following steps:

1. Define a vector of N offset angles $\theta_{ofs} = [\theta_1, \theta_2 \ldots \theta_N]$.
2. Define a vector of N time instants $T = [t_1, t_2, \ldots t_N]$.
3. Predict the azimuth positions of the Sun for the defined time instants: $A_z(T) = [A_z(t_1), A_z(t_2), \ldots A_z(t_N)]$.
4. Compute the scanning positions: $S = A_z + \theta_{ofs}$.
5. Point the antenna to positions S at time instants T and compute the power received for each position.
6. Locate the peak power and update the offset with the new angle value.
7. Repeat for the elevation angle.

The scan should be performed by alternating between azimuth and elevation to avoid the coupling error of moving azimuth and elevation simultaneously and for several iterations until the obtained offsets converge within a given uncertainty. This routine ensures that the calculated positions compensate for the Sun's movement during the scan.

The operator should make sure the timestamps between points are long enough to be able to move the antenna and record the signal at the correct time.

This method should only be used when the beamdwith is larger than the diameter of the Sun. In cases where the beamwidth is larger, the scanning will return the beam pattern of the antenna, and we can confidently say that the maximum noise reading corresponds to the center of the Sun. However, if the beamwidth is smaller than the diameter of the Sun, there will be multiple maximum values when passing inside the solar disk, which will result in a beam pattern with a flat top instead of a peak.

2.5. Signal Model

Let us consider a radar system emitting a pulse with a carrier at frequency f_0, modulated with a complex baseband linear frequency modulation (LFM)

$$u(t) = \text{rect}\left[\frac{t}{T_p}\right] e^{j\pi \frac{B}{T_p} t^2} e^{j(2\pi f_o)}, \tag{3}$$

where T_p is the pulse duration, t is the fast time, B is the signal bandwidth, and $j = \sqrt{-1}$. Now let us assume that the radar is illuminating a target with a constant radial velocity, causing the range to change according to

$$r(t_s) = r_0 + v_r t_s, t_s \in [0, \text{CPI}], \tag{4}$$

where t_s is the slow time, r_0 is the initial range, v_r is the radial velocity, and CPI is the coherent processing interval. The slow time can be written as dependent on the pulse number

$$t_s = nT_r, n \in [0, 1, 2, \ldots, M-1], \tag{5}$$

where T_r is the pulse repetition interval, and M is the number of pulses in a CPI.

If the radar emits a burst of identical pulses, given by (3), the received echoes after carrier demodulation are given by

$$u_r(t, t_s) = A_r \text{rect}\left[\frac{t - 2r(t_s)/c}{T_p}\right] e^{j\pi \frac{B}{T_p}(t - 2r(t_s)/c)^2} \times e^{-j\frac{4\pi r(t_s)}{\lambda}} + n(t), \tag{6}$$

where A_r is the complex amplitude of the echo, c is the speed of light, $\lambda = c/f_0$, and $n(t)$ is the additive noise term. The sources of noise arise from inside and outside a circuit: solar noise, cosmic and galactic noise, interference from other antennas, and instrumental noise. Different noise sources are modeled by different distributions, with the most predominant one being thermal noise [23] (Chapter 1), which is white and follows a normal distribution in amplitude. Given that the final noise signal is a byproduct of interference from potentially many different sources, the Central Limit Theorem can be applied, and it is reasonable to assume additive white Gaussian noise (AWGN) with a zero mean and a variance σ^2: $\mathcal{N}(0, \sigma^2)$.

After passing the signal through a matched filter, the target's echoes are given by

$$u_{pc}(r_i, t_s) = A_{pc} sinc\left[\pi B \frac{2(r_i - r_0 - v_r t_s)}{c}\right] \times e^{-j\frac{4\pi(r_0 + v_r t_s)}{\lambda}} + n(r_i), \tag{7}$$

where A_{pc} is the complex amplitude of the compressed echoes, and $r_i = ct/2$ is the range. Equation (7) contains a *sinc* term [23] (Chapter 20), which is centered at r_0 and migrates during the CPI.

Real examples of the signal model in Equation (7) are presented in Figures 4 and 5 (r_i is the range axis, and t_s is the Illumination time axis). These figures correspond to the squared magnitude of the pulse-compressed echoes received during the observation of two satellites. In both cases, it is possible to see a distinct line that corresponds to range migration during the illumination time.

The complex exponential term at the end of Equation (7) can be developed into

$$e^{-j\frac{4\pi(r_0+v_r t_s)}{\lambda}} = e^{-j\frac{4\pi r_0}{\lambda}} e^{-j2\pi f_D t_s}, \tag{8}$$

where $f_D = \frac{2v_r}{\lambda}$ is the Doppler shift. The sequence of phase-change terms along the slow time forms a complex sinusoid at the Doppler frequency and is sometimes called the spatial Doppler signal.

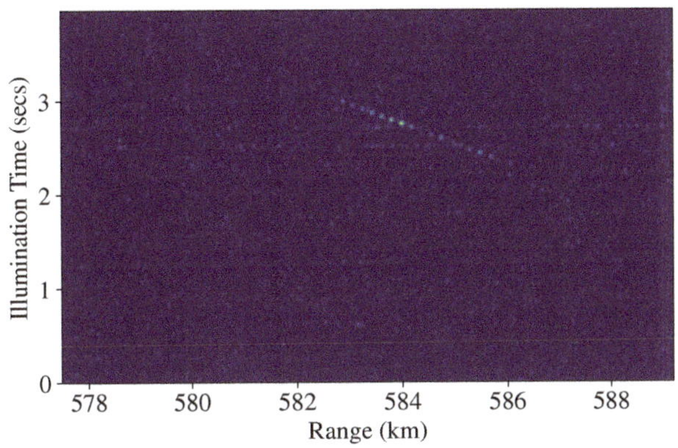

Figure 4. NORAD ID: 25544, ISS. A streak is formed by the echoes received from the sattelite.

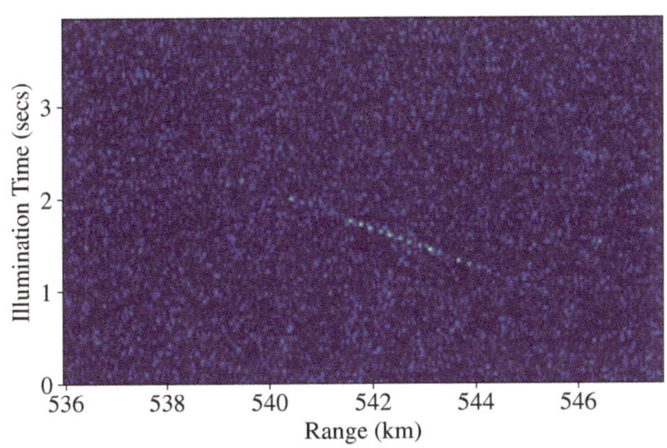

Figure 5. NORAD ID: 54216, CSS (MENGTIAN). A streak is formed by the echoes received from the sattelite.

2.6. Signal Processor

ATLAS uses an AM Chirp waveform with a bandwidth of 2 MHz and emits bursts of 100 pulses at a frequency of 33 Hz. This configuration results in a range resolution of 75 m after pulse compression and a Doppler resolution of 0.009 m/s. The maximum unambiguous range is 4545 km, and the maximum unambiguous velocity is 0.44 m/s. The signal processor is depicted in Figure 6. The raw data acquired by the ADCs for each pulse are stacked in a range-pulse complex matrix [34] (Chapter 3). The complex data are processed by a matched filter [34] (Chapter 4) and then split into two branches.

The first one computes the sample power and passes the signal through a cell-averaging constant false-alarm ratio (CA-CFAR) [34] (Chapter 6) detector with a probability of false alarm of 10^{-9}. At the end of this branch, a list of range detections is obtained. The second branch compensates for range migration and performs pulse-Doppler processing.

Range migration compensation is achieved using the Radon transform (RT) for line detection [35] and envelope cross-correlation for range alignment [36,37]. The RT of a function $f(x,y)$ is defined as [38]

$$R(p,\tau) = \int_{-\infty}^{+\infty} f(x, px+\tau)dx, \tag{9}$$

where p is the slope of the line and τ is its intercept with the y-axis. So, $R(p,\tau)$ is the line integral of $f(x,y)$ over the line defined by the parameters p and τ. By applying the RT across a range of slopes and intercepts with the y-axis, we obtain the sinogram of $f(x,y)$, which contains maximums for the pairs of (p,τ) that define existing lines in the (x,y) domain.

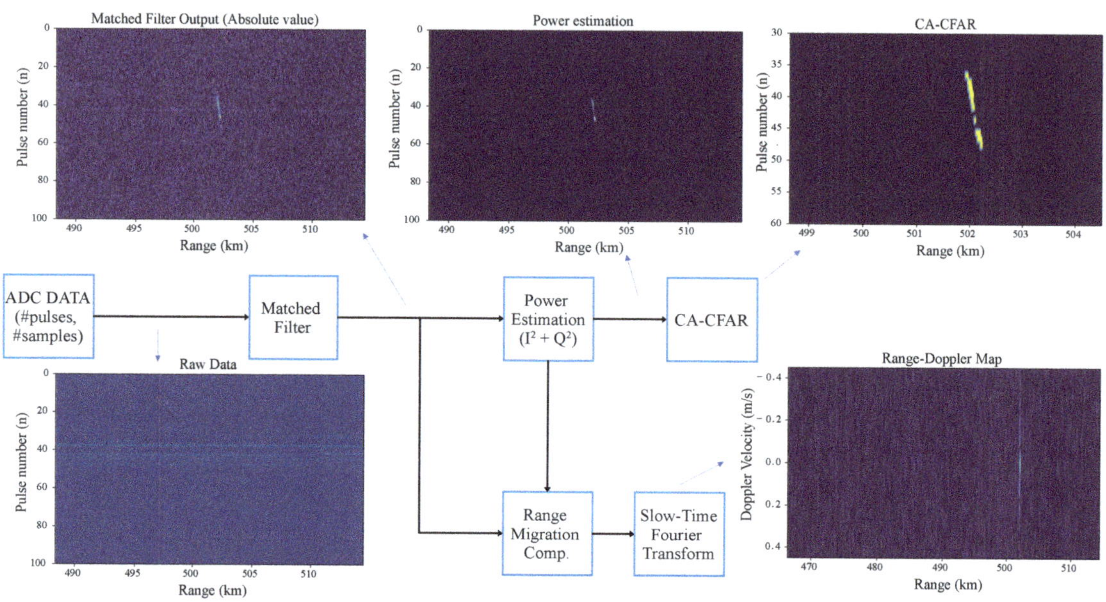

Figure 6. ATLAS signal processor. The output of the signal processor consists of a list of detections containing range and velocity information.

By combining (7) and (9), it is possible to obtain the RT of the received signal in terms of the range r_0 and the radial velocity v_r

$$R(r_0, v_r) = \int_0^{CPI} u_{pc}(r_0 + v_r \frac{\Delta t_s}{\Delta r_i} t_s, t_s)dt_s, r_0 \in [r_{min}, r_{max}], \tag{10}$$

where Δt_s is the pulse repetition interval, and Δr_i is the range resolution.

The radial velocity can also be defined as the anticlockwise angle from the slow time axis to the range walk line

$$v_r = \tan(\theta)\frac{\Delta r_i}{\Delta t_s}, \theta \in [0, \pi]. \tag{11}$$

If there is one line in $R(r_0, v_r)$, it can be detected using

$$r_{RT}, v_{RT} = \max(R(r_0, v_r(\theta)), r_0 \in [r_{min}, r_{max}], \theta \in [0, \pi], \tag{12}$$

where r_{RT} and v_{RT} are estimates for the initial range and radial velocity computed with the RT. After detecting the line, we can filter out the rest of the components in the sinogram (by setting them to 0) and apply the inverse Radon transform (IRT), resulting in the detected line being isolated, denoted as u_{clean}

$$u_{clean} = \mathrm{IRT}\left(\begin{cases} R(r_0, v_r) & \text{if } (r_0, v_r) = (r_{RT}, v_{RT}) \\ 0 & \text{else} \end{cases}\right). \quad (13)$$

Figure 7 showcases the line detection technique applied to the observation shown in Figure 5.

Applying the envelope cross-correlation to u_{clean} estimates the range cell shift between the peaks in the line, which are then used to align the predominant scatterer within the same range bin, as depicted in Figure 8.

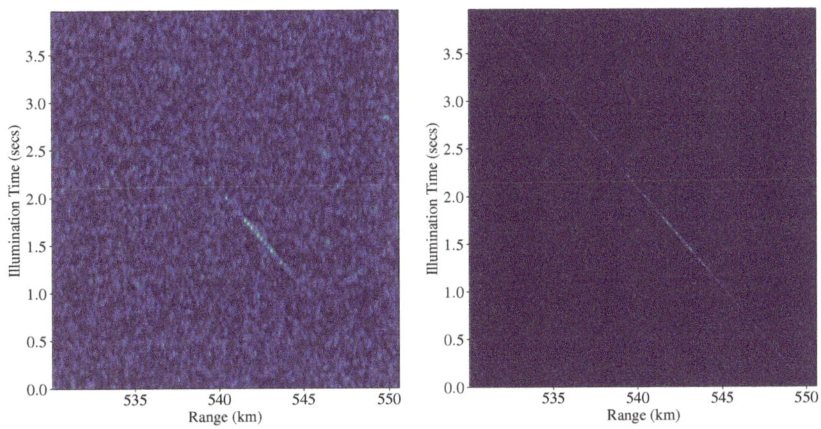

Figure 7. Line detection based on the Radon transform applied to an observation of CSS (MENG-TIAN).

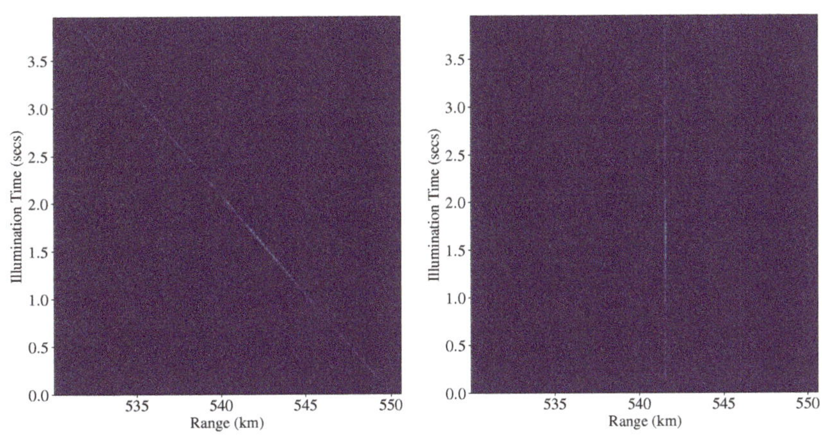

Figure 8. Range alignment of the observation shown in Figure 7.

Now that the range alignment is completed, we need to correct for the Doppler term (Equation (8)). To do this, we can implement a filterbank, with each filter tuned to a different Doppler frequency. This can be obtained by applying the Fourier transform along the slow time, generating a range-Doppler map [34] (Chapter 5). Due to the usage of a low PRF, the Doppler velocity obtained is highly ambiguous. Unambiguous velocity estimation is obtained from the line parameters found with the RT.

3. Results

3.1. Calibration

The results from the last calibration procedure with cross-scanning using the Sun are depicted in Figures 9 and 10. Figure 9 showcases some of the steps of scanning. Initially, an azimuth scan with an 8° searching angle yielded an offset estimate of 6°. Then, an elevation scan with a 4° searching angle provided an initial estimate of 1°. Subsequently, finer-resolution scans were conducted around these initial offsets, and the azimuth offset was corrected to 5.4°, and the elevation offset was corrected to 0.5°. The convergence was obtained after the offset in azimuth stabilized within an uncertainty of 0.2° and the offset in elevation stabilized within an uncertainty of 0.05°. The results in Figure 10 correspond to the last iteration on each axis after convergence was obtained.

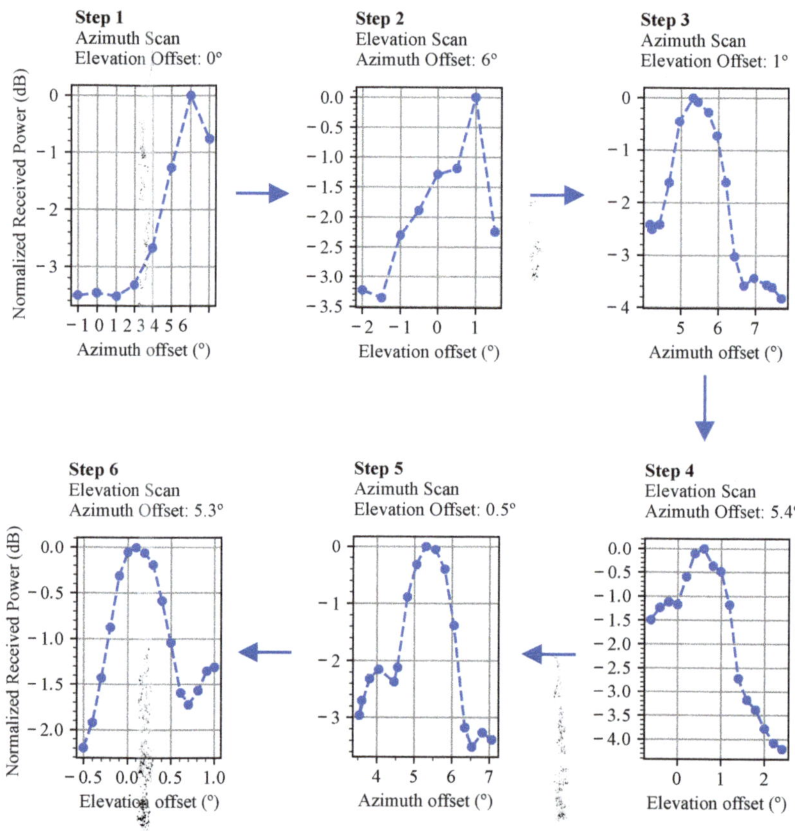

Figure 9. Example of the steps of the calibration procedure. In Step 1, an azimuth scan was performed with an 8°searching angle to obtain an initial guess of the offset, where a peak was obtained at 6°. In Step 2, the 6°offset in azimuth was fixed, and an elevation scan was performed with a 4°searching angle to obtain an initial guess of the offset, where a peak was obtained at 1°. Once an initial guess for both offsets was obtained, the same process was repeated but with finer resolution around the new offsets (Steps 3 and 4), where the azimuth offset was corrected to 5.4° and the elevation offset was corrected to 0.5°. The process converged in Steps 5 and 6, as detailed in Figure 10. The number of scans required until convergence may vary, and for illustration purposes, intermediary scans were omitted.

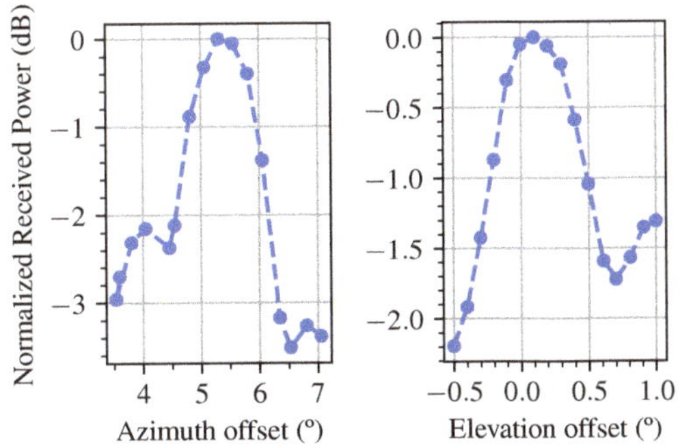

Figure 10. Offsets obtained for azimuth and elevation after convergence in the cross-scanning procedure. The azimuth offset is 5.3 ± 0.2°, where the mainlobe can be identified roughly in the region between 4.9° and 5.7°, the left sidelobe is visible between 3.5° and 4.5°, and the right sidelobe is between 6.5° and 7° (incomplete). The elevation offset is 0.10 ± 0.05°, where the mainlobe can be identified roughly in the region between −0.27° and 0.47°, and the right sidelobe is between 0.6° and 1° (incomplete).

3.2. Space Object Observations

The observations presented in this section resulted from the first campaign of observations performed with ATLAS in July 2023 and a second campaign during the last week of January 2024. The preferred candidate for observation was the ISS, as explained in Section 2.2. Figure 11 showcases one of the observations obtained. After applying the signal processing chain to the raw data, a streak of detections is visible. The object returned reflections from pulses 35 to 48 out of the 100 pulses emitted. Range migration is obvious given that the object moved to other range cells from pulse to pulse. Figure 12 depicts the range-Doppler map obtained for the observation in Figure 11.

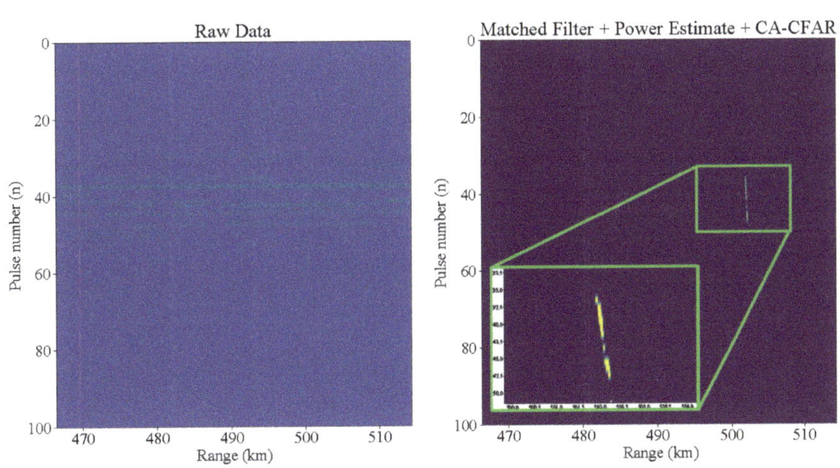

Figure 11. Image on the left shows the absolute value of the complex raw data recorded by the ADCs. The image on the right shows the detections obtained. The detections cluster into a streak, a common pattern when observing space objects due to the effect of range migration.

A total of 35 range detections and 5 velocity detections were obtained from the campaign. Table 3 shows the complete data for some of the observations. The rest of the observations were used to assess the performance of the system, as explained in Section 4.2.

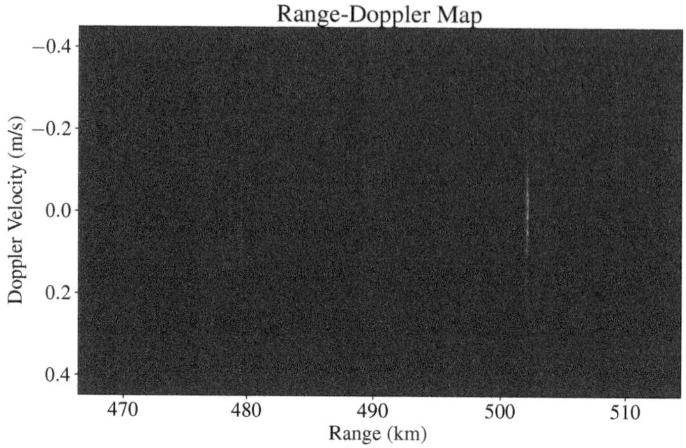

Figure 12. Range-Doppler map with detection at 502.0 km of range and 0.014 m/s of Doppler velocity.

Table 3. Some examples of observations.

NORAD ID	Timestamp	Azimuth (°)	Elevation (°)	Range (km)	Doppler Velocity (m/s)	Velocity (From Line Detection) (m/s)
25544	2023-07-04T06:51:05	91.94	57.37	502.0	0.014	623.115
25544	2023-07-06T23:34:16	258.87	45.27	589.5	−0.105	4407.915
25544	2023-07-13T21:11:20	27.68	46.97	573.5	−0.040	3931.333
54216	2024-01-31T12:40:09	287.22	45.39	541.5	0.052	4653.335
25544	2024-02-02T18:44:46	303.62	45.84	583.9	0.037	4593.325

4. Discussion

4.1. Calibration

Figure 10 shows not only the offset of the main beam but also the shape of the main beam and sidelobes. The sidelobes of the antenna are not symmetric and require re-alignment. This is due to it being an old antenna that was moved and underwent several upgrades and repairs over the years. A 3D point cloud was recently obtained using LIDAR, showing that the feed and the secondary reflector are aligned, but some of the panels of the main reflector are misaligned, compromising the parabolic shape. This issue is currently being addressed.

4.2. Revisiting the Radar Equation

By adapting one of the many forms of the radar equation, the single pulse signal-to-noise ratio (SNR_p) after pulse compression is given by [39]:

$$\mathrm{SNR}_p = \frac{P_{peak} G^2 \lambda^2 \sigma \tau \beta}{4\pi^3 R^4 k_B T N_F B L_s} \tag{14}$$

where P_{peak} is the transmitted peak power (W), G is the antenna gain, λ is the operating wavelength (m), σ is the target radar cross-section (RCS) (m^2), τ is the pulse width (s), β is the bandwidth of the waveform (Hz), R is the distance to the object (m), N_F is the noise factor of the receiver, k is the Boltzmann constant ($\frac{J}{K}$), T is the receiver temperature (K), B is the bandwidth of the receiver (Hz), and L_s accounts for all losses.

The loss term L_s represents all the factors that are difficult to obtain/identify, e.g., system losses and atmospheric losses. While an educated guess can be a useful starting point to evaluate the future performance of a radar system, ideally, this term should be updated by performing measurements with the system. Figure 13 represents the theoretical curves obtained for ATLAS for different RCS values, represented as solid lines. Since we have now made observations of the ISS, we decided to use these to revisit the equation and improve it. To do this, we fixed the equation terms for an RCS of 400 m^2 and fitted the loss term L_s to the observations using a non-linear least-squares method. The fitting resulted in an L_s of 15.94 dB with a variance of 6.09 dB. The fit curves are represented as dashed lines. The variance of the estimate is high due to the dispersion and number of measurements. We can see that for ranges around 500 km, the detections have variations of around 5 dB in the SNR, which can be due to many factors such as the scintillation of the object or variations in the gains of the system during pulse emission/reception. We are currently investigating the cause of these differences in the SNR for the same range and RCS. Currently, the number of observations is limited, and the objective is to gather more observations at different ranges and RCSs to obtain a better estimate of L_s.

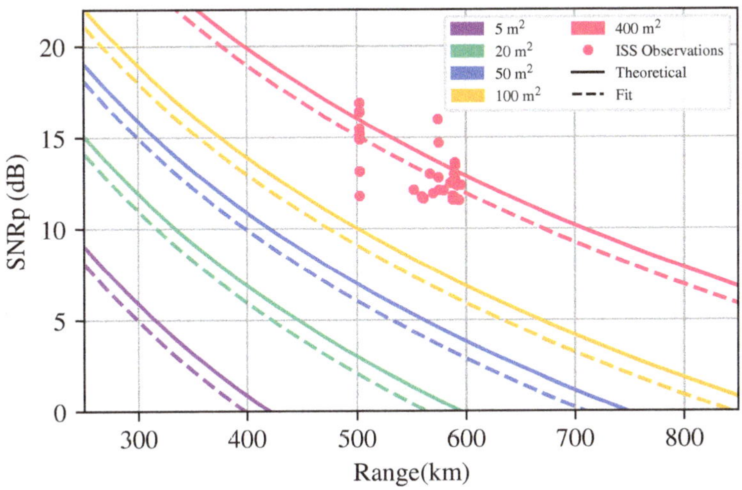

Figure 13. SNR dependence on the range for different RCSs. The theoretical curves (solid lines) were adjusted using the ISS observations to obtain a better estimate of the performance of ATLAS in real scenarios (dashed lines).

5. Conclusions and Future Work

After 4 years of development, testing, and validation, the ATLAS system has successfully achieved its first uncooperative observations of space objects. These results were only possible due to the efforts invested in creating an adequate observation strategy, implementing calibration techniques, and exploring the capabilities of the system in terms of baseband waveform generation and processing.

As of now, ATLAS is evolving to be not only a radar system but a fully-fledged SST ground station with the capacity for orbit prediction, schedule generation, and maintenance of an internal catalog. ATLAS is still a work in progress that is moving closer to being used in an operational scenario. Ongoing and future work comprises the following:

- Obtain more observations of well-known objects at different ranges and RCSs to obtain a better estimate of the radar's performance;
- Compare range and range-rate measurements with precise ephemeris to correct unaccounted bias terms;

- Increase the probability of detection by designing more sophisticated detectors. This includes exploring subspace detector design [40,41], multi-stage adaptive detectors [42,43], and machine learning-based detectors [44,45];
- Address ambiguous Doppler velocity extraction using multiple PRFs [46,47];
- Increase the peak power, bandwidth, and PRF of the system to increase the SNR, range resolution, and Doppler resolution.

Author Contributions: Conceptualization, J.P., M.B. and P.M.; methodology J.P., M.B. and P.M.; software, J.P.; validation, M.B., P.M. and B.C.; formal analysis, J.P., M.B., P.M. and B.C.; investigation, J.P., M.B., P.M. and B.C.; resources, J.P.; data curation, J.P. and B.C.; writing—original draft preparation, J.P.; writing—review and editing, M.B., P.M., B.C., D.B. and M.F.; visualization, J.P.; supervision, P.M. and M.B.; project administration, D.B. All authors have read and agreed to the published version of the manuscript.

Funding: This work was supported by the Fundação para a Ciência e Tecnologia (FCT), Ph.D. grant No. 2022.12341.BDANA.

Data Availability Statement: Data were obtained from the Portuguese Ministry of Defense and are available from the corresponding author with the permission of the Portuguese Ministry of Defense.

Acknowledgments: The team acknowledges the European Commission's Horizon 2020 Programme under the Space Surveillance and Tracking (SST) grant agreement 2-3SST2018-20. Additionally, the team acknowledges LC Technologies for their technical developments in close collaboration with Instituto de Telecomunicações. We also acknowledge the support from the European Regional Development Fund (FEDER) through the Competitiveness and Internationalization Operational Programme (COMPETE 2020) of the Portugal 2020 framework (Project SmartGlow with No. 069733 (POCI-01-0247-FEDER-069733)). The team further acknowledges the support from the ENGAGE-SKA Research Infrastructure, Ref. POCI-01-0145-FEDER-022217, funded by COMPETE 2020 and FCT, Portugal. IT team members acknowledge the support from Projecto Lab. Associado UID/EEA/50008/2019. Finally, the team acknowledges the Portuguese Ministry of Defense for allowing the usage and dissemination of the results.

Conflicts of Interest: Authors M.B. and B.C. are employed by ATLAR Innovation. The remaining authors declare that the research was conducted in the absence of any commercial or financial relationships that could be construed as a potential conflict of interest.

Abbreviations

The following abbreviations are used in this manuscript:

SSA	Space Situational Awareness
SST	Space Surveillance and Tracking
ESA	European Space Agency
NASA	National Aeronautics and Space Administration
EUSST	European Space Surveillance and Tracking
ATLAS	rAdio TeLescope pAmpilhosa Serra
PASO	Pampilhosa da Serra Space Observatory
ACSA	ATLAS Cloud Service API
LEO	Low Earth Orbit
SGP	Simplified General Perturbation
SNR	Signal-to-Noise Ratio
PRF	Pulse Repetition Frequency
TLE	Two-Line Element
RSO	Resident Space Object
ISS	International Space Station
RCS	Radar Cross-Section
MDPI	Multidisciplinary Digital Publishing Institute
LIDAR	Light Detection and Ranging
ADC	Analogue-to-Digital Converter

References

1. *ESA'S Annual Space Environment Report*; Technical report; ESA Space Debris Office: Darmstadt, Germany, 2023
2. United Nations Office for Outer Space Affairs. Treaty on Principles Governing the Activities of States in the Exploration and Use of Outer Space, including the Moon and Other Celestial Bodies. Available online: https://www.unoosa.org/oosa/en/ourwork/spacelaw/treaties/introouterspacetreaty.html (accessed on 8 January 2021).
3. European Space Agency. SSA Programme Overview. Available online: https://www.esa.int/Space_Safety/SSA_Programme_overview (accessed on 12 February 2024).
4. EU SST. Eusst-European Space Surveillance and Tracking Projects. Available online: https://www.eusst.eu (accessed on 12 February 2024).
5. Pandeirada, J.; Bergano, M.; Neves, J.; Marques, P.; Barbosa, D.; Coelho, B.; Ribeiro, V. Development of the First Portuguese Radar Tracking Sensor for Space Debris. *Signals* **2021**, *2*, 122–137. [CrossRef]
6. Coelho, B.; Barbosa, D.; Bergano, M.; Correia, A.; Freitas, J.; Marques, P.; Pandeirada, J.; Ribeiro, V. New SST optical sensor of Pampilhosa da Serra—Studies on image processing algorithms and multi-filter characterization of space debris. In Proceedings of the 8th European Conference on Space Debris (Virtual), Darmstadt, Germany, 20–23 April 2021; ESA Space Debris Office: Darmstadt, Germany 2021; pp. 1–4.
7. Pandeirada, J.; Bergano, M.; Marques, P.; Barbosa, D.; Freitas, J.; Ribeiro, V. Design of Pulsed Waveforms for Space Debris Detection with Atlas. In Proceedings of the 8th European Conference on Space Debris (Virtual), Darmstadt, Germany, 20–23 April 2021.
8. Pandeirada, J.; Bergano, M.; Marques, P.; Barbosa, D.; Coelho, B.; Ribeiro, V.; Freitas, J.; Nunes, D.; Eduardo, J. A Portuguese radar tracking sensor for Space Debris monitoring. *arXiv* **2022**, arXiv:2111.02232. https://doi.org/10.48550/arXiv.2111.02232.
9. Pandeirada, J.; Bergano, M.; Marques, P.; Barbosa, D.; Coelho, B.; Freitas, J.; Nunes, D. ATLAS: Deployment, Control Platform and First RSO Measurements. *arXiv* **2023**, arXiv:2211.03586. https://doi.org/10.48550/arXiv.2211.03586.
10. Gonçalves, L.F.P. Space Data Processing Pipeline for PASO (Pipeline de Análise de Detritos Espaciais Para o PASO). Master's Thesis, University of Aveiro, Aveiro, Portugal, 5 July 2023.
11. Gonçalves, L.F.P.; Sousa, A. CLOWN: A new tool for cloud detection with an All-Sky camera for optimization of space-debris surveys. In Proceedings of the 73rd International Astronautical Congress (IAC), Paris, France, 18–22 September 2022.
12. Coelho, B.; Barbosa, D.; Berganoa, M.; Pandeirada, J.; Marques, P.; Correia, A.C.M.; Freitas, J.M.d. Developing a data fusion concept for radar and optical ground based SST station. In Proceedings of the 73rd International Astronautical Congress (IAC), Paris, France, 18–22 September 2022; International Astronautical Federation: Paris, France; pp. 1–3.
13. Fossa, C.; Raines, R.; Gunsch, G.; Temple, M. An overview of the IRIDIUM (R) low Earth orbit (LEO) satellite system. In Proceedings of the IEEE 1998 National Aerospace and Electronics Conference. NAECON 1998. Celebrating 50 Years (Cat. No.98CH36185), Dayton, OH, USA, 17 July 1998; pp. 152–159, ISSN 0547-3578. [CrossRef]
14. Flohrer, T.; Krag, H.; Klinkrad, H. Assessment and Categorization of TLE Orbit Errors for the US SSN Catalogue. In Proceedings of the Advanced Maui Optical and Space Surveillance Technologies Conference, Maui, HI, USA, 17–19 September 2008.
15. Kelecy, T.; Hall, D.; Hamada, K.; Stocker, M.D. Satellite Maneuver Detection Using Two-line Element (TLE) Data. In Proceedings of the Advanced Maui Optical and Space Surveillance Technologies Conference, Maui, HI, USA, 12–15 September 2007.
16. Qi, X.; Wang, J.; Zhao, L.; Ji, J. Antenna beam angle calibration method via solar electromagnetic radiation scan. *J. Eng.* **2019**, *2019*, 7890–7893. [CrossRef]
17. National Aeronautics and Space Administration Jet Propulsion Laboratory. Horizons System. Available online: https://ssd.jpl.nasa.gov/horizons/ (accessed on 12 February 2024).
18. National Aeronautics and Space Administration Jet Propulsion Laboratory. Spot the Station. Available online: https://spotthestation.nasa.gov/trajectory_data.cfm# (accessed on 7 February 2024).
19. Vallado, D.A.; McClain, W.D. *Fundamentals of Astrodynamics and Applications*, 3rd ed.; Number 21 in Space technology library; Microcosm Press: Hawthorne, CA, USA, 2007.
20. Pearlman, M.R.; Noll, C.E.; Pavlis, E.C.; Lemoine, F.G.; Combrink, L.; Degnan, J.J.; Kirchner, G.; Schreiber, U. The ILRS: Approaching 20 years and planning for the future. *J. Geod.* **2019**, *93*, 2161–2180. [CrossRef]
21. Mochan, J.; Stophel, R.A. Dynamic Calibration of Space Object Tracking Systems. In Proceedings of the Space Congress, Brevard County, FL, USA, 1 April 1968.
22. Krebs, G.D. Starlette/Stella. Available online: https://space.skyrocket.de/doc_sdat/starlette.htm (accessed on 12 February 2024).
23. Richards, M.A.; Scheer, J.A.; Holm, W.A. (Eds.) *Principles of Modern Radar: Basic Principles*; Institution of Engineering and Technology: London, UK, 2010. [CrossRef]
24. Whiton, R.C.; Smith, P.L., Jr.; Harbuck, A.C. Calibration of weather radar systems using the sun as a radio source. In Proceedings of the 17th Conference on Radar Meteorology, Seattle, WA, USA, 26–29 October 1976; pp. 60–65.
25. Mano, K.; Altshuler, E.E. Tropospheric refractive angle and range error corrections utilizing exoatmospheric sources. *Radio Sci.* **1981**, *16*, 191–195. [CrossRef]
26. Huuskonen, A.; Holleman, I. Determining Weather Radar Antenna Pointing Using Signals Detected from the Sun at Low Antenna Elevations. *J. Atmos. Ocean. Technol.* **2007**, *24*, 476–483. [CrossRef]
27. Reimann, J.; Hagen, M. Antenna Pattern Measurements of Weather Radars Using the Sun and a Point Source. *J. Atmos. Ocean. Technol.* **2016**, *33*, 891–898. [CrossRef]

28. Holleman, I.; Huuskonen, A. Analytical formulas for refraction of radiowaves from exoatmospheric sources. *Radio Sci.* **2013**, *48*, 226–231. [CrossRef]
29. Appleton, E.V. Departure of Long-Wave Solar Radiation from Black-Body Intensity. *Nature* **1945**, *156*, 534–535. [CrossRef]
30. Meftah, M.; Corbard, T.; Hauchecorne, A.; Morand, F.; Ikhlef, R.; Chauvineau, B.; Renaud, C.; Sarkissian, A.; Damé, L. Solar radius determined from PICARD/SODISM observations and extremely weak wavelength dependence in the visible and the near-infrared. *Astron. Astrophys.* **2018**, *616*, A64. [CrossRef]
31. Morgan, M. Determination of Earth Station Antenna G/T Using the Sun or the Moon as an RF Source. In Proceedings of the 32nd Annual AIAA/USU Conference on Small Satellites, Logan, UT, USA, 4–9 August 2018.
32. Roy, A.E.; Clarke, D. *Astronomy: Principles and Practice*, 4th ed.; Institute of Physics Pub: Bristol, UK; Philadelphia, PA, USA, 2003.
33. Cakaj, S.; Keim, W.; Malaric, K. Sun noise measurement at low Earth orbiting satellite ground station. In Proceedings of the 47th International Symposium ELMAR, Zadar, Croatia, 8–10 June 2005; pp. 345–348. [CrossRef]
34. Richards, M.A. *Fundamentals of Radar Signal Processing*, 2nd ed.; McGraw-Hill Education: New York, NY, USA, 2014.
35. Xu, J.; Yu, J.; Peng, Y.N.; Xia, X.G. Radon-Fourier Transform for Radar Target Detection, I: Generalized Doppler Filter Bank. *IEEE Trans. Aerosp. Electron. Syst.* **2011**, *47*, 1186–1202. [CrossRef]
36. Uysal, F.; Goodman, N. The effect of moving target on range-doppler map and backprojection algorithm for focusing. In Proceedings of the 2016 IEEE Radar Conference (RadarConf), Philadelphia, PA, USA, 2–6 May 2016; pp. 1–5, ISSN 2375-5318. [CrossRef]
37. Chen, V.C.; Martorella, M. *Inverse Synthetic Aperture Radar Imaging: Principles, Algorithms and Applications*; Institution of Engineering and Technology: London, UK, 2014. [CrossRef]
38. Tian, J.; Xia, X.G.; Cui, W.; Yang, G.; Wu, S.L. A Coherent Integration Method via Radon-NUFrFT for Random PRI Radar. *IEEE Trans. Aerosp. Electron. Syst.* **2017**, *53*, 2101–2109. [CrossRef]
39. Skolnik, M.I. *Introduction to Radar Systems*, 2nd ed.; McGraw-Hill: New York, NY, USA, 1980.
40. Kraut, S.; Scharf, L.; McWhorter, L. Adaptive subspace detectors. *IEEE Trans. Signal Process.* **2001**, *49*, 1–16. [CrossRef]
41. Pulsone, N.; Rader, C. Adaptive beamformer orthogonal rejection test. *IEEE Trans. Signal Process.* **2001**, *49*, 521–529. [CrossRef]
42. De Maio, A.; Orlando, D. Feature article: A survey on two-stage decision schemes for point-like targets in Gaussian interference. *IEEE Aerosp. Electron. Syst. Mag.* **2016**, *31*, 20–29. [CrossRef]
43. Bandiera, F.; Orlando, D.; Ricci, G. A Subspace-Based Adaptive Sidelobe Blanker. *IEEE Trans. Signal Process.* **2008**, *56*, 4141–4151. [CrossRef]
44. Ball, J.E. Low signal-to-noise ratio radar target detection using Linear Support Vector Machines (L-SVM). In Proceedings of the 2014 IEEE Radar Conference, Cincinnati, OH, USA, 19–23 May 2014; IEEE: New York, NY, USA, 2014; pp. 1291–1294. [CrossRef]
45. Coluccia, A.; Fascista, A.; Ricci, G. A k-nearest neighbors approach to the design of radar detectors. *Signal Process.* **2020**, *174*, 107609. [CrossRef]
46. Trunk, G.; Brockett, S. Range and velocity ambiguity resolution. In Proceedings of the The Record of the 1993 IEEE National Radar Conference, Lynnfield, MA, USA, 20–22 April 1993; IEEE: New York, NY, USA, 1993; pp. 146–149. [CrossRef]
47. Lei, W.; Long, T.; Han, Y. Resolution of range and velocity ambiguity for a medium pulse Doppler radar. In Proceedings of the Record of the IEEE 2000 International Radar Conference [Cat. No. 00CH37037], Alexandria, VA, USA, 12 May 2000; IEEE: New York, NY, USA, 2000; pp. 560–564. [CrossRef]

remote sensing MDPI

Article

Joint Power and Bandwidth Allocation with RCS Fluctuation Characteristic for Space Target Tracking

Qingwei Yang, Libing Jiang, Shuyu Zheng, Yingjian Zhao and Zhuang Wang *

National Key Laboratory of Science and Technology on Automatic Target Recognition, College of Electronic Science and Technology, National University of Defense Technolody, Changsha 410073, China; yangqingwei18@nudt.edu.cn (Q.Y.)
* Correspondence: wangzhuang@nudt.edu.cn; Tel.: +86-0139-7516-3062

Abstract: Reasonable allocation of space-based radar resources is a crucial aspect of improving the accuracy of space multi-target tracking and enhancing spatial awareness. The conventional resource allocation algorithm fails to exploit the high dynamic radar cross-section (RCS) characteristics, resulting in poor tracking robustness, tracking divergence, or even loss of tracking. However, the RCS of space targets fluctuates considerably in actual tracking scenarios, which cannot be disregarded for space target tracking tasks. To address this issue, we propose an adaptive allocation method that considers the dynamic RCS fluctuation characteristic for space-based radar tracking assignments. The proposed method exploits the predictable orbital information of space target to calculate the real-time observation angle of radar, and then obtains the multi-target dynamic RCS through the target RCS dataset. By combining the obtained RCS sequence, radar power, and bandwidth, an optimal model for radar tracking accuracy is established based on the multi-target posterior Cramér–Rao lower bound (PCRLB) to evaluate the tracking performance. By resolving the aforementioned multivariance optimization problem, we eventually acquire the results of power and bandwidth pre-allocation for tracking multiple space targets. Simulation results validate that, compared with the traditional methods, the proposed joint dynamic RCS power and bandwidth allocation (JRPBA) method can achieve superior tracking accuracy and minimize instances of missed tracking.

Keywords: space multi-target tracking; joint power and bandwidth resource allocation method; posterior Cramér–Rao lower bound (PCRLB); space-based radar; dynamic radar cross section (RCS)

1. Introduction

1.1. Background and Problem Statement

With the growing exploration of space activities, effective tracking, monitoring, and protection technology for various space targets has brought significant attention to the field of space situational awareness (SSA) region. In [1–3], an analysis is conducted on the number of space targets launched by the United States, Europe, the Soviet Union, and China in recent years, as well as the evolving trend of additional space debris generated.

Therefore, sparing no efforts on space resource development, space security and space control [4,5] is indeed necessary for enhancing the SSA ability of a nation. Many countries and regions have developed ground-based equipment for the SSA task. For instance, the European Incoherent Scatter Scientific Association (EISCAT) has developed three incoherent scatter radar systems, and the operating frequency of the two radars located in northern Scandinavia are 224 MHz and 931 MHz, while the other one in Svalbard operates at 500 MHz. The operating frequency of the Italian Bi-static Radar for LEO Tracking (BIRALET) system ranges from 410 to 415 MHz [6]. The Tracking and Imaging Radar (TIRA) in Germany is a system devoted to detailed studies of satellites and space debris, and contributes to European SSA [7]. The United States Space Surveillance Network (SSN), operated by the United States Strategic Command (USSTRATCOM), consists of a worldwide network of 29 space surveillance radars and optical telescopes [8]. The French air force has operated the GRAVES bistatic radar system since 2005 [9].

However, it should be noted that the space-based radar system is superior to both ground-based equipment and space-based optical systems. Unlike other systems, it can monitor space targets of interest in all weather conditions, regardless of the Earth's curvature or weather constraints. Therefore, we focus on implementing a space-based radar system to track multiple space targets in this paper. It is worth emphasizing that enhancing the accuracy and efficiency of multi-target tracking using space-based radar has been a significant research topic in recent years.

Space-based radars can accomplish simultaneous multi-beam tracking of space targets by carrying a collocated multiple input multiple output (C-MIMO) radar payload. As a new system radar, C-MIMO radar launches different beams to multiple targets by means of transmitting waveform diversity, which can accomplish simultaneous multi-beam tracking of space targets [10–14]. The technology addresses the limitations of traditional phased array radar (PAR) time-sharing tracking and is extensively applied in multi-target detection and tracking. Nevertheless, due to the reliance on solar panels for energy supply and hardware design constraints, the power and bandwidth of space-based radar are more restricted compared to their ground-based counterparts. Space targets can achieve relative velocities ranging from several to tens of Mach, causing rapid changes in the azimuth and elevation angles observed by space-based radar. Moreover, the structures of space targets are complex and different. During the tracking process of space-based radar, the radar cross section (RCS) of the space targets exhibits high dynamic variation characteristics. These factors will result in unstable tracking performance of space-based radar. The above-mentioned problems in the field of space-based C-MIMO radar resource allocation research to allocate the limited total transmit power and bandwidth to each transmit beam reasonably are in urgent need of solving, so as to obtain higher space target tracking accuracy.

At present, most research mainly focuses on the tracking resource scheduling of ground-based radar to airmobile targets. Ref. [15] carried out research on the selection and optimization of MIMO radar parameters for multi-target detection and tracking. Ref. [16], the target detection and tracking under the framework of cognitive radar was studied. Refs. [17–19] established a cost function based on Bayesian Cramér–Rao lower bound, and used the gradient projection algorithm to obtain the optimal power scheduling results of a radar system for multi-target tracking under different motion parameters. A joint strategy of power and bandwidth scheduling for maneuvering multi-target tracking (MMTT) was proposed in [20] and utilized a relaxation method to obtain the optimized allocation scheme by solving the nonconvex optimization problem. Refs. [21–23] studied the resource allocation method for air target tracking for the two institutional configurations of distributed MIMO radar and collocated MIMO radar. Based on the sum of PCRLB of multiple batches of targets, the adaptive allocation scheme of radar resource was realized. Refs. [24–26] studied the differentiated tracking resource scheduling scheme, considering the system constraints and external factors comprehensively, they proposed the joint beam and power scheduling (JBPS) method, and realized the global objective function design and solution. The authors of [27] studied the antenna selection or power allocation problem. Most of the existing algorithms are aimed at radar resource scheduling for tracking air target tasks, while the stable space targets have obvious semi-cooperative characteristics which can be further exploited [28]. For instance, the orbit of the space target can be predicted relatively accurately, and its orientation mode is often fixed. So, it is necessary to study the resource scheduling scheme which is more suitable for space multi-target tracking.

In addition, Ref. [29] studied the power allocation scheme of C-MIMO radar for RCS non-fluctuation models and RCS fluctuated models, and verified the effectiveness of the adaptive power optimal allocation algorithm. Similarly, Ref. [30] aims at the problem of C-MIMO radar resource scheduling, the influence factors of distance, RCS, and motion model error are studied, respectively, and the decision-making effect of each factor on radar resource allocation is discussed comprehensively. Ref. [31] studied the low-Earth-orbit (LEO) satellite detection techniques and analyzed the detection capability of various walker constellation space-based radar deployment schemes for space targets. However, the above works only discuss the detection ability of space-based radar with different walker constellations and ignore the target RCS fluctuation characteristic in the radar resource scheduling tasks. The existing works and algorithms simply construct the target RCS as the non-fluctuation model or give RCS simple time-varying characteristic and these assumptions are inaccurate in actual scenarios, which fail to be eligible for the actual high dynamic RCS space target tracking requirements.

1.2. Problem Analysis and Contributions

The existing works have provided a rich theoretical and technical basis for the ground-based radar resource scheduling problem and improved its comprehensive application efficiency. However, this research rarely considers the radar resource allocation problem of space-based radar tracking multiple space targets and does not comprehensively consider factors such as the orbital motion characteristics or dynamic RCS characteristic of space targets.

Bayesian theory points out that prior information can improve the accuracy of parameter estimation. Inspired by this, we can model the on-orbit space target and obtain its full-angle static RCS database and orbit information, which can be introduced into the resource allocation framework as a priori information to solve the above problems. In this paper, we propose an adaptive joint dynamic RCS power and bandwidth allocation (JRPBA) method for multiple space target tracking via space-based radar. The main contributions of the paper are given as follows.

(1) The space satellite target is different from the aircraft target in the air. It mainly moves around the Earth due to the gravity of the Earth. Considering that the space target orientation model is usually fixed, this means that the motion attitude of the space target can be calculated more accurately. Can this semi-cooperative characteristic of space targets be utilized in space-based radar resource allocation to improve tracking accuracy? For space surveillance systems, the space target information database is often continuously constructed and expanded. In this paper, we introduce the space target information database as a priori information into the resource allocation framework. The purpose is to utilize the target information to adopt appropriate signal processing methods for different satellites and take various characteristics of space targets into account in the actual target tracking process to obtain higher multi-target tracking accuracy.

(2) In the actual radar tracking scene, the radar line of sight (RLOS) has a great influence on the target RCS characteristics. Specifically, the RCS sequence of the target shows high dynamic fluctuation characteristics. However, the existing radar resource allocation method does not take into account the dynamic RCS characteristics of the targets, which leads to the fact that the previous radar resource allocation scheme cannot be fully adapted to the actual tracking scene and eventually leads to tracking divergence or even mistracking. In this paper, an adaptive joint dynamic RCS power and bandwidth allocation (JRPBA) method for space-based C-MIMO radar is proposed. The core idea is to reasonably allocate the limited power and bandwidth resources of C-MIMO radar in multi-target high dynamic RCS tracking scenarios by utilizing the semi-cooperative characteristic of space targets and the predictable characteristics of orbit. The mismatched problem between the radar resource allocation scheme and the actual tracking scene is solved and the multi-target tracking accuracy and efficiency of the space-based radar system are improved.

The rest of this paper is organized as follows. Section 2 presents a space target motion model and space-based radar observation model. Section 3 derives the multi-target tracking PCRLB. Furthermore, the dynamic RCS sequence mapping method and the adaptive joint dynamic RCS power and bandwidth allocation method is given in Section 4. Simulation results and conclusions are given in Sections 5 and 6, respectively.

2. Space Target Observation System Model

2.1. Space Target Motion Model

The schematic diagram of space-based radar observation for space targets tracking is depicted in Figure 1. Note that we assume that space-based radar is able to track space targets in the observation region continuously.

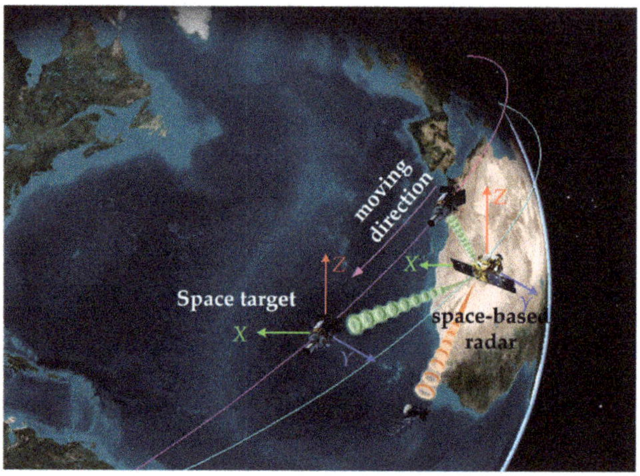

Figure 1. Schematic diagram of space-based radar observation for space targets tracking.

For the sake of simplicity, assuming that there are Q distinguishable space targets in the three-dimensional observation scene and the targets kinemics motion obeys the constant velocity (CV) mode, the motion model of the q-th target at the k-th sample interval is defined as follows:

$$\mathbf{x}_k^q = \mathbf{F}_q \mathbf{x}_{k-1}^q + \mathbf{u}_{k-1}^q \tag{1}$$

where $\mathbf{x}_k^q = [x_{Tk}^q, \dot{x}_{Tk}^q, y_{Tk}^q, \dot{y}_{Tk}^q, z_{Tk}^q, \dot{z}_{Tk}^q]^{\mathrm{T}}$ denotes the target state. $(x_{\mathrm{Tk}}^q, y_{\mathrm{Tk}}^q, z_{\mathrm{Tk}}^q)$ and $(\dot{x}_{\mathrm{Tk}}^q, \dot{y}_{\mathrm{Tk}}^q, \dot{z}_{\mathrm{Tk}}^q)$ represent the range and velocity of the q-th target at the k-th sample interval in the Cartesian coordinate system. $\mathbf{F}_q$ represents the state transition matrix of q-th target.

$$\mathbf{F}_q = \begin{bmatrix} 1 & \Delta T \\ 0 & 1 \end{bmatrix} \otimes \mathbf{I}_3 \tag{2}$$

where $\otimes$ and ΔT represent the Kronecker operator and sample interval of adjacent tracking moments, respectively. The matrix $\mathbf{I}_3$ denotes the 3×3 identity matrix. In the above motion model, the matrix $\mathbf{u}_{k-1}^q$ is modeled as a multidimensional Gaussian distribution with zero-mean and the covariance $\mathbf{Q}_{k-1}^q$, which is used to represent the random error or fluctuation of the target motion state. Its covariance is expressed as [32]:

$$\mathbf{Q}_{k-1}^q = \begin{bmatrix} \int_0^{\Delta T} t \times t dt & \int_0^{\Delta T} 1 \times t dt \\ \int_0^{\Delta T} t \times 1 dt & \int_0^{\Delta T} 1 \times 1\, dt \end{bmatrix} \otimes \mathrm{e}_q \mathbf{I}_3 = \begin{bmatrix} \frac{1}{3}\Delta T^3 & \frac{1}{2}\Delta T^2 \\ \frac{1}{2}\Delta T^2 & \Delta T \end{bmatrix} \otimes \mathrm{e}_q \mathbf{I}_3 \tag{3}$$

where e_q is the process noise coefficient that has the ability to adjust the size of $\mathbf{Q}_{k-1}^q$.

2.2. Space-Based Observation Model

Assume that the C-MIMO radar with coordinates of (x_k, y_k, z_k) tracks multiple space targets with the simultaneous multi-beam mode. After all the radar echo signals are preprocessed by pulse compression and moving target detection, a series of radar measurements are formed. At the k-th sample interval, the radar tracks Q distinguishable space targets. The relation between the nonlinear observation vector of the q-th target $\mathbf{z}_{q,k}$ and the target state vector is as follows:

$$\mathbf{z}_{q,k} = h_{q,k}(\mathbf{x}_k^q) + \mathbf{v}_{q,k} \tag{4}$$

where $\mathbf{v}_{q,k} \sim \mathcal{N}(0,\ \Xi_{q,k})$ is the measurement noise of the system at the k-th sample interval for the q-th target. $h_{q,k}(\bullet)$ denotes the nonlinear transformation between the target state and the radar measurement, which includes distance $R_{q,k}$, azimuth angle $\theta_{q,k}$ and elevation angle $\varphi_{q,k}$.

The three-dimensional measurements are expressed as

$$\begin{cases} R_{q,k}(\mathbf{x}_k^q) = \sqrt{(x_{Tk}^q - x_k)^2 + (y_{Tk}^q - y_k)^2 + (z_{Tk}^q - z_k)^2} \\ \theta_{q,k}(\mathbf{x}_k^q) = \mathrm{arctan2}((y_{Tk}^q - y_k)/(x_{Tk}^q - x_k)) \\ \varphi_{q,k}(\mathbf{x}_k^q) = \mathrm{arctan2}((z_{Tk}^q - z_k)/\sqrt{(x_{Tk}^q - x_k)^2 + (y_{Tk}^q - y_k)^2}) \end{cases} \tag{5}$$

The covariance matrix of measurement noise $\mathbf{v}_{q,k} \sim \mathcal{N}(0, \Xi_{q,k})$ is shown as follows:

$$\Xi_{q,k} = \begin{bmatrix} \Xi_{R_{q,k}}^2 & 0 & 0 \\ 0 & \Xi_{\theta_{q,k}}^2 & 0 \\ 0 & 0 & \Xi_{\varphi_{q,k}}^2 \end{bmatrix} \propto \begin{bmatrix} (SNR_k^q T_k^q \beta_{q,k}^2)^{-1} & 0 & 0 \\ 0 & (SNR_k^q T_k^q / B_w)^{-1} & 0 \\ 0 & 0 & (SNR_k^q T_k^q / B_w)^{-1} \end{bmatrix} \tag{6}$$

where $\Xi_{R_{q,k}}^2, \Xi_{\theta_{q,k}}^2$, and $\Xi_{\varphi_{q,k}}^2$ represent noise covariance of $R_{q,k}$, $\theta_{q,k}$, and $\varphi_{q,k}$, respectively [33].

Equation (6) can be further arranged as

$$\begin{cases} \Xi_{R_{q,k}}^2 \propto (\alpha_{q,k} P_{q,k} |\sigma_k^q|^2 T_k^q \beta_{q,k}^2)^{-1} \\ \Xi_{\theta_{q,k}}^2 \propto (\alpha_{q,k} P_{q,k} |\sigma_k^q|^2 T_k^q / B_w)^{-1} \\ \Xi_{\varphi_{q,k}}^2 \propto (\alpha_{q,k} P_{q,k} |\sigma_k^q|^2 T_k^q / B_w)^{-1} \end{cases} \tag{7}$$

where $\alpha_{q,k}$ is the attenuation coefficient which is inversely proportional to the fourth power of the distance at the k-th sample interval for the q-th target, $P_{q,k}$ is the assigned transmit power, σ_k^q is the RCS, which depends on the characteristics of the target itself and the RLOS, T_k^q is the beam dwell time, and $\beta_{q,k}$ and B_w are the effective bandwidth of the transmit signal and the width of the receiving beam, respectively.

As seen from Equation (7), the space-based radar measurement covariance is related to σ_k^q and the radar transmit parameters in the tracking process. Due to the observation that the angle of space-based radar to the space target changes greatly in the process of space intersection, the RCS of space moving targets shows dramatic fluctuation characteristics. Adding the dynamic RCS sequence to target real-time tracking scenes a priori information can further improve the accuracy of space multi-target tracking.

3. Multi-Target Tracking PCRLB Recursion

Considering the limitation of radar power and bandwidth in the actual space surveillance task, space targets of different orbits fall into different monitoring regions. As a consequence, the dynamic RCS of different targets changes as the RLOS varies, which affects the tracking performance of the space-based radar simultaneously.

According to the measurement model in Equation (4), we constrain the unbiased estimate $\hat{\mathbf{x}}^q_{k|k}\left(\mathbf{z}_{q,k}\right)$ and the target state as follows:

$$\mathcal{F}^{-1}\left(x^q_k\right) \preceq \mathbb{E}_{\mathbf{x}^q_k,\, \mathbf{z}_{q,k}}\left\{[\hat{\mathbf{x}}^q_{k|k}\left(\mathbf{z}_{q,k}\right) - \mathbf{x}^q_k][\hat{\mathbf{x}}^q_{k|k}\left(\mathbf{z}_{q,k}\right) - \mathbf{x}^q_k]^{\mathrm{T}}\right\} \tag{8}$$

where $\mathbf{z}_{q,k}$ is the radar measurement, $\mathbb{E}_{\mathbf{x}^q_k,\, \mathbf{z}_{q,k}}(\bullet)$ represents the expectation operation of the target state and the radar measurement, and $\mathcal{F}^{-1}\left(\mathbf{x}^q_k\right)$ represents the PCRLB.

Then the Bayesian Fisher information matrix (BFIM) $\mathcal{F}\left(\mathbf{x}^q_k\right)$ corresponding to the q-th target can be obtained by

$$\mathcal{F}\left(\mathbf{x}^q_k\right) = -\mathbb{E}_{\mathbf{x}^q_k,\mathbf{z}^q_k}\left[\nabla^{\mathbf{x}^q_k}_{\mathbf{x}^q_k} \ln p(\mathbf{z}^q_k, \mathbf{x}^q_k)\right] \tag{9}$$

where the notation $\nabla^{\mathbf{x}^q_k}_{\mathbf{x}^q_k}$ is the second order partial derivative vectors and ∇^{κ}_{η} is equal to $\nabla_{\eta}\nabla_{\kappa}{}^{\mathrm{T}}$.

Next, the joint probability density function (PDF) $p(\mathbf{z}^q_k, \mathbf{x}^q_k)$ is given by

$$p(\mathbf{z}^q_k, \mathbf{x}^q_k) = p(\mathbf{x}^q_k)p(\mathbf{z}^q_k|\mathbf{x}^q_k) \tag{10}$$

where $p(\mathbf{x}^q_k)$ is the PDF of the space target state and the $p(\mathbf{z}^q_k|\mathbf{x}^q_k)$ is the joint conditional PDF.

The BFIM $\mathcal{F}\left(\mathbf{x}^q_k\right)$ can be factorized into the prior information FIM $\mathcal{F}_P\left(\mathbf{x}^q_k\right)$ and the data FIM $\mathcal{F}_Z\left(\mathbf{x}^q_k\right)$, which can be calculated as

$$\begin{cases} \mathcal{F}_P\left(\mathbf{x}^q_k\right) = -\mathbb{E}_{\mathbf{x}^q_k}\left[\nabla^{\mathbf{x}^q_k}_{\mathbf{x}^q_k} \ln p(\mathbf{x}^q_k)\right] \\ \qquad\qquad = \mathbf{D}^{22}_{k-1} - \mathbf{D}^{21}_{k-1}\left(\mathcal{F}_P\left(\mathbf{x}^q_{k-1}\right) + \mathbf{D}^{11}_{k-1}\right)^{-1}\mathbf{D}^{12}_{k-1} \\ \mathcal{F}_Z\left(\mathbf{x}^q_k\right) = -\mathbb{E}_{\mathbf{x}^q_k,\mathbf{z}^q_k}\left[\nabla^{\mathbf{x}^q_k}_{\mathbf{x}^q_k} \ln p(\mathbf{z}^q_k|\mathbf{x}^q_k)\right] \end{cases} \tag{11}$$

where

$$\begin{cases} \mathbf{D}^{11}_{k-1} = -\mathbb{E}_{\mathbf{x}^q_{k-1},\mathbf{x}^q_k}\left[\nabla^{\mathbf{x}^q_{k-1}}_{\mathbf{x}^q_{k-1}} \ln p(\mathbf{x}^q_k|\mathbf{x}^q_{k-1})\right] \\ \mathbf{D}^{12}_{k-1} = -\mathbb{E}_{\mathbf{x}^q_{k-1},\mathbf{x}^q_k}\left[\nabla^{\mathbf{x}^q_{k-1}}_{\mathbf{x}^q_k} \ln p(\mathbf{x}^q_k|\mathbf{x}^q_{k-1})\right] = \left(\mathbf{D}^{12}_{k-1}\right)^{\mathrm{T}} \\ \mathbf{D}^{22}_{k-1} = -\mathbb{E}_{\mathbf{x}^q_{k-1},\mathbf{x}^q_k}\left[\nabla^{\mathbf{x}^q_k}_{\mathbf{x}^q_k} \ln p(\mathbf{x}^q_k|\mathbf{x}^q_{k-1})\right] \end{cases} \tag{12}$$

By substituting the motion model of targets into (18) and simplifying the expectation calculation, the following can be obtained:

$$\begin{cases} \mathbf{D}^{11}_{k-1} = \mathrm{F}^{\mathrm{T}}\mathrm{Q}^{-1}\mathrm{F} \\ \mathbf{D}^{12}_{k-1} = -\mathrm{F}^{\mathrm{T}}\mathrm{Q}^{-1} \\ \mathbf{D}^{22}_{k-1} = \mathrm{Q}^{-1} \end{cases} \tag{13}$$

Finally, the BFIM $\mathcal{F}\left(\mathbf{x}_k^q\right)$ yields

$$\begin{aligned}\mathcal{F}\left(\mathbf{x}_k^q\right) &= \mathcal{F}_P\left(\mathbf{x}_k^q\right) + \mathcal{F}_Z\left(\mathbf{x}_k^q\right) \\ &= \underbrace{\left[\mathbf{Q}_{k-1}^q + \mathbf{F}_q \mathcal{F}^{-1}\left(\mathbf{x}_{k-1}^q\right)\mathbf{F}_q^{\mathrm{T}}\right]^{-1}}_{\mathcal{F}_P(\mathbf{x}_k^q)} + \underbrace{\mathbb{E}\left[\mathbf{H}_{q,k}^{\mathrm{T}}(\Xi_{q,k})^{-1}\mathbf{H}_{q,k}\right]}_{\mathcal{F}_Z(\mathbf{x}_k^q)} \\ &\approx \left[\mathbf{Q}_{k-1}^q + \mathbf{F}_q \mathcal{F}^{-1}\left(\mathbf{x}_{k-1}^q\right)\mathbf{F}_q^{\mathrm{T}}\right]^{-1} + \mathbf{H}_{q,k}^{\mathrm{T}}(\hat{\Xi}_{q,k})^{-1}\mathbf{H}_{q,k}\end{aligned} \tag{14}$$

where $\mathbf{x}_{k|k-1}^q$ represents the predicted state vector of the q-th target, $\hat{\Xi}_{q,k} = \Xi_{q,k}(P_{q,k}, \beta_{q,k}, \mathbf{x}_k^q, \sigma_{q,k})$ represents the measurement covariance matrix of $\mathbf{x}_{k|k-1}^q$, $\mathbf{H}_{q,k}^{\mathrm{T}}$ is the Jacobian matrix of the measurement function $h_{q,k}(\bullet)$ with respect to the target state $\mathbf{x}_k^q$, and $\mathbf{H}_{q,k}^{\mathrm{T}}$ can be expressed as

$$\mathbf{H}_{q,k}^{\mathrm{T}} = \left[\nabla_{\mathbf{x}_k^q} R_{q,k}, \nabla_{\mathbf{x}_k^q}\theta_{q,k}, \nabla_{\mathbf{x}_k^q}\varphi_{q,k}\right] \tag{15}$$

where

$$\nabla_{\mathbf{x}_k^q} \mathrm{R}_{q,k} = \begin{bmatrix} \frac{x_{Tk}^q - x_k}{\left[(x_{Tk}^q - x_k)^2 + (y_{Tk}^q - y_k)^2 + (z_{Tk}^q - z_k)^2\right]^{1/2}} \\ 0 \\ \frac{y_{Tk}^q - y_k}{\left[(x_{Tk}^q - x_k)^2 + (y_{Tk}^q - y_k)^2 + (z_{Tk}^q - z_k)^2\right]^{1/2}} \\ 0 \\ \frac{z_{Tk}^q - z_k}{\left[(x_{Tk}^q - x_k)^2 + (y_{Tk}^q - y_k)^2 + (z_{Tk}^q - z_k)^2\right]^{1/2}} \\ 0 \end{bmatrix} \tag{16}$$

$$\nabla_{\mathbf{x}_k^q} \theta_{q,k} = \begin{bmatrix} -(y_{Tk}^q - y_k)/\left[(x_{Tk}^q - x_k)^2 + (y_{Tk}^q - y_k)^2\right] \\ 0 \\ (x_{Tk}^q - x_k)/\left[(x_{Tk}^q - x_k)^2 + (y_{Tk}^q - y_k)^2\right] \\ 0 \\ 0 \\ 0 \end{bmatrix} \tag{17}$$

$$\nabla_{\mathbf{x}_k^q} \varphi_{q,k} = \begin{bmatrix} -\frac{(x_{Tk}^q - x_k) \times (z_{Tk}^q - z_k)/\left[(x_{Tk}^q - x_k)^2 + (y_{Tk}^q - y_k)^2\right]^{1/2}}{(x_{Tk}^q - x_k)^2 + (y_{Tk}^q - y_k)^2 + (z_{Tk}^q - z_k)^2} \\ 0 \\ -\frac{(y_{Tk}^q - y_0) \times (z_{Tk}^q - z_k)/\left[(x_{Tk}^q - x_k)^2 + (y_{Tk}^q - y_k)^2\right]^{1/2}}{(x_{Tk}^q - x_k)^2 + (y_{Tk}^q - y_k)^2 + (z_{Tk}^q - z_k)^2} \\ 0 \\ \frac{\left[(x_{Tk}^q - x_k)^2 + (y_{Tk}^q - y_k)^2\right]^{1/2}}{(x_{Tk}^q - x_k)^2 + (y_{Tk}^q - y_k)^2 + (z_{Tk}^q - z_k)^2} \\ 0 \end{bmatrix} \tag{18}$$

where $\nabla_{\mathbf{x}_k^q}(\bullet)$ represents the first order partial derivative vector. By performing the inversion operation on $\mathcal{F}\left(\mathbf{x}_k^q\right)$, the PCRLB matrix of target state $\mathbf{x}_k^q$ can be obtained.

$$M_{PCRLB}(P_{q,k}, \beta_{q,k}, \mathbf{x}_k^q, \sigma_k^q) = \mathcal{F}^{-1}\left(P_{q,k}, \beta_{q,k}, \mathbf{x}_k^q, \sigma_k^q\right) \tag{19}$$

where the diagonal elements of the $M_{PCRLB}(P_{q,k}, \beta_{q,k}, \mathbf{x}_k^q, \sigma_k^q)$ correspond to the lower bound for the unbiased estimation variance of the target state vector.

Considering that $M_{PCRLB}(P_{q,k}, \beta_{q,k}, \mathbf{x}_k^q, \sigma_k^q)$ is a function of radar transmit power, bandwidth, RCS, and target state, the cost function can be given by

$$\mathbb{F}(P_k, \beta_k, \sigma_k) = \max\left(\sqrt{\mathrm{tr}\left\{M_{PCRLB}(P_{q,k}, \beta_{q,k}, \mathbf{x}_k^q, \sigma_k^q)\right\}}_{q=1,2,\ldots,Q}\right) \tag{20}$$

where at the k-th sample interval and for Q independent targets, $P_k = [P_{1,k}, P_{q,k}, \ldots, P_{Q,k}]^{\mathrm{T}}$, $\beta_k = [\beta_{1,k}, \beta_{q,k}, \ldots, \beta_{Q,k}]^{\mathrm{T}}$, and $\sigma_k = [\sigma_k^1, \sigma_k^2, \ldots, \sigma_k^Q]^{\mathrm{T}}$ represent the transmit power allocation scheme, radar bandwidth allocation scheme, and the set of dynamic RCS of all space targets, respectively, and $\mathbb{F}(P_k, \beta_k, \sigma_k)$ denotes the PCRLB of the target with the worst radar tracking accuracy.

4. Joint Dynamic RCS Power and Bandwidth Allocation Method

4.1. Dynamic RCS Sequence Mapping under Space-Based Observation

For the typical space targets in the satellite database, we can establish the target full-angle static RCS database by computer-aided-design (CAD) modeling, grid meshing, scattering characteristic computation, and post processing procedure. The RLOS has a great influence on target RCS characteristics in the actual detection and tracking scenario. In order to obtain the exact dynamic RCS sequence of the space targets through space-based radar, the space target observation geometry transformation is derived in this section, ensuring the correct RLOS.

Considering that the ECI (Earth-centered inertial) coordinate system is an inertial coordinate system, this coordinate system has good performance and adaptability for real-time target tracking, and can track moving targets more conveniently and accurately. Therefore, this paper adopts the ECI coordinate system to describe the relative position relationship between space-based radar and space targets. As illustrated in Figure 2, the ECI coordinate system $O_e - X_eY_eZ_e$ is marked in black lines. The center of the Earth is the origin of the ECI coordinate system. The Z_e axis points to the pole of the agreement along the Earth's rotation axis, and the X_e is located in the equatorial plane and points to the equinox. The Y_e axis conforms to the right-hand Cartesian coordinate system. The satellite-based orbital RTN (radial, tangential, normal) coordinate system $O_t - X_tY_tZ_t$ is marked in red lines. For the space target of the Earth-oriented attitude stabilization mode, the X_t axis points to the Earth's center from the center of mass of the space target, the Y_t axis is in the orbit plane and constrained by the direction of the target velocity, and the Z_t axis is the normal of the orbit plane.

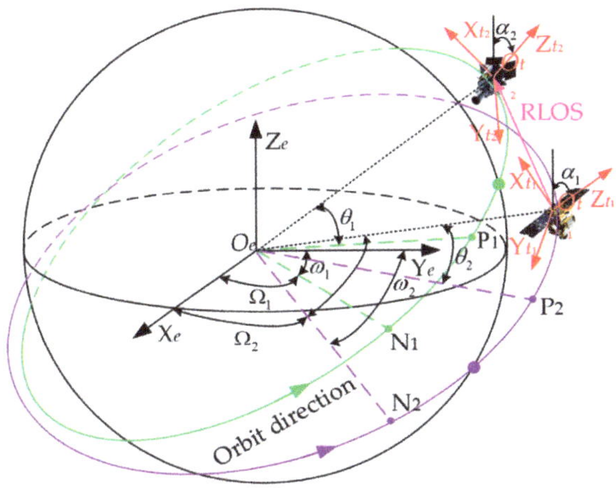

Figure 2. Schematic of the coordination systems transformation.

The space targets orbit can be determined by the six orbital elements deriving from a given two-line element (TLE). It can be expressed as a (semi-major axis), α (inclination), Ω (right ascension of the ascending node), e (eccentricity), ω (argument of perigee), and θ (true anomaly) [34–36]. Thereafter, the accurate RLOS can be obtained by calculating the real-time position of space-based radar in the RTN coordinate system of the space target. The derivation of RLOS is as follows:

(1) The positions of the space-based radar in the ECI coordinate system are given as

$$\begin{aligned}\mathbf{O}_{t_2\mathrm{ECI}} &= \mathbf{R}_Z(\Omega_2)\mathbf{R}_X(\alpha_2)\mathbf{R}_Z(\omega_2)\\ &\times\left[a_2\frac{e_2+\cos\theta_2(k)}{1+e_2\cos\theta_2(k)} \quad a_2\frac{(1-e_2{}^2)\sin\theta_2(k)}{1+e_2\cos\theta_2(k)} \quad 0\right]^{\mathrm{T}}\end{aligned} \tag{21}$$

where $\mathbf{R}_X$ and $\mathbf{R}_Z$ represent the rotation matrix around the X axis and the Z axis, respectively. Ω_2, α_2, ω_2, e_2, and $\theta_2(k)$ are the orbital parameters of space-based radar at the k-th sample interval.

(2) Then the coordinate transformation matrix between the ECI coordinate system and the RTN coordinate system of space target can be expressed as

$$\begin{aligned}M_{\mathrm{ECI-RTN}}(\Omega_1,\alpha_1,\omega_1,\theta_1(k)) &= [\mathbf{R}_Z(\Omega_1)\mathbf{R}_X(\alpha_1)\mathbf{R}_Z(\omega_1+\theta_1(k))]^{-1}\\ &= \left(\begin{bmatrix}\cos\Omega_1 & -\sin\Omega_1 & 0\\ \sin\Omega_1 & \cos\Omega_1 & 0\\ 0 & 0 & 1\end{bmatrix} \times\begin{bmatrix}1 & 0 & 0\\ 0 & \cos\alpha_1 & -\sin\alpha_1\\ 0 & \sin\alpha_1 & \cos\alpha_1\end{bmatrix} \times\begin{bmatrix}\cos(\omega_1+\theta_1(k)) & -\sin(\omega_1+\theta_1(k)) & 0\\ \sin(\omega_1+\theta_1(k)) & \cos(\omega_1+\theta_1(k)) & 0\\ 0 & 0 & 1\end{bmatrix}\right)^{-1}\end{aligned} \tag{22}$$

where Ω_1, α_1, ω_1, and $\theta_1(k)$ are the orbital parameters of the space target at the k-th sample interval.

(3) Thus, the position of space-based radar in the RTN coordinate system of the space target can be derived as

$$\mathbf{O}_{t_2\text{RTN}} = \mathbf{M}_{\text{ECI-RTN}}(\Omega_1, \alpha_1, \omega_1, \theta_1(k))\mathbf{O}_{t_2\text{ECI}} \tag{23}$$

(4) Finally, the RLOS, azimuth angle and elevation angle are calculated as

$$\overrightarrow{RLOS(k)} = \mathbf{O}_{t_2\text{RTN}} = (\mathrm{x}_{RLOS}(k), \mathrm{y}_{RLOS}(k), \mathrm{z}_{RLOS}(k)) \tag{24}$$

$$\theta_{RLOS}(k) = \arctan\left(\mathrm{x}_{RLOS}(k) \Big/ \mathrm{y}_{RLOS}(k)\right) \tag{25}$$

$$\varphi_{RLOS}(k) = \arctan\left(\mathrm{z}_{RLOS}(k) \Big/ \sqrt{\mathrm{x}_{RLOS}^2(k) + \mathrm{y}_{RLOS}^2(k) + \mathrm{z}_{RLOS}^2(k)}\right) \tag{26}$$

where $(\mathrm{x}_{RLOS}(k), \mathrm{y}_{RLOS}(k), \mathrm{z}_{RLOS}(k))$ represents the RLOS in the space target RTN coordinate system at the k-th sample interval. $\theta_{RLOS}(k)$ and $\varphi_{RLOS}(k)$ are the azimuth angle and elevation angle of the $\overrightarrow{RLOS(k)}$, respectively. $\overrightarrow{RLOS(k)}$ can be utilized to calculate $\theta_{RLOS}(k)$ and $\varphi_{RLOS}(k)$ to realize the dynamic RCS sequence mapping.

Considering the true anomaly can uniquely determine that the position of the space target and the orbit of low Earth orbit (LEO) satellite is close to a circle. The true anomaly $\theta_1(k)$ of space target and $\theta_2(k)$ of space-based radar at the k-th sample interval can be calculated by

$$\theta_1(k) = (\theta_1 + k\Delta T \times \sqrt{\mu/a_1}) \tag{27}$$

$$\theta_2(k) = (\theta_2 + k\Delta T \times \sqrt{\mu/a_2}) \tag{28}$$

where $\mu = 3.9861 \times 10^{14}$ $(\mathrm{m}^3/\mathrm{s}^2)$ is the Kepler constant. θ_1 and θ_2 are the initial true anomaly of the space target and space-based radar, respectively.

The dynamic RLOS can be given as

$$\overrightarrow{RLOS(k)} = \mathbf{M}_{\text{ECI-RTN}}(\Omega_1, \alpha_1, \omega_1, \theta_1(k))\mathbf{O}_{t_2\text{ECI}}(\theta_2(k)) \tag{29}$$

Finally, the flowchart of dynamic RCS sequence mapping is shown in Figure 3.

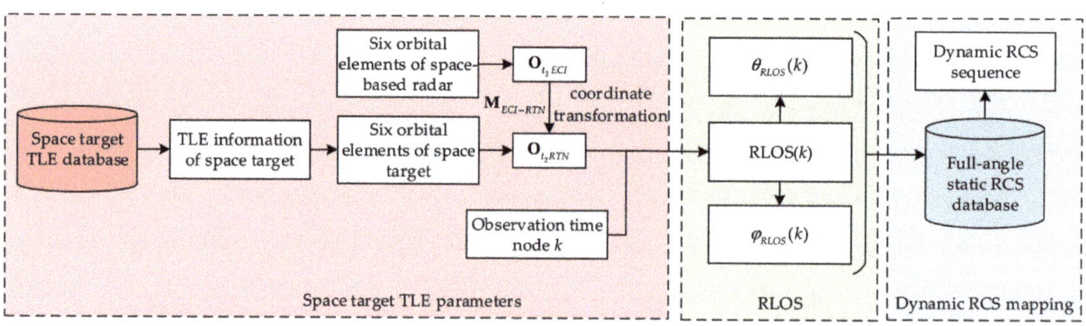

Figure 3. Flowchart of the dynamic RCS sequence mapping.

4.2. Joint Dynamic RCS Power and Bandwidth Allocation Optimization Modeling

In this section, the Min-Max PCRLB (minimum maximum PCRLB) optimization criterion is employed to solve the power and bandwidth allocation problem in limited transmit resources of collocated MIMO radar. By fully utilizing the dynamic target RCS fluctuation characteristics, the Min-Max PCRLB optimization model can be modeled as

$$\begin{aligned} &min[\mathbb{F}(P_k, \beta_k, \sigma_k^q(\theta_{RLOS}(k), \varphi_{RLOS}(k)))] \\ &s.t. \quad \mathbf{1}_Q^T P_{q,k} = P_{total}, q = 1,2,\ldots,Q \\ &\quad \overline{P}_{\min} \leq P_{q,k} \leq \overline{P}_{\max} \\ &\quad \mathbf{1}_Q^T \beta_{q,k} = \beta_{total}, q = 1,2,\ldots,Q \\ &\quad \overline{\beta}_{\min} \leq \beta_{q,k} \leq \overline{\beta}_{\max} \\ &\quad \sigma_k^q(\theta_{RLOS}(k), \varphi_{RLOS}(k)) \in \{\sigma^Q\}, q = 1,2,\ldots,Q \\ &\quad \theta_{RLOS}(k) \in [0, 360^\circ] \\ &\quad \varphi_{RLOS}(k) \in [0, 180^\circ] \end{aligned} \tag{30}$$

where P_{total} and β_{total} are the total transmit power and bandwidth, respectively, $\overline{P}_{\max}$, $\overline{P}_{\min}$, $\overline{\beta}_{\max}$, and $\overline{\beta}_{\min}$ are the maximum power, the minimum power, the maximum bandwidth, and the minimum bandwidth of a single tracking beam, respectively, 1_Q^T is the vector of size $1 \times Q$ whose total elements is 1, σ^Q represents the multi-target full-angle static RCS database, and $\theta_{RLOS}(k)$ and $\varphi_{RLOS}(k)$ are the azimuth angle and elevation angle of the RLOS, respectively.

4.3. Joint Dynamic RCS Power and Bandwidth Allocation Method

It can be seen from Equations (7) and (20) that the power and bandwidth variables of space-based radar are coupled in the optimized objective function. Since the bandwidth is applied to the optimization function by squaring, the adaptive joint power and bandwidth allocation problem in this paper is a typical non-convex optimization problem. In order to solve the above problem, the C-MIMO radar bandwidth is first allocated equally, and the sub-optimal scheme of power allocation is obtained by using the convex optimization algorithm. Then the bandwidth constraint is relaxed, and finally the cyclic minimization method is used to obtain the optimal allocation results to solve the problem. The illustration of the joint dynamic RCS power and bandwidth allocation (JRPBA) method is shown in Figure 4. Additionally, the detailed process of the resource allocation method is shown in Table 1.

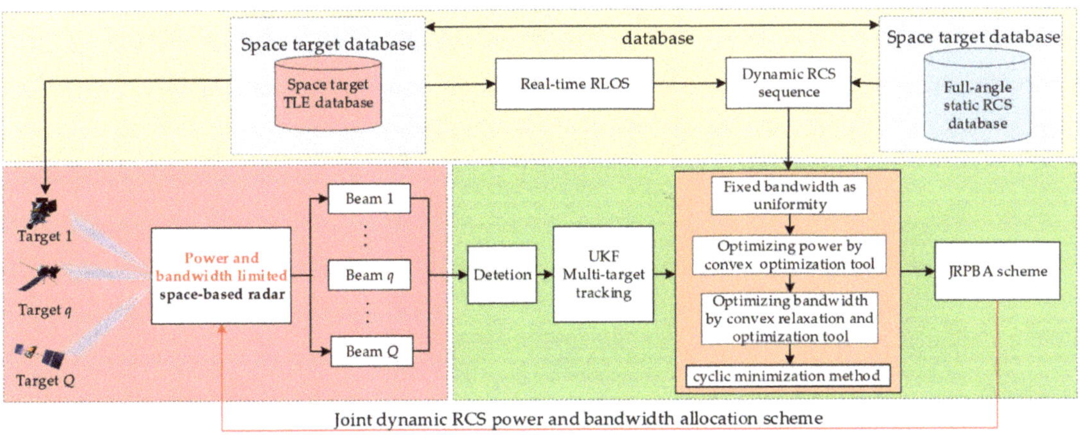

Figure 4. Illustration of the proposed JRPBA method.

Table 1. JRPBA method steps.

Algorithmic Flowchart:		
Step 1:	Initializes $k = 1$, $\beta_{k,opt} = \beta_0$, β_0 is the bandwidth uniformity allocation scheme. Initializes $P_{k,opt} = P_0$, P_0 is the power uniformity allocation scheme.	
Step 2:	According to the $P_{k,opt}$, the radar obtains the observation $\mathbf{z}_{q,k}(P_{k,opt})$ and estimates the target distance $R_{q,k}$, azimuth angle $\theta_{q,k}$, and elevation angle $\varphi_{q,k}$.	
Step 3:	The dynamic RCS $\sigma_k^q(\theta_{RLOS}(k), \varphi_{RLOS}(k))$ of the target is obtained by the one-step predicted motion state, and the variances $\Xi^2_{R_{q,k}}$, $\Xi^2_{\theta_{q,k}}$, and $\Xi^2_{\varphi_{q,k}}$ of each observation are calculated according to Equation (7).	
Step 4:	Constructing the resource allocation optimization model according to Equation (30).	
Step 5:	Step 5-1:	Calculating the power allocation scheme $P_{k,opt}$ at the k-th sample interval. Relaxing the constraints of bandwidth.
	Step 5-2:	Introducing $\gamma_k = [\gamma_k^1, \gamma_k^2, \ldots, \gamma_k^Q]^{\mathrm{T}} = [\lvert\beta_k^1\rvert^2, \lvert\beta_k^2\rvert^2, \ldots, \lvert\beta_k^Q\rvert^2]^{\mathrm{T}}$, λ_1, and λ_2. Transforming the bandwidth optimization problem into a convex optimization problem.
	Step 5-3:	Calculating the bandwidth sub-optimal scheme $\beta_{k,sub-opt}$ at the k-th sample interval.
	Step 5-4:	$\beta_{k,sub-opt}$ is utilized as the initial solution of the exact search algorithm (cyclic minimization method) to obtain the optimal solution $\beta_{k,opt}$. The algorithm flow of the cyclic minimization method is as follows: Step 5-4-1: Set $\beta_{l_0} = \beta_{k,sub-opt}$, set the step size $\triangle\beta$ and algorithm termination condition ε. Step 5-4-2: $\begin{cases} \beta_l = \mathrm{argmax}\{\mathbb{F}(P_k, \beta^+{}_{l-1}, \sigma_k^q(\theta_{RLOS}(k), \varphi_{RLOS}(k))) - \mathbb{F}(P_k, \beta_{l-1}, \sigma_k^q(\theta_{RLOS}(k), \varphi_{RLOS}(k)))\} \\ s.t. \bar{\beta}_{\min} \le \beta_k^q \le \bar{\beta}_{\max} \\ \triangle\beta_l = \triangle\beta_{l-1}(\mathbf{1}_Q^{\mathrm{T}}\beta_l / \mathbf{1}_Q^{\mathrm{T}}\beta_{l-1}) \end{cases}$ Step 5-4-3: while $\mathbb{F}(P_k, \beta_{l-1}, \sigma_k^q(\theta_{RLOS}(k), \varphi_{RLOS}(k))) - \mathbb{F}(P_k, \beta_l, \sigma_k^q(\theta_{RLOS}(k), \varphi_{RLOS}(k))) > \varepsilon$ do $\beta_{MLE} = \beta_{l-1}$, end
	Step 5-5:	Feedback the allocation results $\beta_{k,opt}$ and $P_{k,opt}$ to form the allocation scheme.
Step 6:	Let $k = k + 1$, return to Step2.	
Step 7:	Achieve the JRPBA.	

5. Simulation and Performance Evaluation

5.1. Simulation Scenario and Parameters

In the given simulation scenario, Figure 5 represents a typical multi-target tracking scenario with four space targets within the surveillance region. The space-based satellite in this scenario utilizes a C-MIMO radar system to track these four space targets. The TLE (two-line element) information [37] for the selected four space targets is depicted in Table 2. Table 3 displays the six orbital elements of the space-based radar, and Table 4 provides the radar system parameters in detail.

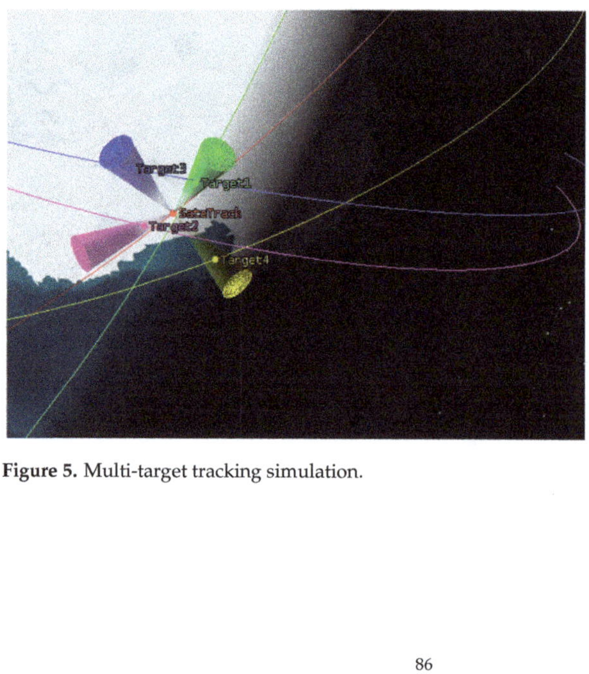

Figure 5. Multi-target tracking simulation.

Table 2. TLE information for the four space targets.

Targets	TLE Information
Target 1	1 54109U 22137A 22313.89342876 .00018851 00000+0 32060-3 0 9998 2 54109 97.0764 33.6544 0008642 190.3376 169.7701 15.51428531 2950
Target 2	1 54148U 22138AM 22313.91668981 -.00038156 00000+0 -38369-2 0 9992 2 54148 87.3610 211.0069 0008503 225.5845 43.1975 14.91526080 1304
Target 3	1 54040U 22132F 22314.20664134 .00056903 00000+0 30279-2 0 9996 2 54040 97.1602 317.8748 0122535 126.4581 234.7794 15.08704769 4257
Target 4	1 37348U 11002A 22175.00479290 .00018492 00000-0 15645-3 0 9997 2 37348 97.8853 286.4946 0526835 125.0700 234.9893 14.81374348 3501

Table 3. Six orbital elements of space-based radar.

Orbital Elements	Value
Semimajor Axis	6678.1 km
Inclination	98.5°
Eccentricity	0
RAAN	−20°
Argument of perigee	0°
True anomaly	245°

Table 4. Radar system parameters.

Parameters	λ	P_{total}	$\overline{P}_{\min}$	$\overline{P}_{\max}$	β_{total}	$\overline{\beta}_{\min}$	$\overline{\beta}_{\max}$
Value	0.3 m	3000 W	300 W	2100 W	100 MHz	10 MHz	70 MHz

In the simulation experiments, the UKF (unscented Kalman filter) recursion is employed for tracking space targets using one hundred frames of radar raw data [38], with a data update rate of 1 s and 100 Monte Carlo trails. To provide a more intuitive comparison of multi-target tracking performance and tracking accuracy, the root mean square error (RMSE) of location and velocity are utilized and presented in this section.

$$RMSE_{position}^{k,q} = \sqrt{\frac{1}{MC}\sum_{i=1}^{MC}\left[\left(x_{Tk}^{q} - \hat{x}_{Tk}^{q,i}\right)^2 + \left(y_{Tk}^{q} - \hat{y}_{Tk}^{q,i}\right)^2 + \left(z_{Tk}^{q} - \hat{z}_{Tk}^{q,i}\right)^2\right]} \tag{31}$$

$$RMSE_{velocity}^{k,q} = \sqrt{\frac{1}{MC}\sum_{i=1}^{MC}\left[\left(\dot{x}_{Tk}^{q} - \hat{\dot{x}}_{Tk}^{q,i}\right)^2 + \left(\dot{y}_{Tk}^{q} - \hat{\dot{y}}_{Tk}^{q,i}\right)^2 + \left(\dot{z}_{Tk}^{q} - \hat{\dot{z}}_{Tk}^{q,i}\right)^2\right]} \tag{32}$$

where MC represents the total Monte Carlo trials and $(\hat{x}_{Tk}^{q,i}, \hat{y}_{Tk}^{q,i}, \hat{z}_{Tk}^{q,i})$ and $(\hat{\dot{x}}_{Tk}^{q,i}, \hat{\dot{y}}_{Tk}^{q,i}, \hat{\dot{z}}_{Tk}^{q,i})$ are the estimated position and velocity of the i-th Monte Carlo simulation, respectively.

5.2. Dynamic RCS Sequence Mapping

In this section, the monostatic static RCS database of the space targets is calculated using electromagnetic computing software, and the simulation results are depicted in Figure 6a–d. To obtain the RLOS parameter sequences in practical applications, different methods can be employed. Here, we determine the position of satellite targets using TLE and calculate their relative position to the radar station using the SGP4 (simplified general perturbations 4) model through aerospace orbit calculation software. The RLOS parameters can then be computed using Equation (29). Finally, the dynamic RCS sequence results can be obtained by combining the static full-angle RCS and the RLOS parameters.

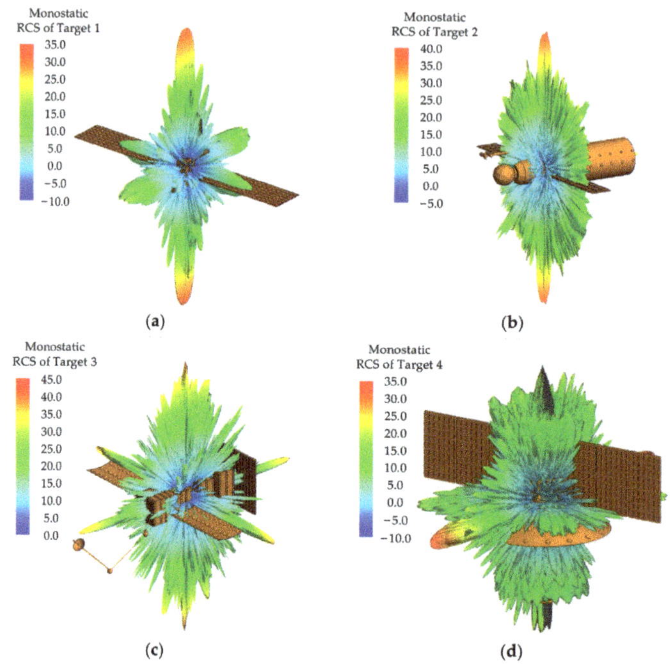

Figure 6. Monostatic RCS of four typical space targets. (**a**) Static RCS of target 1. (**b**) Static RCS of target 2. (**c**) Static RCS of target 3. (**d**) Static RCS of target 4.

Figures 7 and 8 depict the full-angle monostatic RCS of four typical space targets and the RLOS parameters of the four space targets, respectively. Figure 9 displays the dynamic RCS sequence obtained through dynamic RLOS during tracking. Upon examining Figure 7, it is evident that the fluctuation characteristics of the monostatic RCS differ among the four typical space targets as the azimuth angle varies.

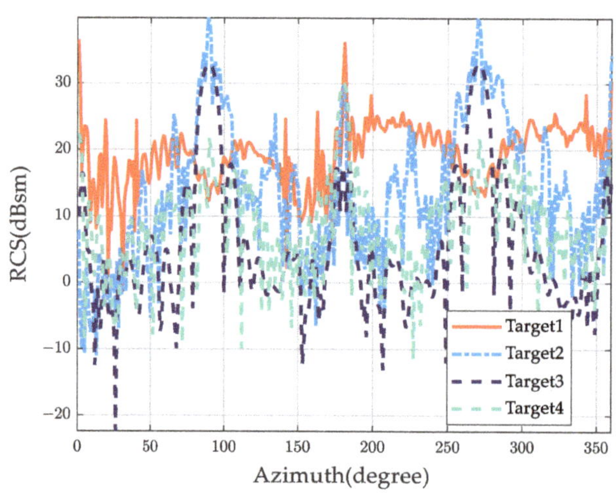

Figure 7. Monostatic RCS of four typical space targets (azimuth degree varies from 0° to 360°, elevation degree is fixed at 90°).

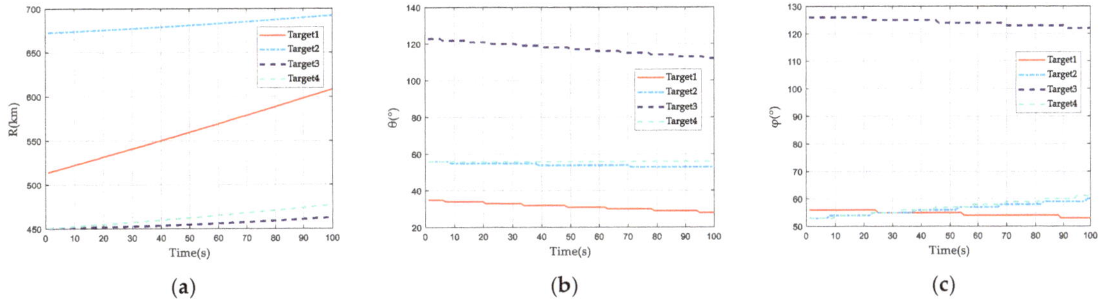

Figure 8. Dynamic RLOS of four typical space targets. (**a**) Range of RLOS. (**b**) θ of RLOS. (**c**) φ of RLOS.

Notice that from Figure 9 it can be observed that the dynamic RCS of different targets varies to a significant degree. The dynamic RCS of target 1 and target 2 fluctuates in a large dynamic range, while target 3 generally has the smallest RCS among the four targets. In addition, the dynamic RCS of target 1, target 2, and target 4 overlapped with each other. Therefore, the real-time dynamic RCS results can be effectively utilized to enhance the accuracy of multi-target tracking.

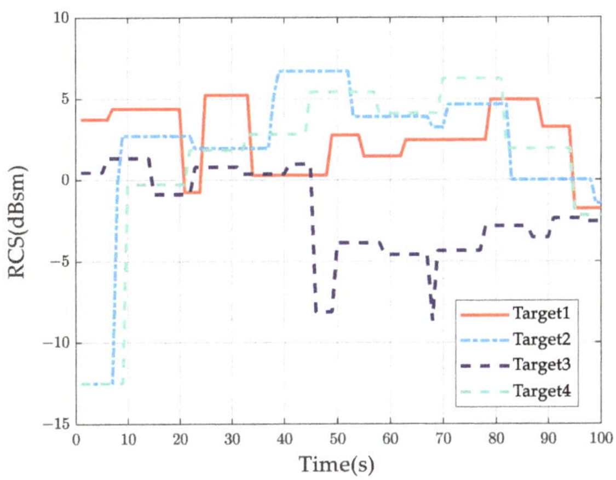

Figure 9. Dynamic RCS sequence results.

5.3. Joint Dynamic RCS Power and Bandwidth Allocation Method Results

Figure 10a gives the multi-target tracking result of the space targets monitored by the space-based radar. As can be seen from Figure 10b–e, the C-MIMO radar achieves multiple target tracking with simultaneous multi-beam mode. The image indicated by the arrow in Figure 10 b–e is a zoomed-in view of the region inside the red box.

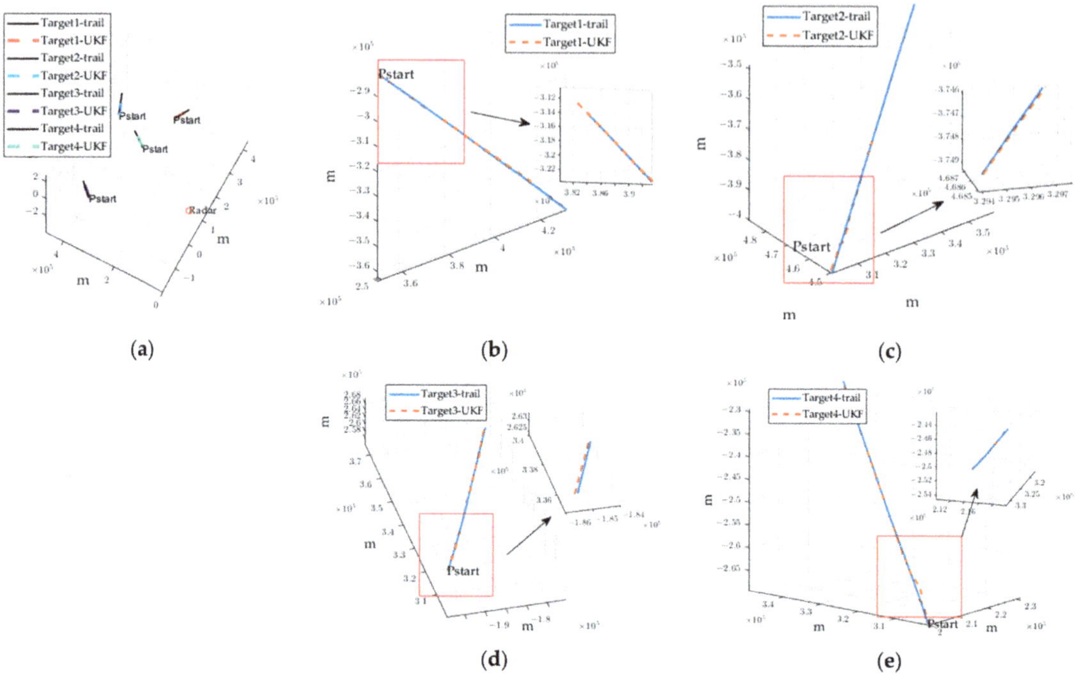

Figure 10. Space targets tracking results. (**a**) Multi-target tracking. (**b**) Tracking result of target 1. (**c**) Tracking result of target 2. (**d**) Tracking result of target 3. (**e**) Tracking result of target 4.

The Min-Max PCRLB criterion is employed in this paper to dynamically allocate the power and bandwidth resource of the space-based radar. The core idea of the Min-Max PCRLB criterion is to optimize the target with the worst tracking accuracy. By solving Equation (30), the proposed method calculates the joint allocation result of power and bandwidth, as shown in Figure 11.

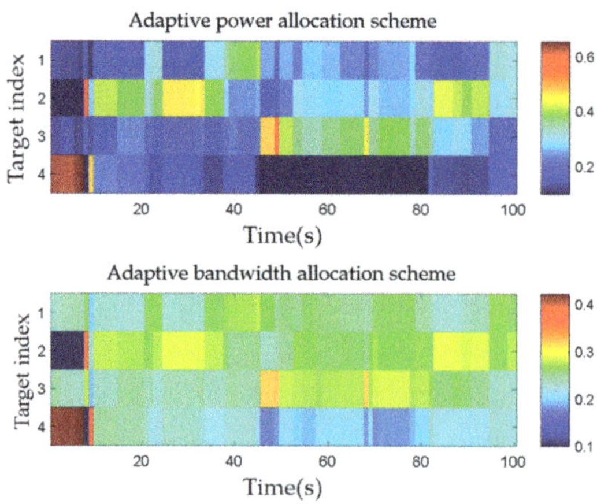

Figure 11. Adaptive joint power and bandwidth allocation result.

As can be seen from Figure 11, the dynamic RCS of four space targets is relatively close at the tenth frame and the later tracking stage. The allocation of power and bandwidth resources in the space-based radar is influenced by the distance between the targets and the radar. As a result, targets that are located farther from the radar receive a higher allocation of radar resources. From Figure 11, it can be seen that target 2 is the farthest from the radar and the target 3 and target 4 have the closest distance. So, when the RCS of the four targets is close, the largest power bandwidth resources is allocated to target 2 and the least radar resources are allocated to target 3 and target 4.

Moreover, as the dynamic RCS of the space targets rapidly declines, the tracking accuracy of the space-based radar on these targets decreases. In order to maintain the original tracking accuracy, the proposed method allocates more system resources to the specific space target. Specifically, space target 3 receives the largest allocation of radar system resources. When tracking the 46th frame, the real-time dynamic RCS of target 3 is the smallest and to guarantee the tracking accuracy, more radar power and bandwidth resources are allocated to target 3. When tracking the 46th frame to the 82nd frame, the distance between the space-based radar and target 4 is the smallest and the dynamic RCS of target 4 is the largest, and the power and bandwidth resources allocated to target 4 are minimal. By analyzing the above simulation results, it can be seen that the distances and observation angles between the space targets play vital roles in influencing the performance of joint power and bandwidth allocation.

Figures 12 and 13 depict the RMSE curves and PCRLB of position and velocity of 100 Monte Carlo experiments during the whole tracking scenario. It can be seen from Figures 12 and 13 that both the ideal PCRLB for position tracking accuracy and the ideal PCRLB for velocity tracking accuracy experience a significant decrease starting from the tenth frame. Considering that the Min-Max PCRLB criterion is utilized for multi-target joint power bandwidth allocation, the PCRLB of position and the PCRLB of velocity for four space targets have the same tendency. Since the Min-Max PCRLB criterion is employed for the multi-target joint power band-width allocation, the PCRLBs for position and velocity of all four space targets exhibit a similar trend.

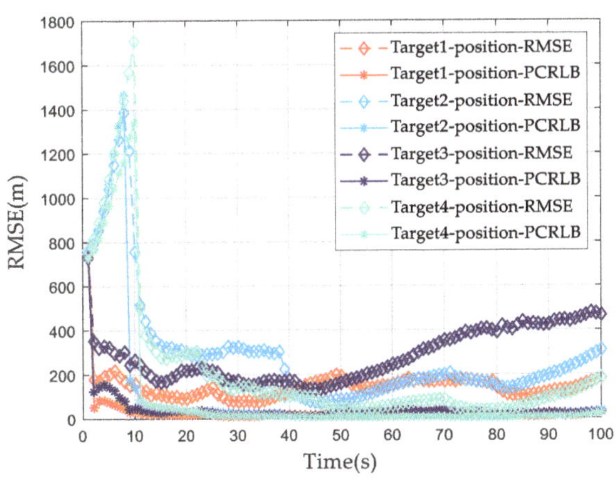

Figure 12. The position PCRLB and RMSE of the adaptive allocation scheme.

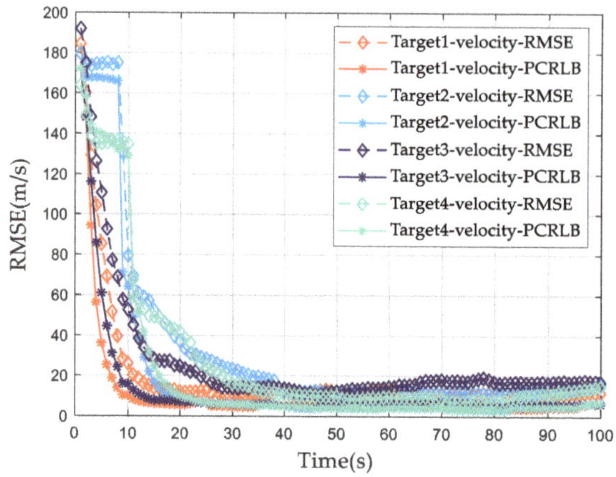

Figure 13. The velocity PCRLB and RMSE of the adaptive allocation scheme.

It can also be seen from Figures 12 and 13 that the RMSE of position and the RMSE of velocity are affected by the dynamic target RCS. Notably, when the RCS of a target decreases rapidly within a frame, the corresponding RMSE of the target's position increases. For instance, in the 45th frame, the dynamic RCS of target 3 decreases and its corresponding tracking error increases. By optimizing the tracking performance of the target with the worst tracking accuracy, the radar power and bandwidth resources are increased to be allocated for target 3, so that the RMSE of the position estimation of target 3 decreases in the later tracking process.

5.4. Joint Power and Bandwidth Allocation Scheme Mismatch Results

To compare the multi-target tracking accuracy under different RCS models, tracking results of the traditional RCS fluctuation model and tracking results of the high dynamic RCS fluctuation adaptive joint power and bandwidth allocation are compared in this paper.

Figure 14 displays the traditional RCS fluctuation model, while Figure 15 showcases the joint power and bandwidth allocation results for multi-target tracking using this mod-el. Notice that the overall tendency of the power bandwidth allocated to the four space targets is mainly influenced by the distance between the space target and the space-based radar. Since target 2 has the greatest distance from the radar, it receives the highest allocation of radar resources. Furthermore, the RCS of four space targets is intertwined, and the radar resources obtained by the four space targets change at different times. However, the change trend is irregular and the optimal allocation of radar power and bandwidth is not formed according to the dynamic RCS fluctuation trend of space targets, which leads to the divergence of tracking error or target missing problems.

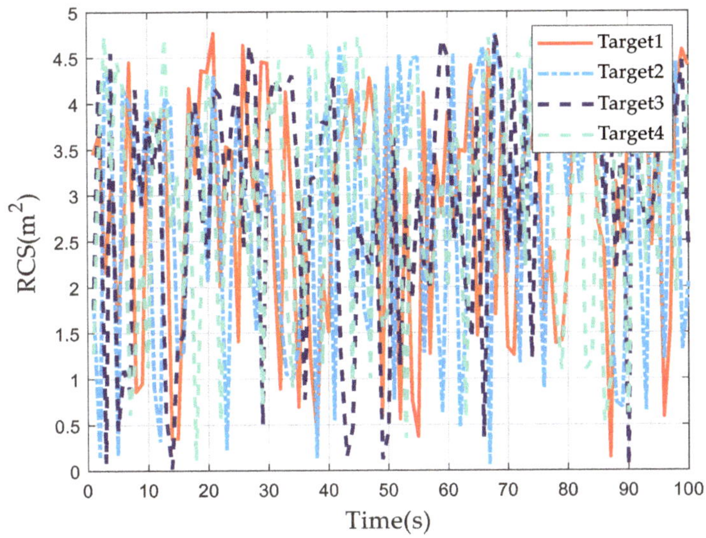

Figure 14. The mismatched multi-target RCS.

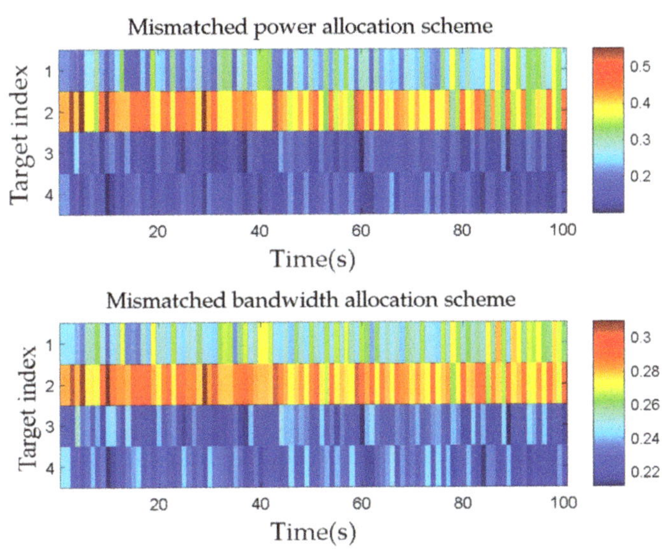

Figure 15. Mismatched power and bandwidth allocation result.

Figures 16 and 17 show the RMSE curves and PCRLBs of position and velocity with the traditional RCS fluctuation model of 100 Monte Carlo experiments, respectively.

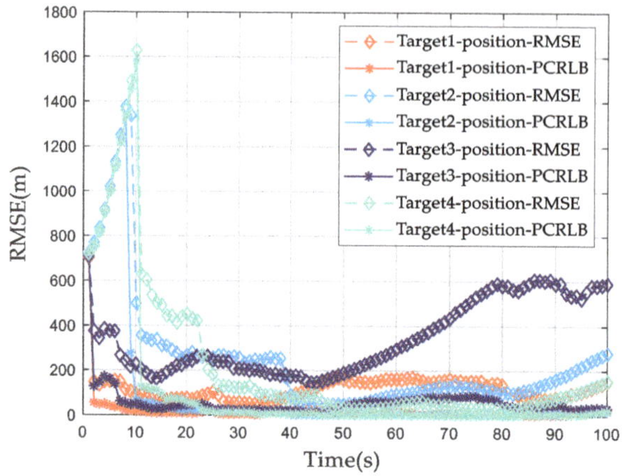

Figure 16. The position PCRLB and RMSE of mismatched allocation result.

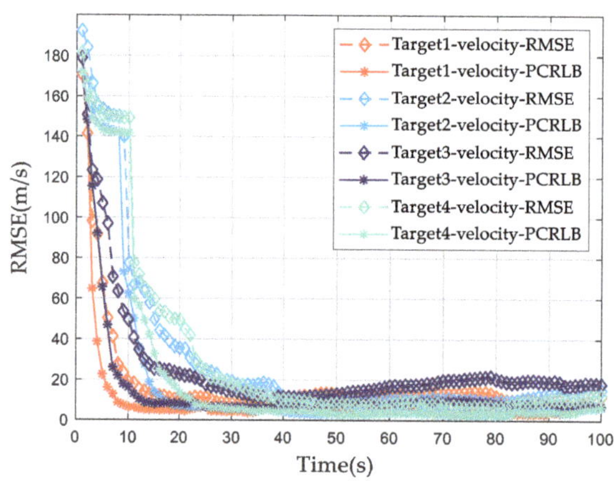

Figure 17. The velocity PCRLB and RMSE of mismatched allocation result.

It can be clearly seen from Figures 16 and 17 that while the dynamic RCS of target 3 decreases significantly at the 47th frame, the resource allocation scheme with the traditional RCS model does not change the power and bandwidth allocation ratio simultaneously, which leads to larger tracking errors for targets with smaller RCS. Under this condition, the RMSE of the position and the RMSE of the speed of target 3 are much larger than the other two targets. Consequently, it can be concluded that the traditional RCS model fails to effectively incorporate the dynamic changes in multi-target RCSs, leading to suboptimal radar power and bandwidth allocation schemes.

To show the effectiveness of the proposed method more intuitively, multi-target tracking results are compared with the high dynamic RCS model and the traditional RCS model. The comparison outcomes regarding the position RMSE and velocity RMSE are displayed in Figures 18 and 19, respectively. From Figure 18, the position RMSE of target 3 is the largest, reaching 466.2 m in the 100th frame. As for the traditional RCS model, target 3 also demonstrates the largest position tracking error, showing an RMSE of 590.7 m. The proposed method shows an improved position estimation accuracy of 21.1% when compared to the traditional RCS model.

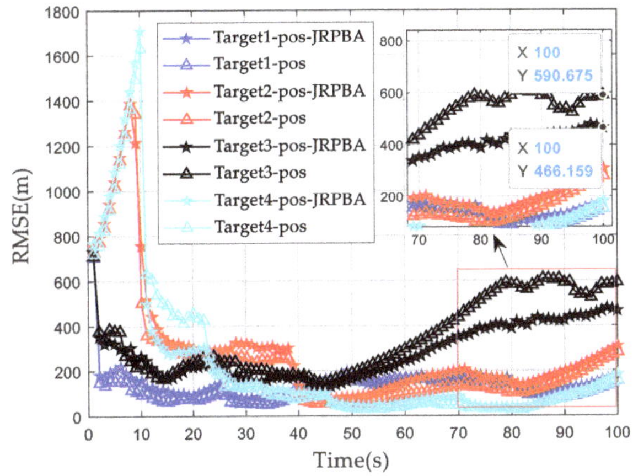

Figure 18. Multi-target tracking position accuracy of different scheme.

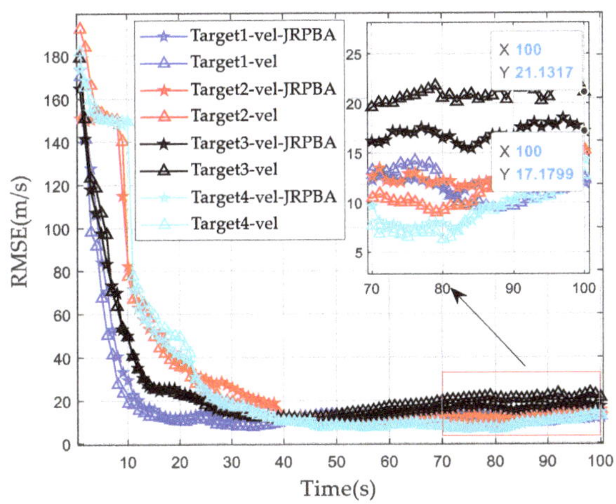

Figure 19. Multi-target tracking velocity accuracy of different scheme.

It can be seen from Figure 19 that the traditional RCS model algorithm has the largest RMSE for target 3, reaching 21.1 m/s in the 100th frame. The RMSE of velocity obtained by the proposed method is the largest in the later tracking stage of target 3 and the corresponding value is about 17.2 m/s. The velocity estimation accuracy is improved by 18.4%. In conclusion, under the criterion of Min-Max PCRLB, the proposed method can effectively combine the real-time RCS dynamic characteristics of multiple targets in the tracking process and improve the multi-target tracking performance of the radar system.

6. Conclusions

In this paper, an adaptive joint dynamic RCS power and bandwidth allocation (JRPBA) method is proposed to solve the problem of radar tracking resources allocation in a space multi-target intersection scenario. The proposed approach establishes a mapping method between the space target motion model and the full-angle static RCS database, enabling accurate retrieval of the dynamic RCS sequence. The high dynamic characteristics of RCS are considered as prior information in the radar resource allocation framework. The optimization model, considering space multi-target RCS, power, and bandwidth, is formulated and solved using a convex relaxation method to obtain the optimal allocation scheme suitable for the actual tracking scenario. Simulation results demonstrate that compared with the traditional RCS model radar resource allocation method, the proposed JRPBA method enhances the space multi-target tracking performance and effectively resolves mismatches between the allocation scheme and the actual tracking scenario. Further research will focus on resource allocation in multi-target observation scenarios of space-based networking radar.

Author Contributions: Q.Y. proposed the conceptualization and methodology; S.Z. wrote the draft manuscript; Z.W. and L.J. supervised the experimental analysis and revised the manuscript; Y.Z. proofread and revised the first draft. All authors have read and agreed to the published version of the manuscript.

Funding: This work was funded by the National Defense Basic Scientific Research program of China, grant number WDZC20225250203.

Data Availability Statement: Not applicable.

Acknowledgments: The authors would like to thank the editors and all the reviewers for their very valuable and insightful comments during the revision of this work.

Conflicts of Interest: The authors declare no conflict of interest.

References

1. Klinkrad, H. The Current Space Debris Environment and its Sources. In *Space Debris: Models Risk Analysis*; Springer: Berlin/Heidelberg, Germany, 2006; pp. 5–58. [CrossRef]
2. Weeden, B.; Cefola, P.; Sankaran, J. Global space situational awareness sensors. In Proceedings of the Advanced Maui Optical and Space Surveillance Technologies Conference, Maui, HI, USA, 14–17 September 2010; pp. 1–10.
3. Vallado, D.A.; Griesbach, J.D. Simulating Space Surveillance Networks. In Proceedings of the AAS 11-580 the AAS/AIAA Astrodynamics Specialist Conference, Napa, CA, USA, 31 July 2011; pp. 1–21.
4. Delande, E.; Frueh, C.; Franco, J.; Houssineau, J.; Clark, D. Novel Multi-Object Filtering Approach for Space Situational Awareness. *J. Guid. Control. Dyn.* **2017**, *41*, 1–15. [CrossRef]
5. Du, Y.; Jiang, Y.; Wang, Y.; Zhou, W.; Liu, Z. ISAR Imaging for Low-Earth-Orbit Target Based on Coherent Integrated Smoothed Generalized Cubic Phase Function. *IEEE Trans. Geosci. Remote Sens.* **2019**, *58*, 1205–1220. [CrossRef]
6. Schirru, L.; Pisanu, T.; Podda, A. The Ad Hoc Back-End of the BIRALET Radar to Measure Slant-Range and Doppler Shift of Resident Space Objects. *Electronics* **2021**, *10*, 577. [CrossRef]
7. Karamanavis, V.; Dirks, H.; Fuhrmann, L.; Schlichthaber, F.; Egli, N.; Patzelt, T.; Klare, J. Characterization of deorbiting satellites and space debris with radar. *Adv. Space Res.* **2023**. [CrossRef]
8. Ender, J.; Leushacke, L.; Brenner, L.; Wilden, H. RaWAdar Techniques for Space Situational Awareness. In Proceedings of the 12th International Radar Symposium (IRS), Dresden, Germany, 9–21 June 2011; pp. 1–6.
9. Console, A. *Command and Control of a Multinational Space Surveillance and Tracking Network*; Joint Air Power Competence Centre (JAPCC): Kalkar, Germany, 2019.
10. Li, J.; Stoica, P. MIMO Radar with Colocated Antennas. *IEEE Signal Process Mag.* **2007**, *24*, 106–114. [CrossRef]
11. Haimovich, A.M.; Blum, R.S.; Cimini, L.J. MIMO radar with widely separated antennas. *IEEE Signal Process Mag.* **2008**, *25*, 116–129. [CrossRef]
12. Cheng, Z.; Shi, S.; Tang, L.; He, Z.; Liao, B. Waveform Design for Collocated MIMO Radar With High-Mix-Low-Resolution ADCs. *IEEE Trans. Signal Process* **2021**, *69*, 28–41. [CrossRef]
13. Zhang, W.W.; Shi, C.G.; Salous, S.; Zhou, J.J.; Yan, J.K. Convex optimization-based power allocation strategies for target localization in distributed Hybrid non-coherent active-passive radar networks. *IEEE Trans. Signal Process.* **2022**, *70*, 2476–2488. [CrossRef]
14. Xu, L.; Li, J.; Stoica, P. Target detection and parameter estimation for MIMO radar systems. *IEEE Trans. Aerosp. Electron. Syst.* **2008**, *44*, 927–939. [CrossRef]
15. Garcia, N.; Haimovich, A.M.; Coulon, M. Resource allocation in MIMO radar with multiple targets for non-coherent locali-zation. *IEEE Trans. Signal Process* **2014**, *62*, 2656–2666. [CrossRef]
16. Bell, K.L.; Baker, C.J.; Smith, G.E.; Johnson, J.T.; Rangaswamy, M. Cognitive Radar Framework for Target Detection and Tracking. *IEEE J. Sel. Top. Signal Process* **2015**, *9*, 1427–1439. [CrossRef]
17. Yan, J.K.; Liu, H.W.; Jiu, B.; Bao, Z. Power Allocation Algorithm for Target Tracking in Unmodulated Continuous Wave Radar Network. *IEEE Sensors J.* **2015**, *15*, 1098–1108. [CrossRef]
18. Yan, J.K.; Pu, W.Q.; Zhou, S.H.; Liu, H.W.; Bao, Z. Collaborative detection and power allocation framework for target tracking in multiple radar system. *Inf. Fusion* **2020**, *55*, 173–183. [CrossRef]

19. Yan, J.K.; Jiu, B.; Liu, H.W.; Chen, B.; Bao, Z. Prior knowledge-based simultaneous multibeam power allocation algorithm for cog-nitive multiple targets tracking in clutter. *IEEE Trans. Signal Process* **2015**, *63*, 512–527. [CrossRef]
20. Li, Z.J.; Xie, J.W.J.; Liu, W.; Zhang, H.W.; Xiang, H.H. Joint Strategy of Power and Bandwidth Allocation for Multiple Maneuvering Target Tracking in Cognitive MIMO Radar With Collocated Antennas. *IEEE Trans. Veh. Technol.* **2023**, *72*, 190–204. [CrossRef]
21. Zhang, H.W.; Liu, W.J.; Zhang, Z.J.; Lu, W.L.; Xie, J.W. Joint Target Assignment and Power Allocation in Multiple Distributed MIMO Radar Networks. *IEEE Syst. J.* **2021**, *15*, 694–704. [CrossRef]
22. Zhang, H.W.; Liu, W.J.; Xie, J.W.; Zhang, Z.J.; Lu, W.L. Joint Subarray Selection and Power Allocation for Cognitive Target Tracking in Large-Scale MIMO Radar Networks. *IEEE Syst. J.* **2020**, *14*, 2569–2580. [CrossRef]
23. Zhang, H.W.; Shi, J.; Zhang, Q.; Zong, B.; Xie, J.W. Antenna Selection for Target Tracking in Collocated MIMO Radar. *IEEE Trans. Aerosp. Electron. Syst.* **2021**, *57*, 423–436. [CrossRef]
24. Yi, W.; Yuan, Y.; Hoseinnezhad, R.; Kong, L. Resource Scheduling for Distributed Multi-Target Tracking in Netted Colocated MIMO Radar Systems. *IEEE Trans. Signal Process* **2020**, *68*, 1602–1617. [CrossRef]
25. Tichavsky, P.; Muravchik, C.H.; Nehorai, A. Posterior Cramer-Rao bounds for discrete-time nonlinear filtering. *IEEE Trans. Signal Process* **1998**, *46*, 1386–1396. [CrossRef]
26. Yuan, Y.; Yi, W.; Kirubarajan, T.; Kong, L.J. Scaled accuracy based power allocation for multi-target tracking with colocated MIMO radars. *Signal Process* **2019**, *158*, 227–240. [CrossRef]
27. Xie, M.C.; Yi, W.; Kirubarajan, T.; Kong, L.-J. Joint Node Selection and Power Allocation Strategy for Multitarget Tracking in Decentralized Radar Networks. *IEEE Trans. Signal Process* **2018**, *66*, 729–743. [CrossRef]
28. Grey, J.P.; Mann, I.R.; Fleischauer, M.D.; Elliott, D.G. Analytic Model for Low Earth Orbit Satellite Solar Power. *IEEE Trans. Aerosp. Electron. Syst.* **2020**, *56*, 3349–3359. [CrossRef]
29. Li, Z.J.; Xie, J.W.; Ge, J.A. Power allocation optimization algorithm based on collocated MIMO radar. In Proceedings of the International Conference on Electronic Information Technology and Computer Engineering (EITCE), Xiamen, China, 18–20 October 2019; pp. 1096–1100. [CrossRef]
30. Zhang, H.W.; Zong, B.; Xie, J.W. Power and Bandwidth Allocation for Multi-Target Tracking in Collocated MIMO Radar. *IEEE Trans. Veh. Technol.* **2020**, *69*, 9795–9806. [CrossRef]
31. Chang, J.Y.; Fu, X.; Cheng, G.J.; Fang, G.Q.; Liu, S.L. Low-earth-orbit object detection by spaceborne netted radars. In Proceedings of the 2015 12th International Bhurban Conference on Applied Sciences and Technology (IBCAST), Islamabad, Pakistan, 13–17 January 2015.
32. Challa, S.; Morelande, M.R.; Mušicki, D.; Evans, R.J. *Fundamentals of Object Tracking*; University Press: Cambridge, UK, 2011. [CrossRef]
33. Shi, C.; Fei, W.; Zhou, J. Resource management for target tracking in distributed radar network system. In Proceedings of the IEEE International Conference on Signal Processing, Communications and Computing (ICSPCC), Zhejiang, China, 5–8 August 2015.
34. Chen, J.; Fu, T.J.; Chen, D.F.J.; Gao, M.G.J. Observation angle and plane characterisation for ISAR imaging of LEO space objects. *Adv. Space Res.* **2016**, *58*, 30–44. [CrossRef]
35. Ma, J.-T.; Gao, M.-G.; Guo, B.-F.; Dong, J.; Xiong, D.; Feng, Q. High resolution inverse synthetic aperture radar imaging of three-axis-stabilized space target by exploiting orbital and sparse priors. *Chin. Phys. B* **2017**, *26*, 108401. [CrossRef]
36. Yu, X.; Wang, Z.; Du, X.; Jiang, L. Multipass Interferometric ISAR for Three-Dimensional Space Target Reconstruction. *IEEE Trans. Geosci. Remote Sens.* **2022**, *60*, 5110317. [CrossRef]
37. Chae, H.; Kim, H.; Kwon, E.; Ryu, K.; Seong, Y.B.; Lee, J. Development of a real-time satellite orbit propagation system using TLE databases. *Adv. Space Res.* **2020**, *66*, 410–421.
38. Ning, X.L.; Wang, F.; Fang, J.C. An Implicit UKF for Satellite Stellar Refraction Navigation System. *IEEE Trans. Aerosp. Electron. Syst.* **2017**, *53*, 1489–1503. [CrossRef]

remote sensing

MDPI

Review

Overview of High-Power and Wideband Radar Technology Development at MIT Lincoln Laboratory

Michael MacDonald *, Mohamed Abouzahra and Justin Stambaugh

MIT Lincoln Laboratory, Lexington, MA 02421, USA; abouza@ll.mit.edu (M.A.); stambaugh@ll.mit.edu (J.S.)
* Correspondence: mmacdon@ll.mit.edu

Abstract: This paper summarizes over 60 years of radar system development at MIT Lincoln Laboratory, from early research on satellite tracking and planetary radar to the present ability to perform the centimeter-resolution imaging of resident space objects and future plans to extend this capability to geosynchronous range.

Keywords: radar; RCS; ISAR; coherent; integration; Doppler; imaging; tracking; debris; space

1. Introduction

The history of Lincoln Laboratory is intimately tied to that of radar system technology. The Massachusetts Institute of Technology (MIT) established the Radiation Laboratory in 1940 [1] to exploit the British invention of the first practical high-power microwave source, the magnetron, to develop pioneering radar systems for the Allied effort in World War 2. This required the invention of a wide array of microwave technologies, and the Radiation Laboratory ultimately grew to nearly 4000 staff and fielded over 100 different land-, sea- and air-based radar systems which were a decisive factor in achieving Allied victory. When the War ended, the facilities of the Radiation Laboratory were absorbed by MIT.

The establishment of Project Lincoln at MIT in 1951 [2], prompted by Soviet demonstration of nuclear weapons and the capability to deliver them at long range, followed a similar initial history and was co-located at the MIT campus but eventually moved to Hanscom Air Force Base upon the decision to make Lincoln Laboratory an ongoing part of MIT.

Lincoln Laboratory initially fielded radar technology in support of USA efforts to provide an early warning capability against the threat of Soviet nuclear attack via aircraft by means of the Distant Early Warning (DEW) line of modest-sized radars. These radars were capable of tracking aircraft to ~300 km. The MITRE corporation was eventually split from the Laboratory to pursue operational development of these systems.

Later, as the Soviet threat migrated to weapon delivery via intercontinental ballistic missile (ICBM), the Laboratory took a lead role in developing the Ballistic Missile Early Warning System (BMEWS), which required large-aperture, high-power radar systems. Approximately 60 dB greater sensitivity was needed to detect and track warheads having a small radar cross-section (RCS) at ranges of 4500 km.

As a Federally Funded Research and Development Center (FFRDC), Lincoln Laboratory prioritizes the prototyping of advanced technology for national security. That priority is manifested in the history of the Laboratory's radars. The focus of this review will be on the progress made in the area of large-aperture, high-power radar systems from the development of the Millstone Hill Radar (MHR) as a UHF prototype for the BMEWS to the present capability at the Lincoln Space Surveillance Center (LSSC) in Westford, Massachusetts, USA, and the Reagan Test Site (RTS) at Kwajalein Atoll in the Republic of the Marshall Islands. This includes radars operating from UHF (435 MHz) to the W band (96 GHz) with narrowband metric tracking and, in the C band and above, wideband range profiling and inverse synthetic-aperture radar (ISAR) imaging modes of operation.

The missions of these radars are continuously evolving and are often intertwined, with ballistic missile defense and space surveillance having similar needs in regard to sensitivity and bandwidth. There have also been complementary developments in scientific research and defense applications owing to the capability of these radars to provide insights into fundamental physical processes. Lincoln Laboratory maintains emphasis on employing these radars as instruments to investigate target physics and phenomenology rather than serving some narrowly defined operational function.

Figure 1 shows a timeline of these radar developments along with some milestones in resident space objects. A more detailed account of technology development in these radars is given in the next section.

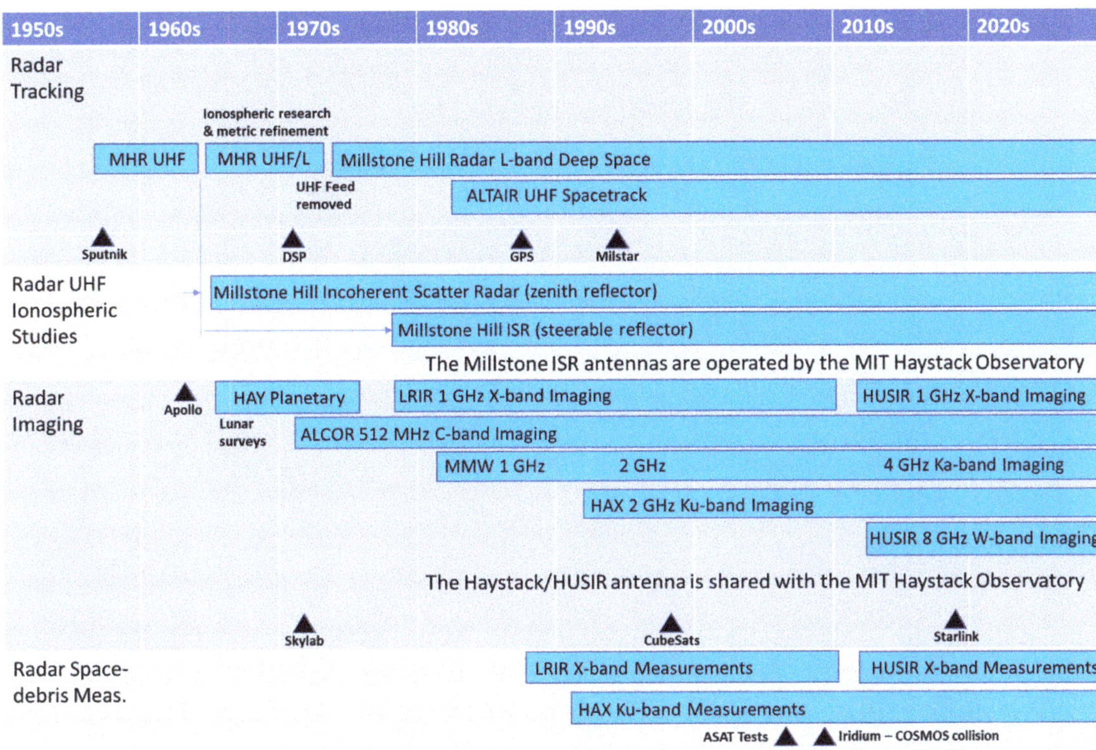

Figure 1. Timeline of radar developments at Lincoln Laboratory.

2. Progress in High-Power and Wideband Radar Technology at MIT Lincoln Laboratory

2.1. The Millstone Hill Radar (MHR; Constructed in 1956, Reconfigured in 1965)

The Millstone Hill Radar was first constructed as a UHF (440 MHz) radar in 1956 as a prototype of operational BMEWS radar systems that were fielded at Fylingdales Moor in the UK, at Clear Alaska in the USA and at Thule in Greenland. MHR represented a spectacular advance in radar system performance, with 60 dB higher sensitivity than the other radars of its time, which was made possible by its 25.6 m diameter antenna aperture and 150 kW average transmit power. Its value as an asset for detecting and tracking resident space objects was demonstrated upon launch of the first man-made satellite, Sputnik, in 1957. MHR acquired and collected data on Sputnik, foreshadowing its present space surveillance mission.

In 1965, MHR was reconfigured as an L-band (1.295 GHz) radar with a new transmitter and a 12-horn tracking feed [3] it retains to this day. The simultaneous L-band and UHF tracking of satellites led to order-of-magnitude refinements in metric calibration at MHR, which have since been implemented in other radars. The MHR antenna is pictured in Figure 2. The RF front end of the radar is housed in a "doghouse" visible atop the tower. The microwave configuration of the antenna has changed little since its conversion to the L band in the mid-1960s, although the UHF feed has been removed.

Figure 2. The Millstone Hill Radar (MHR) antenna.

The UHF transmitter now serves two different antennas (one fixed at the zenith and a steerable one), which are operated [4] by the MIT Haystack Observatory Atmospheric Sciences Group. This radar operates in Thomson (incoherent) scatter mode to measure properties of charged particles in the ionosphere, a technique pioneered [5] at Lincoln Laboratory and one requiring a very-large-power–aperture product. This UHF radar is denoted

Millstone Incoherent Scatter Radar, and although its name (and history) is co-mingled with that of MHR, they operate today as distinct sensors in separate frequency bands.

MHR pioneered many modern radar capabilities [6,7], such as coherent pulse-to-pulse integration, polarimetric analysis of returns, computer control of the antenna and pulse-by-pulse measurement of target range, antenna-to-target angle error and target velocity in real time.

MHR also serves [8] as a transmit site for multistatic radar demonstrations, most recently with three European receive sites.

A transmitter upgrade, described in Section 3, has been completed and is currently being tested. This will replace the existing obsolescent klystron with a higher average power version with redundant power supply and modulator.

2.2. The Haystack Planetary Radar (Constructed in 1964, Antenna Replaced in 2014)

The Haystack antenna was constructed less than a mile from Millstone in 1964 to serve the needs of space communication experiments conducted at the Laboratory and to enable radar measurements with higher range resolution. It was designed for highly efficient operation in the X band (7.84 GHz). This required an exceptionally accurate antenna reflector surface [9]. A 36.6 m diameter antenna was constructed within a 45.7 m radome (a BMEWS spare itself) and configured to a surface accuracy of 10 ppm. The resulting antenna gain was 66 dB, and the associated pencil beam required the antenna pointing to be accurate to within 17 microradians. All of this represented a great leap in the state of the art [10].

Haystack was instrumental in supporting many notable and pioneering science experiments, primary among which was support for the Apollo program, which led to the development of techniques to map the lunar surface. This exploited a technique known as delay-Doppler imaging [11], suitable for rotating objects. The fourth test of general relativity was conducted [12] at Haystack by measuring the time of flight of pulses reflected from Venus as it passed behind the sun to quantify the effect of the solar gravitational field.

In 1970, the MIT Haystack Observatory was created to pursue astronomical research separate from the defense-oriented mission of Lincoln Laboratory. The Haystack facility continued to support both mission areas via its ability to exchange RF boxes through a hoist system which allowed both modalities to be coupled to the antenna, as seen in a cut-away view in Figure 3.

2.3. The ARPA-Lincoln C-Band Observables Radar (ALCOR; 1970)

Lincoln Laboratory was tasked [13], in the late 1960s, by the Advanced Research Projects Agency (ARPA) to construct a wideband (512 MHz) C-band (5.67 GHz) radar at the Kiernan Re-entry Measurements System [14] at Kwajalein Atoll. The project was motivated by the need to investigate the physics of ICBM re-entry vehicles (RVs) and the plasma wake they create upon descent through the atmosphere. This radar, ALCOR, required the development of signal processing methods [15] exploiting a "stretch processing" technique [16] by correlation mixing the target return with a replica of the transmit pulse to process it in a much smaller intermediate-frequency (IF) bandwidth.

At the time of its design, it was recognized [17] that some form of time–bandwidth exchange was necessary to process the 512 MHz radar bandwidth in a modest IF bandwidth, and this was accomplished by limiting the range extent, or "window", of data collection. This was in keeping with ALCOR's mission to obtain high-resolution signatures on small hardbodies. The need to collect larger datasets motivated the development of a series of analog pulse compressors [18], initially using a cumbersome lumped-element bridged-T network but, in the early 1970s, progressing to an elegant solution via surface acoustic wave (SAW) transversal filters. This implementation proved highly effective and remained in use for 30 years before being supplanted by digital pulse compression techniques during the Kwajalein modernization and remoting (KMAR) campaign, which implemented the radar open system architecture (ROSA) in the RTS radars. Figure 4 shows the ALCOR antenna inside its radome.

Figure 3. The Haystack antenna as it stood from 1964 to 2014 in a cut-away view. The figures pictured near the subreflector provide a human scale. An RF box can be seen being hoisted into place for insertion at the antenna feed.

ALCOR pioneered inverse synthetic-aperture radar (ISAR) techniques, in which the aspect change of objects as they pass overhead furnishes the Doppler (cross-range) information used to image them. ALCOR obtained the first range-Doppler images [19] of Earth-orbiting satellites.

Figure 5, adapted from [20], shows how an inverse-synthetic-aperture radar (ISAR) image is generated from a time sequence of radar cross-section (RCS), range and Doppler information on individual scatterers. Here, a "dumbbell" of two scatterers is used to illustrate the technique, but it can be extended to the constellation of scatterers which lie within the range "window" over which the radar collects data. As the target changes aspect with respect to the radar, the RCS data are shown in false colors on range–time–intensity (RTI) and Doppler–time–intensity (DTI) plots illustrating how Doppler information resolves the target in the cross-range direction. The integration time is typically chosen to make the resolution in this dimension the same as that in the range dimension. This figure also illustrates that the image plane is the plane in which the target aspect varies.

Figure 4. The ARPA-Lincoln C-band Observables Radar (ALCOR) antenna, pictured during its construction in 1968.

A simulated ALCOR image of the Skylab space station resulting from this process is shown in Figure 6. The development of ISAR imaging pioneered at ALCOR has been leveraged by wideband radars described in this paper, such as LRIR, MMW, HAX and HUSIR, as well as other USA assets, and those fielded by other nations, such as TIRA.

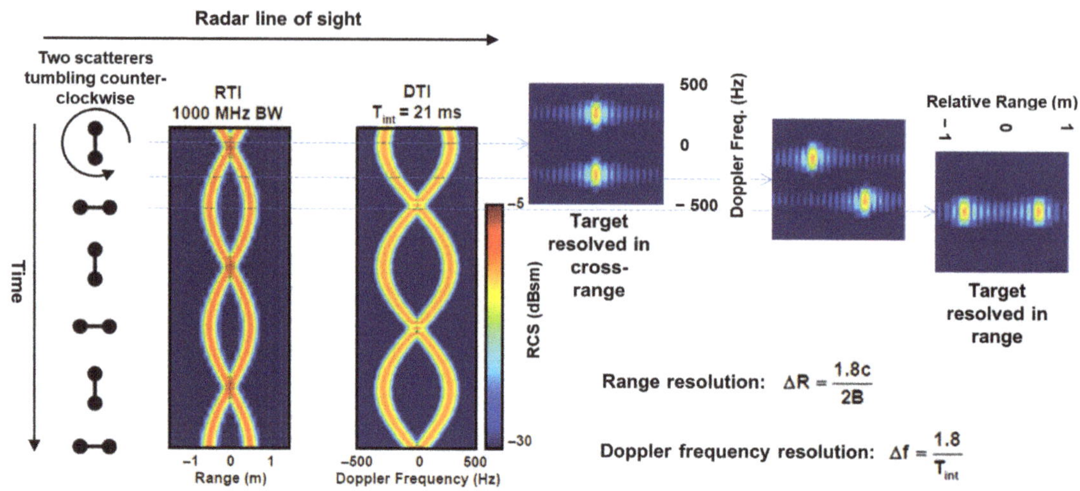

Figure 5. The generation of a range-Doppler image from a time series of range, Doppler and radar cross-section (RCS) measurements of individual scatterers (a dumbbell of two scatterers here).

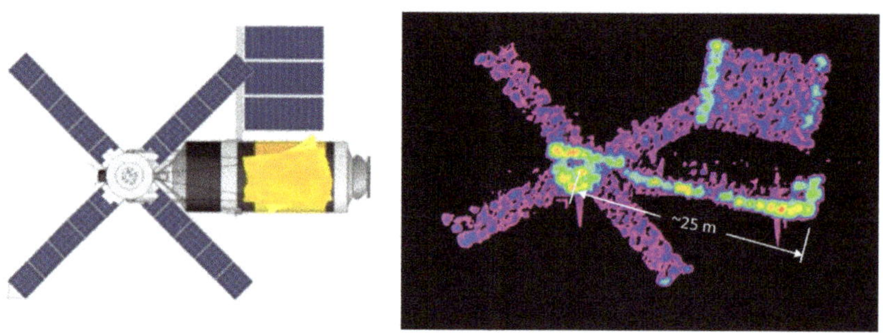

Figure 6. The US Skylab space station, in orbit 1973–1979 (**left**). A simulated ALCOR image [19] of the Skylab space station illustrating a range-Doppler image (**right**).

2.4. The Long-Range Imaging Radar (LRIR (1978)/HUSIR-X (2014))

Following the success of ALCOR, the need to extend this satellite imaging capability to geosynchronous range (approximately 40,000 km) became evident, and the Laboratory was again tasked by the ARPA to achieve this. With the large efficient aperture available at Haystack, and the ability to swap the RF box at the antenna which had been a feature of its design, a wideband (1 GHz) high-power and (by necessity) very compact upgrade was implemented in 1978, termed the Haystack Long-Range Imaging Radar [21]. LRIR required technology advances in high-power coupled-cavity traveling wave tube (TWT) technology [22], monopulse tracking feed [23], receiver protection [24] and waveform generation [25]. LRIR also served as a testbed for the deployment [26] of real-time digital processing and control, which was later improved upon and distributed to other radars.

The LRIR feed is illustrated in Figure 7. When the new HUSIR antenna was installed in 2014, its shape, while more accurate, was identical to that of the original Haystack antenna so that no change to the X-band feed was needed. The RF box behind the feed hosts the four X-band coupled-cavity TWT final power amplifiers and their pulse modulator, as well as the four receiver channels (principal-polarized return, orthogonal-polarized return, elevation error and traverse error). The high-voltage power and intermediate-frequency receive chain are routed down the pedestal via cable wraps to a high-voltage power supply and signal processing equipment in the Haystack facility.

Figure 7. A view of the LRIR feed showing the X-band horn aperture, the support structure and the water-cooled WR28 waveguide.

LRIR has now operated for 45 years and, apart from periodic change-outs of the TWTs (which are lifetime-limited by effects such as occasional high-voltage arcs and, ultimately, depletion of the cathodes), has been a highly reliable sensor. The TWTs developed for LRIR have been employed in other X-band radars since their development.

LRIR has had sufficient sensitivity since its inception to perform the range-Doppler imaging of satellites at geosynchronous range. An upgrade described later plans to bring the W-band capability of HUSIR to this same range. The X-band radar (now referred to as HUSIR-X rather than LRIR, although it is the same sensor) also serves to acquire targets in angle prior to handover to the exceedingly narrow W-band antenna beam.

2.5. The Millimeter-Wave Radar (MMW; Constructed in 1983, Upgraded in 1993 and 2012)

The need to characterize ICBM RVs at millimeter wavelengths at Kwajalein Atoll led to the development of the Millimeter-Wave Radar (MMW) [14]. MMW, whose antenna is shown in Figure 8, was completed in 1983 to gather ballistic missile re-entry signatures in the Ka (35 GHz) and W (95 GHz) bands [27]. The study of re-entry phenomenology at these frequencies was intended to support development of RF seeker technology for ballistic missile defense [28]. Although originally slated to be an adjunct to ALCOR, MMW had sufficient sensitivity to independently track relevant targets with its 25 kW peak-power final-stage Ka-band TWT and monopulse feed, and 1 GHz bandwidth in both the Ka and W bands.

Due to its high operating frequency, the 13.7 m diameter MMW antenna was made with less than 0.1 mm RMS surface tolerance and was also made extremely stiff to support the more than 10 degrees/s angular motion rates needed to track RVs through re-entry at Kwajalein. The antenna was built by Electromagnetic Space Structures Co. (ESSCO; now part of Communications and Power Industries), Ayer, MA USA, which had been spun off from Lincoln Laboratory in the 1960s. The antenna and its 20.7 m diameter radome were placed atop a 12 m high cylinder to elevate it above the treetops and to mitigate the effects of salt spray, which is present in the marine environment of Kwajalein Atoll.

Figure 8. The Millimeter-Wave Radar (MMW) antenna is seen in the foreground as a new GoreTex® radome was installed in 2003.

MMW began providing satellite imagery for the US Space Command in 1988. The ISAR capability at ALCOR and MMW led [29] to the initiation of the Space Object Identification (SOI) mission area at Lincoln Laboratory, which continues to this day. As the highest-resolution radar on Kwajalein Atoll, MMW's role quickly expanded from re-entry signature studies to the ISAR imaging of re-entering and orbiting objects.

The increased demand for high-resolution MMW data motivated numerous major system upgrades throughout the radar's history. At the time of its initial construction, MMW employed a conventional antenna feed structure with an extensive network of water-cooled WR28 and WR10 rectangular waveguides. A drawing of the original microwave system is shown in Figure 9. MMW's Ka-band system underwent a series of major upgrades in 1988–1994, which culminated in the maximum bandwidth being doubled to 2 GHz. These upgrades also increased the radar sensitivity by over 10 dB, tripling the tracking range [27]. These enhancements were due in part to advancements in computing technology and algorithms to improve its real-time coherent tracking and the redesign of the high-power amplifier by Varian (now Communication and Power Industries), doubling the peak-power output of the TWT to 50 kW, as well as doubling its bandwidth.

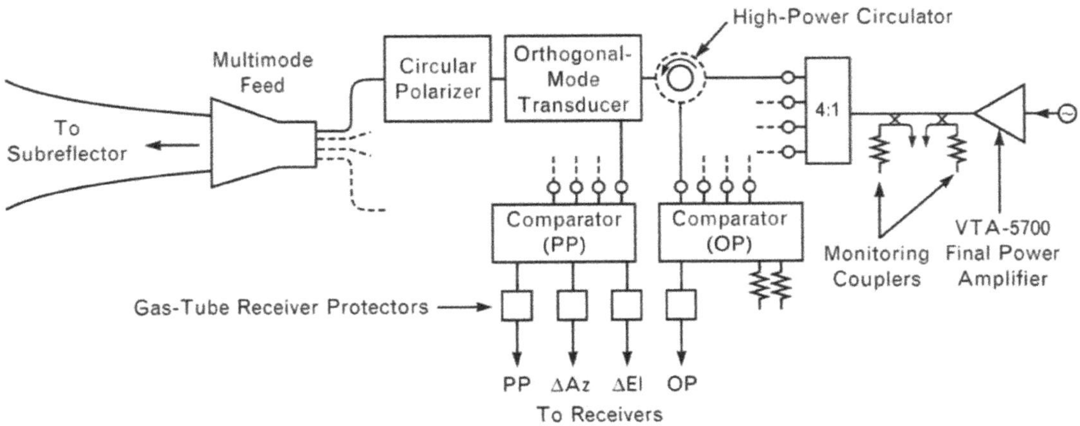

Figure 9. A schematic of the original MMW microwave system (1983).

The high loss of rectangular waveguide at millimeter wavelengths severely limited the achievable sensitivity of the original radar. In 1993, a novel quasi-optical beam waveguide feed was implemented [30], dramatically reducing resistive loss in the transmit feed while enabling the free-space combining of two TWTs with low loss. This effort built upon work that the Laboratory had conducted on submillimeter-wavelength technology employing off-axis paraboloid mirrors turned on a lathe [31]. These mirrors proved suitable for constructing Gaussian beam waveguides [32], in which optical elements such as Faraday rotators [33] could be added to separate the principal-polarized (PP) and orthogonal-polarized (OP) returns. The use of mirrors rather than lenses in the beam waveguide enables high-power operation. An additional advance was the use of a symmetric "clamshell" mirror topology, which effectively cancels the cross-polarization generated [30] by the amplitude asymmetry of a single off-axis paraboloidal mirror. An order-of-magnitude increase in the sensitivity of MMW was achieved while doubling its imaging resolution. The upgraded quasi-optical beam waveguide is shown in Figure 10.

The wideband high-power quasi-optical front end implemented at 35 GHz for MMW was later redesigned for 16.7 GHz for the HAX radar and for 95 GHz for MMW [34] and the Naval Research Laboratory WARLOC radar [35], for which Lincoln Laboratory built the duplexer unit.

The MMW TWT design formulated in 1992 proved difficult to reliably build for Communications and Power Industries (CPI), Palo Alto, CA USA, and the radar sensitivity degraded with time. This, along with the need for improved image resolution, motivated a new coupled-cavity TWT development at CPI, which yielded a new tube having more power than the 1992 design, and an additional two-fold increase in bandwidth. Lincoln Laboratory made an extensive effort [36] to leverage this improvement to upgrade MMW to 4 GHz bandwidth. The linear-FM waveforms employed the existing waveform generator in a dual-chirp configuration to double the bandwidth. The receiver hardware was also extensively modified. The Faraday rotator was redesigned to obtain a seven-fold improvement in its thermal performance, enabling higher-power operation. The HAX polarizer, which functioned over a 12% fractional RF bandwidth, was adapted to 35 GHz to support a similarly wide bandwidth for MMW. This upgrade doubled the image resolution of the radar. Sensitivity enhancements doubled the tracking range of the radar. Improvements in receiver electronics more than tripled the range "window" over which target data are collected. Isolation between the principal- and orthogonal-polarized returns was improved by 16 dB, and the range sidelobe level was improved by 13 dB.

The MMW feed is compatible with the use of quasi-optical quadrature hybrids [32] for TWT power combining [37] by means of a fused quartz disk. Replacing the disk with a metal plate, or omitting it, permits single-TWT operation. The same technique was employed for frequency diplexing, with a perforated metal plate being used for transmission in the W band and reflection in the Ka band for this purpose. Ka-band-only operation could be implemented by means of a flat metal plate. This provides a highly flexible means of adapting the radar to different configurations.

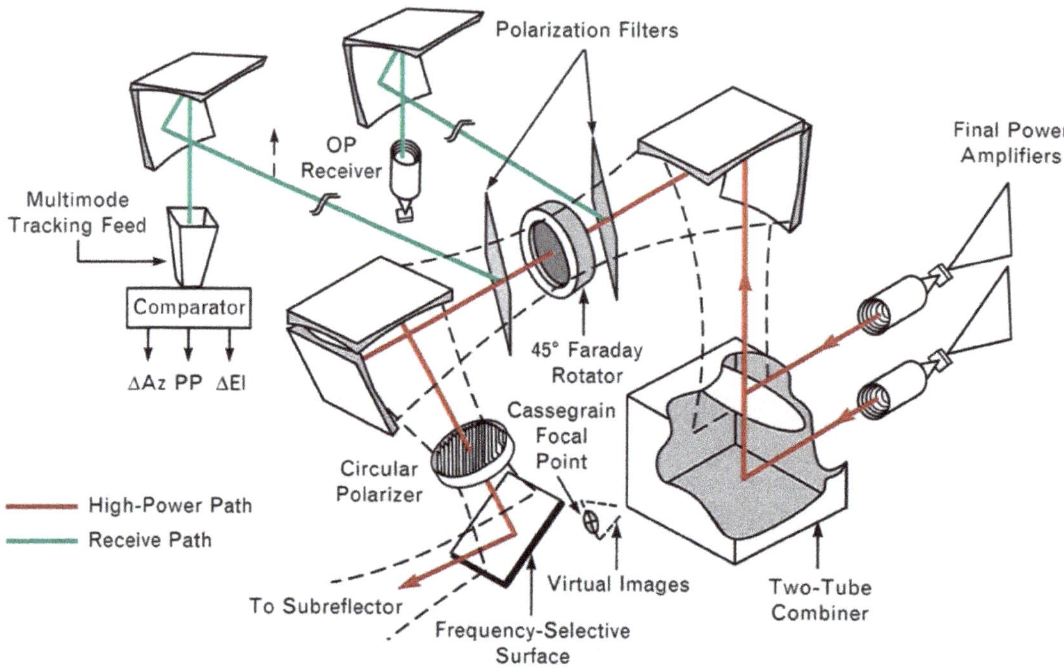

Figure 10. A schematic of the quasi-optical MMW microwave system (1992).

2.6. The Haystack Auxiliary Radar (HAX; 1993)

The Haystack Auxiliary Radar (HAX) was constructed in 1993 to reduce the operational demand on LRIR, which continued to share the Haystack antenna with ongoing radio astronomy activities [38]. HAX complemented LRIR in providing Earth orbiting debris data to NASA [7,39]. The HAX transmitter [40] achieved a 12% fractional bandwidth in the Ku band (16.7 GHz) with the first deployment of the re-entrant double-staggered ladder TWT circuit being invented [41] by Varian (now CPI).

HAX has performed these tasks admirably in terms of satellite imaging and debris characterization. The HAX antenna, pictured in Figure 11, was adapted [42] from an MSC-46 satellite communications antenna originally manufactured by Hughes Aircraft Company. This antenna was extensively modified to improve its pointing and tracking performance and placed in a 20.7 m diameter radome. A quasi-optical beam waveguide similar to that developed earlier for MMW (described in the previous section) was also put in use at HAX to support its 2 GHz bandwidth and high transmit power. A recent redesign of the HAX coupled-cavity TWT by CPI has improved its manufacturability and reliability.

Figure 11. The Haystack Auxiliary Radar (HAX) antenna.

HAX and HUSIR-X continue to jointly be the primary source of the NASA Orbital Debris Program Office [43], with HAX providing shorter-wavelength operation and a wider field of view than HUSIR-X.

The Ku-band vacuum electronics technology developed by CPI for HAX has also found use at the Fraunhofer Institute for High Frequency Physics and Radar Techniques (FHR) Tracking and Imaging Radar (TIRA), which has been extensively used [44,45] for the ISAR imaging of space objects, an example of which is seen in Figure 12.

2.7. The Cobra Gemini Radar (1996) and the XTR-1 Radar (2012)

A need for a transportable dual-band (S and X) radar led to the development of the Cobra Gemini radar [46] at the Laboratory in 1996. The radar open system architecture (ROSA) developed for Cobra Gemini [47] was later distributed to other USA radars at Kwajalein Atoll and to Lincoln Laboratory radars at the LSSC as a common real-time processing architecture, greatly streamlining operations and maintenance needs. It has been further extended [48] to support net-centricity for networked Department of Defense (DOD) needs. Cobra Gemini incorporated motion-compensation techniques, allowing it to be deployed on a T-AGOS class ship. This successful development program led to that of the X-band Transportable Radar (XTR-1) in 2012 [49]. XTR-1 has a significantly larger antenna (11 m diameter) [50] than Cobra Gemini (4.5 m diameter).

2.8. Haystack Ultra-Wideband Satellite Imaging Radar (HUSIR-W; 2014)

Following the previously described and highly successful developments, an ambitious program to implement 8 GHz bandwidth for high-resolution ISAR satellite imaging was achieved in 2014 with the commissioning of the Haystack Ultra-Wideband Satellite Imaging Radar (HUSIR) [51]. The LRIR X-band capability was retained, while the Haystack 37 m reflector was replaced [52], and its shape holographically optimized [53] to accommodate operation in the 92–100 GHz band with a new transmitter [54] and signal processing [55]. HUSIR also retains the ability for the MIT Haystack Observatory to perform astronomical observations with the new antenna [56] and supports continued orbital debris measurements for NASA [57]. It has also found use as a high-power source for terrestrial RF power beaming demonstrations [58]. In 2014, the project was awarded an "R&D 100" prize from R&D Magazine.

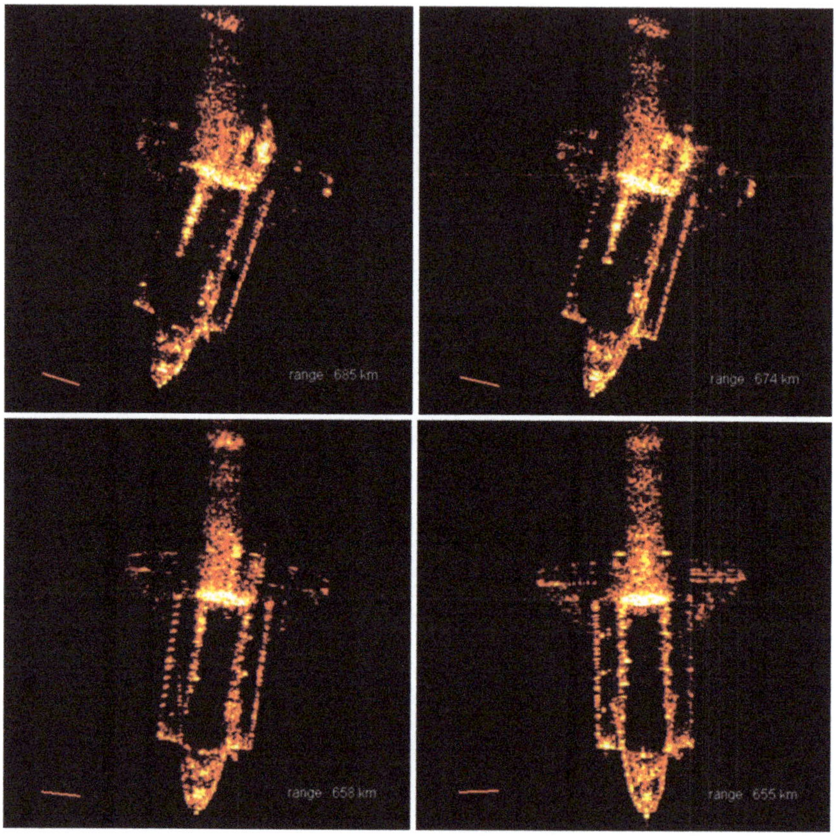

Figure 12. Wideband radar images of the USA Space Shuttle obtained by the TIRA (images courtesy of Fraunhofer FHR, Wachtberg, Germany).

There were many technological advancements enabling the development of HUSIR. A primary one was that of gyrotron amplifiers developed [59] by CPI. These gyrotrons employ superconducting magnets to generate an intense magnetic field in an oversized cylindrical interaction circuit, thus avoiding the electron beam clearance issues with the RF interaction region when attempting to extend coupled-cavity TWT designs to the W band. The W-band transmitter required a low-loss waveguide approximately 100 m long to convey RF power to the feed because the LRIR X-band transmitter and receiver fully occupied the RF box, which is located at the Cassegrain feed of the antenna. This was provided by corrugated over-moded waveguides [60] fabricated by General Atomics (GA), San Diego, CA USA. The power handling of this waveguide is also sufficient to handle the RF power upgrades described later without arcing. While microwave power tube transmitters have historically employed many other tubes in their high-voltage electronics (electron beam-switching tubes, thyratron crowbar triggers and regulator tubes), HUSIR provided a forum for the deployment of all-solid-state high-voltage power supplies and modulators [61] by Diversified Technologies (DTI), Bedford, MA USA, for the gyrotron final power amplifier, and this arrangement has worked flawlessly for the past decade. The HUSIR receiver required development of cryogenically cooled low-noise amplifiers (LNAs), and these were provided [62] by the NASA Jet Propulsion Laboratory (JPL), Pasadena, CA USA. The ability of Lincoln Laboratory to partner with these industry and government colleagues was critical to the success of HUSIR and remains so as plans for further upgrades in capability move forward.

All this technological progress would have been hamstrung without a more capable antenna. The original Haystack antenna was a remarkable achievement for its time, and the RMS surface error (defined as half-path-length error) had been reduced from 0.64 mm RMS at the time of construction [52] to 0.21 mm RMS via a deformable subreflector [38]. This surface error was highly susceptible to deformation from the challenging thermal environment in the radome [63] and gravitational deformations, which are more severe in satellite tracking (due to the range of elevation involved) than celestial observations. A new antenna was required.

The new antenna, pictured in Figure 13, re-used the existing pedestal and hydrostatic azimuth bearing of the 1964 Haystack antenna, as well as the Y-shaped structure supporting the elevation bearings. The structural engineering was performed [64] by Simpson, Gumpertz and Heger (SGH), Waltham, MA USA. The deformations were quantified [65] to permit planning for necessary rigging and measurement campaigns upon construction. All antenna structures from the elevation bearings upward, including the 37 m diameter reflector, were replaced, along with drives and controls [66]. The engineering effort involved [52] was extraordinary. The space frame of the radome was retained, and the skin was replaced [67] to optimize transmission in the W band.

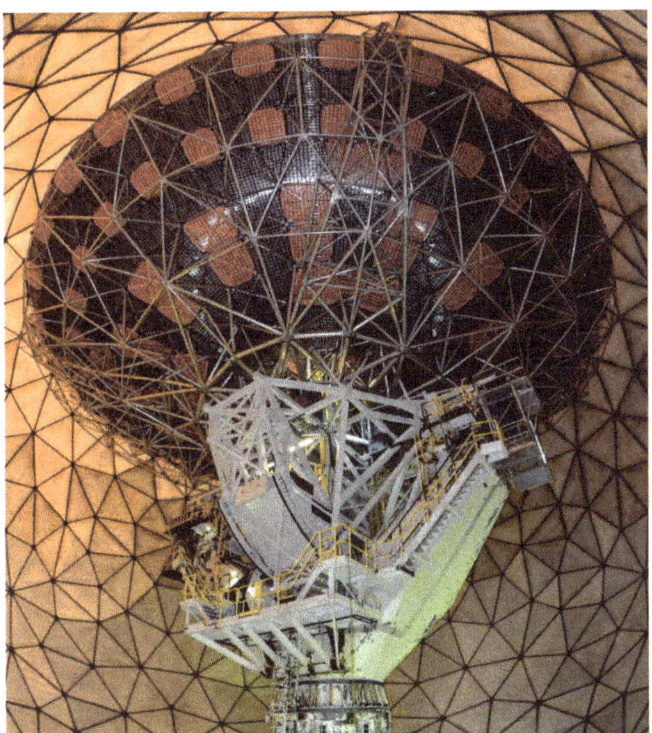

Figure 13. *Cont.*

Figure 13. The Haystack Ultra-Wideband Satellite Imaging Radar (HUSIR) antenna. The backstructure and drives are seen at the top, while at the bottom, a "fisheye" lens photo provides a view of the reflector surface.

The demonstrated performance of the new antenna showed that the RMS surface error, after optimization [53], was 0.07 mm. Calibration procedures on Earth-orbiting satellites, with the repositioning of the subreflector, demonstrate radar sensitivity consistent with an antenna gain of 89 dB [52], the highest for a radar antenna in the world.

3. Upgrades in Progress and Future Plans

A brief summary of key parameters of the radars discussed in this paper is shown in Table 1. The radars described in this overview are often leveraged or used as technology testbeds in addition to their primary operational role. As is evident from the progress in capability over the past six decades, these radars are constantly evolving and being improved upon in modest and major ways. Some ongoing and upcoming efforts are summarized here.

Table 1. A summary of the parameters of the radars described here, in order of frequency.

Radar *	Construction	RF Parameters	Observation Parameters
MHR	25.6 m dia. antenna aperture 12-horn monopulse feed 42.6°N, 71.4°W	1.3 GHz center freq. 20 MHz bandwidth 3.0 MW peak power 300 kW average power *	50 dB reference SNR ** Deep space capable
Cobra Gemini-S	4.5 m dia. antenna aperture Compatible with radome Transportable	S band 300 MHz bandwidth 50 kW avg. power	0.8 m range accuracy
ALCOR	12.2 m dia. antenna aperture 4-horn monopulse feed 20.7 m dia. radome 9.4°N, 167.5°E	5.67 GHz center freq. 512 MHz bandwidth 3 MW peak power 6.0 kW average power	0.4 m range accuracy 100 urad angle accuracy 50 cm image resolution 23 dB reference SNR **
Cobra Gemini-X	4.5 m dia. antenna aperture Compatible with radome Transportable	X band 1 GHz bandwidth 35 kW avg. power	0.25 m range accuracy
LRIR/HUSIR-X	36.6 m dia. antenna aperture 4-horn monopulse feed 45.7 m dia. radome 42.6°N, 71.4°W	10.0 GHz center freq. 1 GHz bandwidth 400 kW peak power 120 kW average power	25 cm image resolution 53 dB reference SNR ** Deep space capable

Table 1. *Cont.*

Radar *	Construction	RF Parameters	Observation Parameters
HAX	12.2 m dia. antenna aperture 4 horn monopulse feed 20.7 m dia. radome 42.6°N, 71.4°W	16.7 GHz center freq. 2 GHz bandwidth 40 kW peak power	12 cm image resolution 36 dB reference SNR **
MMW	13.7 m dia. antenna aperture 4 horn monopulse feed 20.7 m dia. radome 9.4°N, 167.5°E	35 GHz center freq. 4 GHz bandwidth 60 kW peak power	40 urad angle accuracy 6 cm image resolution 26 dB reference SNR **
HUSIR-W	36.6 m dia. antenna aperture 4-horn monopulse feed 45.7 m dia. radome 42.6°N, 71.4°W	96 GHz center freq. 8 GHz bandwidth 1 kW peak power 400 W average power (50 kW Ppk in future *)	3 cm image resolution 34 dB reference SNR ** (Deep space cap. in future *)

* Asterisks indicate upgrades initiated at MIT Lincoln Laboratory. ** This SNR applies to a 1 m^2 (0 dBsm) RCS target at 1000 km range and can be scaled accordingly.

A recent upgrade at the Millstone Hill Radar [68] has constructed a completely new transmitter for the radar. Figures 14 and 15 show the new facility housing this transmitter, and one (of two) CPI VKL-7796 klystron and RF hardware. Two independent transmitters provide "hot spare" redundancy for improved sensor availability. A coolant plant providing 1.5 MW of heat-sinking capacity was included in the new facility.

Figure 14. The MHR antenna with the facility housing the new transmitter in the foreground.

High-voltage power and pulse modulation are provided by 92 series-connected solid-state rectifier and switch modules built by Ampegon Power Electronics AG, Baden, Switzerland. This architecture provides fault tolerance via bypass of any failed module and precise regulation of the pulse voltage. Fault currents in the event of an arc in the klystron are limited to less than 300 A to protect the transistors in the modules. This also limits the energy deposited in the klystron to a safe value.

The VKL-7796 doubles the average power of the previous klystron used at Millstone and provides increased RF bandwidth. It also provides commonality with other L-band radar transmitters to simplify procurement of klystron amplifiers.

In 2020, Lincoln Laboratory tasked CPI to develop a higher-power gyrotron [69] to serve as a final power amplifier for a 50 kW peak-power HUSIR W-band upgrade to extend the radar capability to geosynchronous range (40,000 km). This effort [70] also involves the redesign of the HUSIR W-band feed [71] to handle the increased power. Additionally, greater drive power from advanced solid-state amplifiers is needed [72] for the gyrotron to achieve its designed output.

Figure 15. The interior of the new transmitter facility with one of two klystrons (**left**) and its output waveguide and isolator (**right**). Coolant water is provided from the mezzanine level above.

The redesigned HUSIR feed, pictured in Figure 16, is being tested with the existing X-band feed. The integration of the X-band front end and W-band feed on the removable box allows planned maintenance to be conducted on the ground level with X-band radiation off a "splash plate" reflector oriented to send the radiation skyward during testing prior to re-insertion in the antenna. The ability to radiate in the W band simultaneously on the ground will be implemented in the near future upon addition of an RF switch and corrugated waveguide run to the test dock.

Development of the 50 kW GyroTWT at CPI is progressing, and the initial prototype unit is being tested as pictured in Figure 17. Early results obtained with W-band oscillators to drive the GyroTWT are in line with the predicted performance. More comprehensive testing with the tube operated in saturation at high-pulse duty has motivated the development of high-power solid-state amplifiers with sufficient gain to take the milliwatt power output of the HUSIR waveform generation hardware to 50 W or more, which is enough to saturate the GyroTWT across its full band. At the same time, these solid-state devices have to be protected from the possibility of RF burnout from an unintended oscillation of the tube. An interim drive capability has been brought to CPI in the form of a 20 W peak-power module fabricated by Raytheon Technologies using wideband MMIC devices developed [72] for HUSIR. These are combined in a "puck" architecture demonstrated by

Raytheon [73] over a narrower band. Isolation is provided by operating two HXI circulators developed for the HUSIR band in parallel to share the ~7 W average power. Raytheon has further combined eight such pucks [74] for what will ultimately be a 100 W HUSIR driver, providing sufficient power to saturate the tube through the 4 dB loss of the WARLOC duplexer which has been repurposed as an isolator.

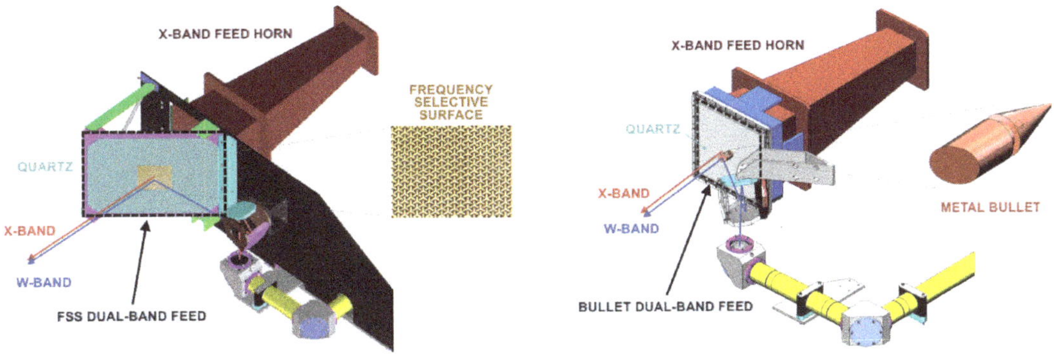

Figure 16. The HUSIR feed (**left**) and the new design for increased power (**right**).

Figure 17. The 50 kW peak-power HUSIR gyroTWT mounted in a test stand at CPI.

Development of improved receiver protection technology is also underway at the Laboratory. The doubling of the average power of the Millstone transmitter motivated a new design for the reflective ionized-gas switch used in that radar as a duplexer. On transmit, a waveguide quadrature hybrid delivers high-power RF to two identical gas-filled sections of waveguide which are ionized by the transmit pulse and reflect the power onward to the antenna. On receive, the switches are no longer ionized, and RF passes through to the receiver.

A different scheme is used at HUSIR, where a wire grid polarizer is used to separate the transmit and receive paths, and this is followed by a three-stage latching circulator, illustrated in Figure 18. The HUSIR receiver is cooled to 20 K, and this profoundly affects the design and testing of the latching circulators. These junction circulators employ a Y-shaped ferrite which is extremely small and difficult to fabricate and impedance-match. Improvements to the design, which will tolerate the 50-fold increase in HUSIR transmit power, are ongoing. Alternate methods such as high-speed switches are also being considered.

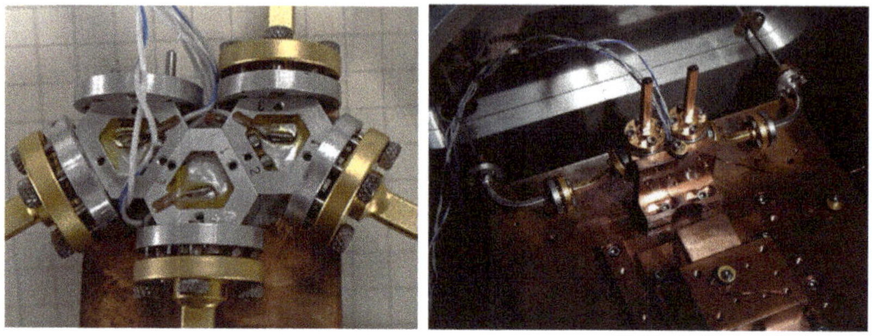

Figure 18. The cryogenic latching-circulator receiver protector for HUSIR (**left**) and its mounting in the 20 K dewar (**right**). The picture on the (**right**) shows mounting provisions for W-band receiver hardware, which is not installed in this picture.

Consideration is being given to further expanding the bandwidth of the MMW radar using frequency-multiplexing techniques developed at the Laboratory for other programs. Analyses and spectrum management considerations are in progress.

4. Discussion

The high-power radars that MIT Lincoln Laboratory has constructed and upgraded over nearly 70 years represent a significant national asset for the United States. Maintaining a commitment to continually improving their performance ensures that the observations they make will be supported in the long term, while new concepts provide a means of inserting advanced technology into this portfolio of sensors.

Figure 19, adapted from [36,51], illustrates the progress made in leveraging increased operating frequency and bandwidth to obtain progressively higher image resolution. As small payloads proliferate in space, and the debris population continues to grow, the need for long-range and wideband imaging sensors is becoming operationally critical for the US Space Force.

Figure 20 illustrates the challenge of keeping tabs on resident space objects as their population (including debris generated from collisions in orbit, unintentional and deliberate) continues to escalate dramatically, as seen in Figure 21. Over 12,000 spacecraft have been placed in orbit since 1957, of which more than 8000 remain. In 2014, NASA estimated the overall debris population to be over 500,000 (for sizes > 1 cm). In mid-2022, NASA also estimated [75] that there were over 25,000 objects on-orbit that are 10 cm or larger. In a 2020 report, the European Space Agency (ESA) estimated the 1 cm to 10 cm sized debris population at 100,000 and the 1 mm to 1 cm sized debris population at 130 million! Wideband radar sensors for space object imaging are complementary to optical sensors: at low Earth orbits, synoptic radar sensors (i.e., the US Space Force "Space Fence") detect and track, while optical sensors can provide images. At long ranges, the roles are reversed, and optical sensors such as the Space Surveillance Telescope [76] provide synoptic data, while the range independence of radar image resolution (provided that a sufficient SNR exists to track the target) permits detailed target characterization in addition to target trajectory.

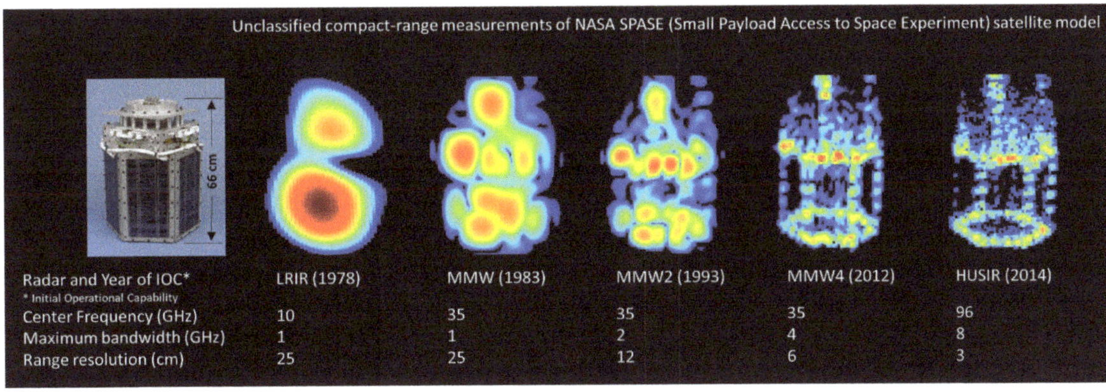

Figure 19. The progressive improvement in image resolution obtained at Lincoln Laboratory. Although ALCOR could image large objects, as seen in Figure 6, it is omitted from this chart, which focuses on the more recent need to image small orbiting payloads. Improved image resolution is achieved via increased radar waveform bandwidth, achieved both through increased center frequency with constant fractional bandwidth (e.g., HUSIR) and increased fractional bandwidth (e.g., MMW).

Figure 20. An illustration of the spatial distribution of man-made resident space objects about Earth (courtesy of AstriaGraph, University of Texas at Austin). Active satellites are color-coded in orange (the geostationary belt is clearly visible as an ellipse), inactive satellites in light blue, rocket bodies in violet and uncategorized objects in pink.

While this review has focused on activities at MIT Lincoln Laboratory, the broader radar community is actively meeting challenging mission scenarios in space. The need to improve space domain awareness, with emphasis on the geostationary orbit, motivated the Deep Space Advanced Radar Capability (DARC) technology demonstration [77] at Johns Hopkins Applied Physics Laboratory. This paves the way for an operational capability covering the entire geostationary belt jointly announced by the United States, Australia and the United Kingdom [78]. With NASA's commitment to return to the Moon, work has been undertaken [79] to use the NASA Deep Space Network radars to extend radar capability to the cislunar region. These radars are also being exploited [80] for the characterization of asteroids. Lincoln Laboratory is also actively working in these areas, and the developments recounted here will continue to be extended to surmount the challenges of radar data collection for progressively smaller objects at ranges exceeding 40,000 km.

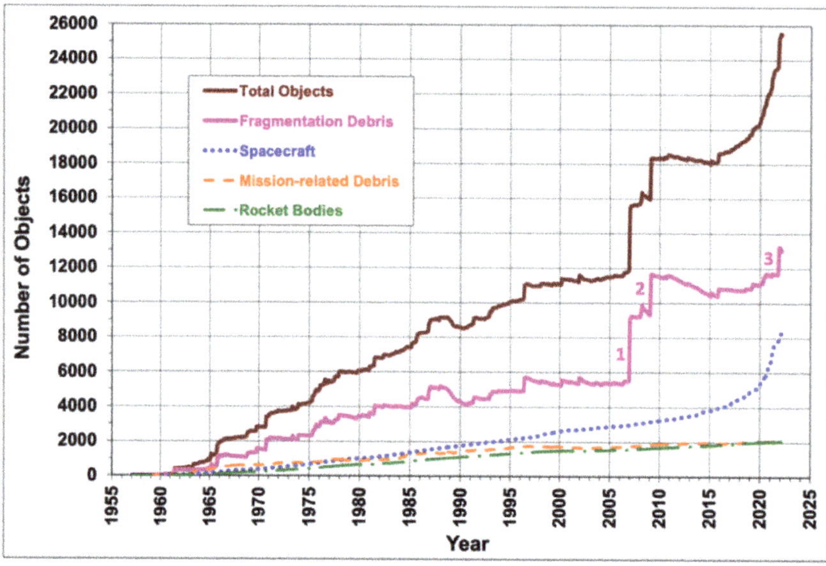

Figure 21. Reproduced from [75], the population of man-made resident space objects as a function of time. The numbered annotations show significant fragmentation events due to collisions in space, these being the 2007 Chinese anti-satellite (ASAT) test (1), the 2009 collision of Cosmos 2251 and Iridium 33 (2) and the 2013 Russian ASAT test (3).

Author Contributions: Resources and data curation, M.M., M.A. and J.S.; writing—original draft preparation, M.M., M.A. and J.S.; writing—review and editing, M.M. All authors have read and agreed to the published version of the manuscript.

Funding: This material is based on work supported by the Department of the Air Force under Air Force Contract No. FA8702-15-D-0001. Any opinions, findings, conclusions or recommendations expressed in this material are those of the author(s) and do not necessarily reflect the views of the Department of the Air Force.

Acknowledgments: In summarizing the breadth of work conducted at MIT Lincoln Laboratory over more than half a century, it is impossible to adequately acknowledge those who achieved these milestones. We have tried to capture as many contributors as possible, but much of the work is not in the open literature. In addition to the more than 80 Lincoln Laboratory authors cited here, we acknowledge the essential contributions of Gary Ahlgren, Gerry Banner, Bob Bergemann, Tom Clark, John Harris, V. Alexander Nedzel, Kurt Schwan, Paul Sebring, Grant Stokes, Henry Thomas and Lee Upton, as well as the financial support of the Defense Advanced Research Projects Agency, and the United States Army, Air Force and Space Force.

Conflicts of Interest: The authors declare no conflicts of interest.

References

1. Massachusetts Institute of Technology. *Radiation Laboratory Series (28 Volumes)*; Ridenour, L., Collins, G., Eds.; McGraw-Hill: New York, NY, USA, 2020; pp. 1948–1951. Available online: https://archive.org/details/mit-rad-lab-series-version-2 (accessed on 9 March 2024).
2. *Technology in Support of National Security*; Grometstein, A. (Ed.) MIT Lincoln Laboratory: Lexington, MA, USA, 2011. Available online: https://www.ll.mit.edu/sites/default/files/other/doc/2018-04/MIT_Lincoln_Laboratory_history_book.pdf (accessed on 9 March 2024).
3. Ricardi, L.; Niro, L. Design of a 12-horn monopulse feed. In Proceedings of the 1958 IRE International Convention Record, New York, NY, USA, 21–25 March 1966; pp. 49–56. [CrossRef]
4. Millstone Hill Incoherent Scatter Radar. Available online: https://www.haystack.mit.edu/about/haystack-telescopes-and-facilities/millstone-hill-incoherent-scatter-radar/ (accessed on 9 March 2024).
5. Evans, J. Theory and practice of ionosphere study by Thomson scatter radar. *Proc. IEEE* **1969**, *57*, 496–530. [CrossRef]
6. Pensa, A. Historical Perspective. In *Perspectives in Space Surveillance*; Sridharan, R., Pensa, A., Eds.; MIT Press: Cambridge, MA, USA, 2017; pp. 1–13.
7. Sridharan, R.; Kupiec, I. Technology Development for Space Surveillance with Narrow-beam Radars. In *Perspectives in Space Surveillance*; Sridharan, R., Pensa, A., Eds.; MIT Press: Cambridge, MA, USA, 2017; pp. 15–73.
8. Welch, S.; Hogan, G.; Morrison, R.; Bassa, C.; Pisano, T. Long baseline radar bistatic measurements of geostationary satellites. In Proceedings of the 2022 19th European Radar Conference (EuRAD), Milan, Italy, 28–30 September 2022; pp. 17–20. [CrossRef]
9. Ruze, J. Antenna tolerance theory-a review. *Proc. IEEE* **1965**, *54*, 633–640. [CrossRef]
10. Weiss, H. The Millstone and Haystack radars. *IEEE Trans. Aerosp. Elect. Syst.* **2001**, *37*, 365–379. [CrossRef]
11. Whitney, A.; Lonsdale, C.; Fish, V. Insights into the Universe: Astronomy with Haystack's radio telescope. *Linc. Lab. J.* **2014**, *21*, 8–27. Available online: https://archive.ll.mit.edu/publications/journal/pdf/vol21_no1/21_1_2_Whitney.pdf (accessed on 9 March 2024).
12. Shapiro, I.; Pettengill, G.; Ash, M.; Stone, M.; Smith, W.; Ingalls, R.; Brockelman, R. Fourth test of General Relativity: Preliminary results. *Phys. Rev. Lett.* **1968**, *20*, 1265–1269. [CrossRef]
13. Delaney, W.; Ward, W. Radar development at Lincoln Laboratory: An overview of the first fifty years. *Linc. Lab. J.* **2000**, *12*, 147–166. Available online: https://archive.ll.mit.edu/publications/journal/pdf/vol12_no2/12_2radardevelopment.pdf (accessed on 9 March 2024).
14. Roth, K.; Austin, M.; Frediani DKnittel, G.; Mrstik, A. The Kiernan re-entry measurements system at Kwajalein Atoll. *Linc. Lab. J.* **1989**, *2*, 247–276. Available online: https://archive.ll.mit.edu/publications/journal/pdf/vol02_no2/2.2.7.kiernanreentry.pdf (accessed on 9 March 2024).
15. Avent, R.; Shelton, J.; Brown, P. The ALCOR C-band imaging radar. *IEEE Antennas Prop. Mag.* **1996**, *38*, 16–27. [CrossRef]
16. Skolnik, M. (Ed.) *Introduction to Radar Systems*, 2nd ed.; McGraw-Hill: New York, NY, USA, 1980.
17. Axelbank, M.; Camp, W.; Lynn, V.; Margolin, J. ALCOR: A high-sensitivity radar with one-half meter range resolution. In Proceedings of the 1971 IEEE International Convention Digest, New York, NY, USA, 22–25 March 1971; pp. 112–113.
18. Purdy, R.; Blankenship, P.; Muehe, C.; Rader, C.; Stern, E.; Williamson, R. Radar signal processing. *Linc. Lab. J.* **2000**, *12*, 297–320. Available online: https://archive.ll.mit.edu/publications/journal/pdf/vol12_no2/12_2radarsignalprocessing.pdf (accessed on 9 March 2024).
19. Camp, W.; Mayhan, J.; O'Donnell, R. Wideband Radar for Ballistic Missile Defense and Range-Doppler Imaging of Satellites. *Linc. Lab. J.* **2000**, *12*, 267–280. Available online: https://archive.ll.mit.edu/publications/journal/pdf/vol12_no2/12_2widebandradar.pdf (accessed on 9 March 2024).
20. Solodyna, C. Overview of Wideband Imaging Radar Technology. In *Perspectives in Space Surveillance*; Sridharan, R., Pensa., A., Eds.; MIT Press: Cambridge, MA, USA, 2017; pp. 75–117.
21. Brown, W.; Pensa, A. History of Haystack. *Linc. Lab. J.* **2014**, *21*, 4–7. Available online: https://archive.ll.mit.edu/publications/journal/pdf/vol21_no1/21_1_1_Brown.pdf (accessed on 9 March 2024).
22. James, B.; Zitelli, L. A 100 kW peak 50 kW average coupled cavity TWT with 10% bandwidth centered at 10 GHz. In Proceedings of the IEEE Electron Devices Meeting, Washington, DC, USA, 1–2 December 1975; p. 133. [CrossRef]
23. Goudey, K.; Sciambi, A. High power X-band monopulse tracking feed for the Lincoln Laboratory Long Range Imaging Radar. *IEEE Trans. Microw. Theory Tech.* **1978**, *26*, 326–332. [CrossRef]
24. Goldie, H. A 2-kilowatt average power X-band receiver protector. *IEEE Trans. Electron. Dev.* **1977**, *24*, 53–55. [CrossRef]
25. Bromaghim, D.; Perry, J. A wideband FM linear ramp generator for the Long-Range Imaging Radar. *IEEE Trans. Microw. Theory Tech.* **1978**, *26*, 322–325. [CrossRef]
26. Sangiolo, T.; Spence, L. PACS: A processing and control system for the Haystack Long-range Imaging Radar. In Proceedings of the 1990 IEEE International Conference on Radar, Arlington, VA, USA, 7–10 May 1990; pp. 480–485. [CrossRef]
27. Abouzahra, M.; Avent, R. The 100-kW millimeter wave radar at the Kwajalein Atoll. *IEEE Antennas Prop. Mag.* **1994**, *36*, 7–19. [CrossRef]
28. Ingwersen, P.; Lemnios, W. Radars for ballistic missile defense research. *Linc. Lab. J.* **2000**, *12*, 245–266. Available online: https://archive.ll.mit.edu/publications/journal/pdf/vol12_no2/12_2ballisticmissiledefense.pdf (accessed on 9 March 2024).

29. Hall, T.; Duff, G.; Maciel, L. The space mission at Kwajalein. *Linc. Lab. J.* **2012**, *19*, 48–63. Available online: https://archive.ll.mit.edu/publications/journal/pdf/vol19_no2/19_2_3_Hall.pdf (accessed on 9 March 2024).
30. Fitzgerald, W. A 35 GHz beam waveguide system for the Millimeter Wave Radar. *Linc. Lab. J.* **1992**, *5*, 245–272. Available online: https://archive.ll.mit.edu/publications/journal/pdf/vol05_no2/5.2.4.35ghzwaveguide.pdf (accessed on 9 March 2024).
31. Dionne, G. Laboratory-scale mirrors for submillimeter wavelengths. *Int. J. Infrared Millim. Waves* **1982**, *3*, 417–423. [CrossRef]
32. Goldsmith, P. *Quasioptical Systems*; IEEE Press: New York, NY, USA, 1998.
33. Dionne, G.; Weiss, A.; Allen, G. Non-reciprocal magneto-optics for millimeter waves. *IEEE Trans. Magn.* **1988**, *24*, 2817–2819. [CrossRef]
34. McHarg, J.; Abouzahra, M.; Lucey, R. 95-GHz Millimeter-wave Radar. In *Millimeter and Submillimeter Waves Appl. III, Proceedings of the 1996 SPIE International Symposium on Optical Science, Engineering and Instrumentation, Denver, CO, USA, 5–9 August 1996*; SPIE: Denver, CO, USA, 1996; Volume 2842, pp. 494–500. [CrossRef]
35. Linde, G.; Ngo, M.; Danly, B.; Cheung, W.; Gregers-Hansen, V. WARLOC: A high-power coherent 94 GHz radar. *IEEE Trans. Aerosp. Elect. Sys.* **2008**, *44*, 1102–1117. [CrossRef]
36. Stambaugh, J.; Lee, R.; Cantrell, W. The 4 GHz bandwidth Millimeter Wave Radar. *Linc. Lab. J.* **2012**, *19*, 64–76. Available online: https://archive.ll.mit.edu/publications/journal/pdf/vol19_no2/19_2_4_Stambaugh.pdf (accessed on 9 March 2024).
37. Jones, G.; Abouzahra, M. Major improvements at the KREMS Millimeter Wave Radar. In Proceedings of the 1992 Space Surveillance Workshop, Lexington, MA, USA, 7–9 April 1992; pp. 239–244.
38. Ingalls, R.; Antebi, J.; Ball, J.; Barvainis, R.; Cannon, J.; Carter, J.; Charpentier, P.; Corey, B.; Crowley, J.; Dudevoir, K.; et al. Upgrading the Haystack radio telescope for operation at 115 GHz. *Proc. IEEE* **1994**, *82*, 742–755. [CrossRef]
39. Stansbery, E.; Settecerri, T. A Comparison of Haystack and HAX measurements of the orbital debris environment. In Proceedings of the Second European Conference on Space Debris, Organised by ESA, Darmstadt, Germany, 17–19 March 1997; ESOC: Darmstadt, Germany, 1997. ESA-SP 393. p. 59.
40. Robinson, P.; North, W. Compact floating-deck modulator for the HAX transmitter. In Proceedings of the Twentieth Conference Record on Power Modulator Symposium, Myrtle Beach, SC, USA, 23–25 June 1992; pp. 41–44. [CrossRef]
41. James, B. Re-Entrant Double-Staggered Ladder Circuit. U.S. Patent 4866343A, 12 September 1989.
42. Corbett, E.; Rossini, P.; Gallon, D.; Hotz, A. Pointing and tracking servo for the Haystack Auxiliary antenna (HAX). In Proceedings of the First IEEE Conference on Control Applications, Dayton, OH, USA, 13–16 September 1992; Volume 2, pp. 1134–1139. [CrossRef]
43. Available online: https://orbitaldebris.jsc.nasa.gov/measurements/radar.html (accessed on 9 March 2024).
44. Ruiz, G.; Patzelt, T.; Leuschake, L.; Loffeld, O. Autonomous tracking of space objects with the FGAN tracking and imaging radar. In Proceedings of the Informatik 2006—Informatik fur Menschen, Band 1, Bonn: Gesellschaft fur Informatik e.V., Dresden, Germany, 2–6 October 2006; pp. 349–353. Available online: https://dl.gi.de/handle/20.500.12116/23700 (accessed on 9 March 2024).
45. Karamanavis, V.; Dirks, H.; Fuhrmann, L.; Schlichthaber, F.; Egli, N.; Patzelt, T.; Klare, J. Characterization of deorbiting satellites and space debris with radar. *Adv. Sp. Res.* **2023**, *72*, 3269–3281. [CrossRef]
46. Mayhan, J.; O'Donnell, W.; Willner, D. The Cobra Gemini radar. In Proceedings of the 1996 IEEE National Radar Conference, Ann Arbor, MI, USA, 13–16 May 1996; pp. 380–385. [CrossRef]
47. Rejto, S. Radar open systems architecture and applications. In Proceedings of the IEEE 2000 International Radar Conference, Alexandria, VA, USA, 12 May 2000; pp. 654–659. [CrossRef]
48. Nelson, J. Radar open system architecture provides net centricity. *IEEE Aerosp. Elect. Syst. Mag.* **2010**, *25*, 17–20. [CrossRef]
49. MIT Lincoln Laboratory. 2012 Annual Report, Page 17. Available online: https://archive.ll.mit.edu/publications/Annual_Report_2012.pdf (accessed on 9 March 2024).
50. Missile Defense Agency. Flight Test Infrastructure Data Sheet, SS Pacific Tracker, 16-MDA-8889, Attachment 5. 14 October 2016. Available online: https://govtribe.com/file/government-file/mda17dtrfi01-attachment-5-xtr-1-tts-2-fact-sheet-dot-pdf (accessed on 9 March 2024).
51. Czerwinski, M.; Usoff, J. Development of the Haystack Ultrawideband Satellite Imaging Radar. *Linc. Lab. J.* **2014**, *21*, 28–44. Available online: https://archive.ll.mit.edu/publications/journal/pdf/vol21_no1/21_1_3_Czerwinski.pdf (accessed on 9 March 2024).
52. Waggener, N. Construction of the HUSIR antenna. *Linc. Lab. J.* **2014**, *21*, 45–82. Available online: https://archive.ll.mit.edu/publications/journal/pdf/vol21_no1/21_1_5_Waggener.pdf (accessed on 9 March 2024).
53. Usoff, J.; Clarke, M.; Liu, C.; Silver, M. Optimizing the HUSIR antenna surface. *Linc. Lab. J.* **2014**, *21*, 83–105. Available online: https://archive.ll.mit.edu/publications/journal/pdf/vol21_no1/21_1_6_Usoff.pdf (accessed on 9 March 2024).
54. MacDonald, M.; Anderson, J.; Lee, R.; Gordon, D.; McGrew, G. The HUSIR W-band transmitter. *Linc. Lab. J.* **2014**, *21*, 106–114. Available online: https://archive.ll.mit.edu/publications/journal/pdf/vol21_no1/21_1_4_MacDonald.pdf (accessed on 9 March 2024).
55. Eshbaugh, J.; Morrison, R.; Hoen, E.; Hiett, T.; Benitz, G. HUSIR signal processing. *Linc. Lab. J.* **2014**, *21*, 115–133. Available online: https://archive.ll.mit.edu/publications/journal/pdf/vol21_no1/21_1_7_Eshbaugh.pdf (accessed on 9 March 2024).
56. Kauffman, J.; Rajagopalan, G.; Akiyama, K.; Fish, V.; Lonsdale, C.; Matthews, L.; Pillai, T. The Haystack telescope as an astronomical instrument. *Galaxies* **2023**, *11*, 9. [CrossRef]

57. Jacobs, J.; Kennedy, T. *Haystack Ultra-Wideband Satellite Imaging Radar Measurements of the Orbital Debris Environment: 2018*; NASA/TP-2020-5006606; National Aeronautics and Space Administration: Washington, DC, USA, 2018.
58. Rodenbeck, C.; Tierney, B.; Park, J.; Parent, M.; Depuma, C.; Bauder, C.; Pizzillo, T.; Jaffe, P.; Simakauskas, B.; Mayhan, T. Terrestrial microwave power beaming. *IEEE J. Microw.* **2021**, *2*, 28–43. [CrossRef]
59. Blank, M.; Borchard, P.; Cauffman, S.; Felch, K. Design and demonstration of W-band gyrotron amplifiers for radar applications. In Proceedings of the 2007 Joint 32nd International Conference on Infrared and Millimeter Waves, Cardiff, UK, 2–9 September 2007; pp. 358–361. [CrossRef]
60. Doane, J. Propagation and Mode Coupling in Corrugated and Smooth-wall Circular Waveguides, Ch. 5. In *Infrared and Millimeter Waves, Vol. 13, Millimeter Components and Techniques, Part IV*; Button, K., Ed.; Academic Press: Orlando, FL, USA, 1985.
61. Kempkes, M.; Hawkey, T.; Gaudreau, M.; Phillips, R. W-band transmitter upgrade for the Haystack Ultra-wideband Satellite Imaging Radar. In Proceedings of the 2006 IEEE International Vacuum Electronics Conference held Jointly with 2006 IEEE International Vacuum Electron Sources, Monterey, CA, USA, 25–27 April 2006; pp. 551–552. [CrossRef]
62. Maiwald, F.; Lin, R.; Dengler, R.; Smith, S.; Pearson, J.; Mehdi, I.; Gaier, T.; Crowley, J. High output power low noise amplifier chains at 100 GHz. In Proceedings of the 18th International Symposium on Space Terahertz Technology, Pasadena, CA, USA, 21–23 March 2007; pp. 131–134.
63. MacDonald, M. Measured thermal dynamics of the HUSIR antenna and Haystack Radome. *IEEE Trans. Antennas Prop.* **2013**, *61*, 2441–2448. [CrossRef]
64. Antebi, J.; Kan, F. Replacement of elevation structure to upgrade Haystack 37-m radio telescope. In Proceedings of the Astronomical Structures and Mechanisms Technology, Glasgow, UK, 21–22 June 2004; Volume 5495. [CrossRef]
65. Doyle, K.; Brenner, M.; Antebi, J.; Kan, F.; Valentine, D. RF-mechanical performance for the Haystack radio telescope. In Proceedings of the SPIE Optical Engineering + Applications, San Diego, CA, USA, 24 September 2011; Volume 8125. [CrossRef]
66. Eshbaugh, J.; Clarke, M.; Maglathlin, G. Haystack Antenna Control System Design Document. Project Report HUSIR-8, MIT Lincoln Laboratory, Defense Technical Information Center Report ESC-TR-2010-071. 7 December 2010. Available online: https://apps.dtic.mil/sti/tr/pdf/ADA533478.pdf (accessed on 9 March 2024).
67. MacDonald, M. An overview of radomes for large ground-based antennas. *IEEE Aerosp. Elect. Syst. Mag.* **2019**, *34*, 36–43. [CrossRef]
68. Beals, D.; Abouzahra, M.; Schoenfeld, M.; MacGibbon, P. Millstone Hill Radar L-band sensitivity upgrade. In Proceedings of the 2023 International Applied Computational Electromagnetics Society Symposium (ACES), Monterey/Seaside, CA, USA, 26–30 March 2023; pp. 1–2. [CrossRef]
69. Grimes, D.; MacDonald, M. Enabling deep space radar with EM simulation. In Proceedings of the 2023 International Applied Computational Electromagnetics Society Symposium (ACES), Monterey/Seaside CA, USA, 26–30 March 2023; pp. 1–2. [CrossRef]
70. Abouzahra, M.; MacDonald, M.; Lee, R.; Grimes, D.; Simakauskas, B.; Lopez, N.; Eckert, C.; Usoff, J. Upgrading the HUSIR radar for deep-space satellite imaging. In Proceedings of the 2022 IEEE/MTT-S International Microwave Symposium—IMS 2022, Denver, CO, USA, 19–24 June 2022; pp. 429–432. [CrossRef]
71. Lee, R.; Simakauskas, B.; Eckert, C.; Abouzahra, M. Development of the HUSIR dual-band feed. In Proceedings of the 2022 19th European Radar Conference (EuRAD), Milan, Italy, 28–30 September 2022; pp. 29–32. [CrossRef]
72. Lopez, N.; MacDonald, M.; Abouzahra, M. LSSC Solid-state high-power amplifiers. In Proceedings of the 2022 19th European Radar Conference (EuRAD), Milan, Italy, 28–30 September 2022; pp. 25–28. [CrossRef]
73. Soric, J.; Kolias, N.; Saunders, J.; Kotce, J.; Brown, A.; Rodenbeck, C.; Gyurcsik, R. A 100-Watt GaN SSPA. *IEEE Microw. Wirel. Comp. Lett.* **2022**, *32*, 712–715. [CrossRef]
74. Soric, J.; Kolias, N.; Saunders, J.; Kotce, J.; Gyurcsik, R.; Abouzhara, M.; MacDonald, M. A 92-100 GHz 100 W SSPA for the HUSIR Deep-Space Upgrade. In Proceedings of the 54th European Microwave Conference (EuMC), Paris, France, 22–27 September 2024. (In press).
75. Available online: https://orbitaldebris.jsc.nasa.gov/quarterly-news/pdfs/odqnv26i1.pdf (accessed on 9 March 2024).
76. Available online: https://spacenews.com/space-surveillance-telescope-developed-by-the-u-s-begins-operations-in-australia/ (accessed on 9 March 2024).
77. Available online: https://www.jhuapl.edu/work/projects/deep-space-advanced-radar-capability-darc-technology-demonstration (accessed on 9 March 2024).
78. Available online: https://www.spaceforce.mil/News/Article-Display/Article/3604036/ (accessed on 9 March 2024).
79. Lee, C.; Rodriguez-Alvarez, N.; Giorgini, J.D.; Jao, J.S.; Majid, W.; Naudet, C.J.; Oudrhiri, K.; Yang, Y.-M. Cis-lunar space debris radar capability and feasibility. In Proceedings of the 2023 IEEE Aerospace Conference, Big Sky, MT, USA, 4–11 March 2023. [CrossRef]
80. Available online: http://echo.jpl.nasa.gov/ (accessed on 9 March 2024).

remote sensing

MDPI

Article

Inverse Synthetic Aperture Radar Imaging of Space Targets Using Wideband Pseudo-Noise Signals with Low Peak-to-Average Power Ratio

Simon Anger *, Matthias Jirousek, Stephan Dill and Markus Peichl

German Aerospace Center (DLR), Microwaves and Radar Institute, Muenchenerstrasse 20, 82234 Weßling, Germany; matthias.jirousek@dlr.de (M.J.); stephan.dill@dlr.de (S.D.); markus.peichl@dlr.de (M.P.)
* Correspondence: simon.anger@dlr.de

Abstract: With the number of new satellites increasing dramatically, comprehensive space surveillance is becoming increasingly important. Therefore, high-resolution inverse synthetic aperture radar (ISAR) imaging of satellites can provide an in-situ assessment of the satellites. This paper demonstrates that pseudo-noise signals can also be used for satellite imaging, in addition to classical linear frequency-modulated chirp signals. Pseudo-noise transmission signals offer the advantage of very low cross-correlation values. This, for instance, enables the possibility of a system with multiple channels transmitting instantaneously. Furthermore, it can significantly reduce signal interference with other systems operating in the same frequency spectrum, which is of particular interest for high-bandwidth, high-power systems such as satellite imaging radars. A new routine has been introduced to generate a wideband pseudo-noise signal with a peak-to-average power ratio (PAPR) similar to that of a chirp signal. This is essential for applications where the transmit signal power budget is sharply limited by the high-power amplifier. The paper presents both theoretical descriptions and analysis of the generated pseudo-noise signal as well as the results of an imaging measurement of a real space target using the introduced pseudo-noise signals.

Keywords: inverse synthetic aperture radar (ISAR); noise radar; pseudo noise; radar imaging; space targets; satellites; space surveillance

1. Introduction

Imaging radars widely use linear frequency-modulated chirp signals to provide broadband characteristics and gather fine range resolution. Chirp signals have an inherently low peak-to-average power ratio (PAPR), which enables power-efficient long-range radar imaging with limited peak transmit power. This is achieved by using a matched filter for range compression, which maximizes the signal-to-noise ratio (SNR). Linear frequency modulations can be either increasing (up-chirp) or decreasing (down-chirp), while non-linear frequency modulations include exponential or hyperbolic modulations [1,2]. In noise radar technology (NRT), random waveforms are transmitted instead of the classical, often sophisticated, deterministic radar signals [3–6]. Compared to deterministic signals, the non-deterministic nature of noise-like transmitted signals can provide advantages in radar imaging. Noise signals and their spread spectrum properties can reduce mutual interference with other radars and communication systems operating in the same frequency spectrum. Due to their low cross-correlation values and quasi-orthogonality, it would be possible to operate several transmitters simultaneously using different noise signals [7,8]. Another well-known advantage is the Low Probability of Intercept (LPI) of noise-like signals, which reduces their visibility to electronic support measures (ESM) and electronic intelligence systems (ELINT). They are more resistant to interference and adversarial attacks such as jamming and signal exploitation (LPE) [9–11].

Nowadays, it is possible to generate analog transmit signals with a random characteristic using an arbitrary waveform generator (AWG). The AWG is capable of converting a digitally available arbitrary signal characteristic into an analog transmit signal using a sample rate of several giga samples per second (GS/s) and analog bandwidths of up to several GHz. The actual characteristic of the signal can be generated by means of software algorithms. The electronics available today allow the receiver to sample the signal with an analog-to-digital converter (ADC) at up to several GS/s with a subsequent fully digital signal processing, enabling comprehensive signal calibrations and error corrections for arbitrary signals used. A digitally matched filter can be used to perform range compression and maximize the SNR. Just like with deterministic signals, the achievable range resolution depends not on the time domain signal characteristic but is determined by the signal bandwidth or, more precisely, the spectral shape of the noise signal [10,12,13].

The German Aerospace Center (DLR) operates an experimental satellite imaging radar called IoSiS (Imaging of Satellites in Space) using a high degree of digital technology able to generate arbitrary transmit signals [14,15]. IoSiS contains a pulse radar system working around the X band with an instantaneous bandwidth of up to 4.4 GHz [16,17]. In the transmitter, a traveling-wave tube amplifier (TWTA) is used to generate the high output pulse power. Due to the limited output peak power of amplifiers, the pulse transmit signal should exhibit a low peak-to-average power ratio (PAPR). This achieves the TWTA's maximum achievable average power, ultimately resulting in a good SNR in the radar image [11]. Especially in the field of space surveillance, tracking or imaging radars, where small targets or low radar cross sections (RCS) have to be tracked or imaged, efficient utilization of the available power is of crucial importance. Hence, using a Gaussian noise signal with its high PAPR as a transmit signal leads to a loss of SNR. A special pseudo-noise signal is required that has the signal characteristics of a noise signal together with a low PAPR like a chirp signal. In addition, LPI and LPE properties as well as the quasi-orthogonality must be retained. This paper outlines the iterative process for creating a pseudo-noise signal with the aforementioned properties. It also includes the theoretical analysis of the signal and its generation based on the IoSiS system parameters for experimental verification measurements. One focus is on investigating the experimental results and the signal's applicability with IoSiS components. Finally, an ISAR measurement of the International Space Station (ISS) using IoSiS is described to demonstrate that the pseudo-noise signal is capable of high-resolution phase-coherent imaging.

2. Pseudo-Noise Signals as Transmit Signal

In radar remote sensing, the power budget of the radar system is critical for achieving a sufficient signal-to-noise ratio (SNR) in the measurement results when detecting or imaging small targets with low radar cross sections (RCS). Therefore, a crucial requirement is to maximize the energy of the transmitted radar signal. With pulse radar technology and a given duration of the transmit pulse, an increase in the energy of the radar pulse can only be achieved by increasing the average power of the pulse. Together with the fact that every high power amplifier (HPA) that supplies the power for the transmit signal has a certain saturation point at which the maximum output power is limited, it results in that the ratio between the average pulse power and the peak pulse power should be as small as possible to maximize the SNR. The PAPR of any time discrete signal characteristic $s(n)$ of length N can be calculated as follows:

$$PAPR = \frac{max|s(n)|^2}{\frac{1}{N}\sum_{n=1}^{N}|s(n)|^2} = \left(\frac{|\hat{s}|}{s_{RMS}}\right)^2 \quad (1)$$

In any case, where the average power is less than the peak power of the signal, this will result in a loss of SNR, which can be calculated as follows:

$$SNR_{loss} = -10\ log_{10}(PAPR) \quad (2)$$

If we consider as an example a real-valued frequency modulated chirp signal with a peak amplitude of $\hat{s}$, and further $s_{RMS} = \hat{s}/\sqrt{2}$ the root mean square (RMS) of a sinusoidal signal, the PAPR can be calculated with $\hat{s}^2/{s_{RMS}}^2 = 2$. Chirp signals with their low PAPR can drive the HPA around its saturation point without changing the signal characteristic in its correlation behavior due to clipping or nonlinearity effects and hence constitute a very power-efficient signal. The main objective of the present work was to generate and experimentally validate a pseudo-noise signal (PNS) that has a comparable power efficiency to a chirp signal, while retaining its fundamental noise characteristics, such as the spread-spectrum properties, LPI, and low cross-correlation values. In the next section, the generation procedure and the basic signal characteristics of the pseudo-noise signal are described.

2.1. Generation Procedure of Pseudo-Noise Signal

With regard to the experimental verification measurements with the IoSiS radar system, the design of the PNS focused on the basic system parameters of IoSiS. These parameters included a system bandwidth of 2.8 GHz, a pulse duration of about 40 µs, and a signal sampling rate of 10 GS/s. Furthermore, the signal should have a peak-to-peak amplitude of $u_{pp} = 2$ V, corresponding to an average power of 10 dBm considering a chirp like PAPR. In addition, in view of the experimental measurement for which the signals are to be designed, only real-valued signals are considered in the following.

The iterative generation procedure is shown in Figure 1. The initial signal used here is a Gaussian noise signal $s_N(t)$ with a standard deviation significantly larger than the intended signal amplitude of the PNS. It should hold $\sigma(s_N(t)) > a \cdot u_{pp}$ with $a > 1000$. In two steps, the signal is shaped. First, in the time domain, the peak amplitude is limited by clipping at u_{pp}. Therefore, the values greater than u_p or less than $-u_p$ will be set to u_p or $-u_p$, respectively. Secondly, with $S(f) = \mathfrak{F}(s(t))$, the signal is shaped in the frequency domain to the intended bandwidth characteristic $H_{BP}(f)$. In the following work, $H_{BP}(f)$ equals a rectangular function. By repeating these shaping steps a minimum of 15 times, a PNS signal with a chirp-like PAPR will be generated.

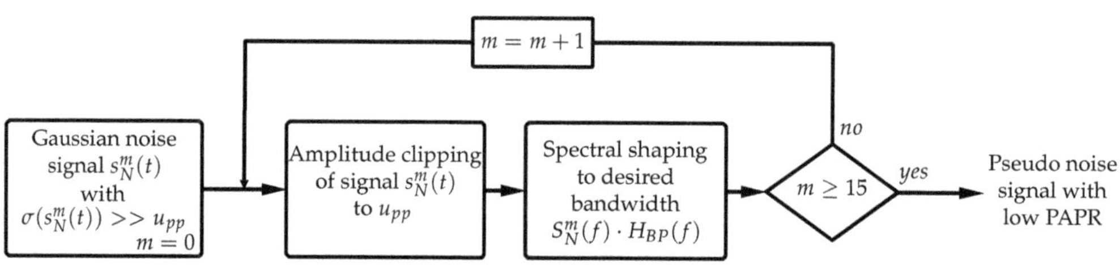

Figure 1. Iterative generation of an amplitude and bandwidth-limited pseudo-noise signal with a low peak-to-average power ratio (PAPR). The generated pseudo-noise signal is limited to the peak-to-peak amplitude u_{pp} in the time domain and to the transfer function $H_{BP}(f)$ in the frequency domain.

The time domain signal of the generated PNS together with a Gaussian noise and a classical chirp signal is shown in Figure 2. In addition, the probability density function (PDF) of the amplitude in the time domain is depicted. All signals shown have an average pulse power of 10 dBm considering an impedance of 50 Ω. Furthermore, the signals exhibit a bandwidth of 2.8 GHz with the cutoff frequencies 200 MHz and 3 GHz. The Gaussian noise signal can be described with the PDF as follows:

$$F_{PDF}(a) = \frac{1}{\sqrt{2\pi\sigma^2}} \exp\left(\frac{a^2}{2\sigma^2}\right) \quad (3)$$

with the variance of the signal of length $t_2 - t_1$:

$$\sigma^2 = \frac{1}{t_2 - t_1} \int_{t_1}^{t_2} s^2(t)\, dt \tag{4}$$

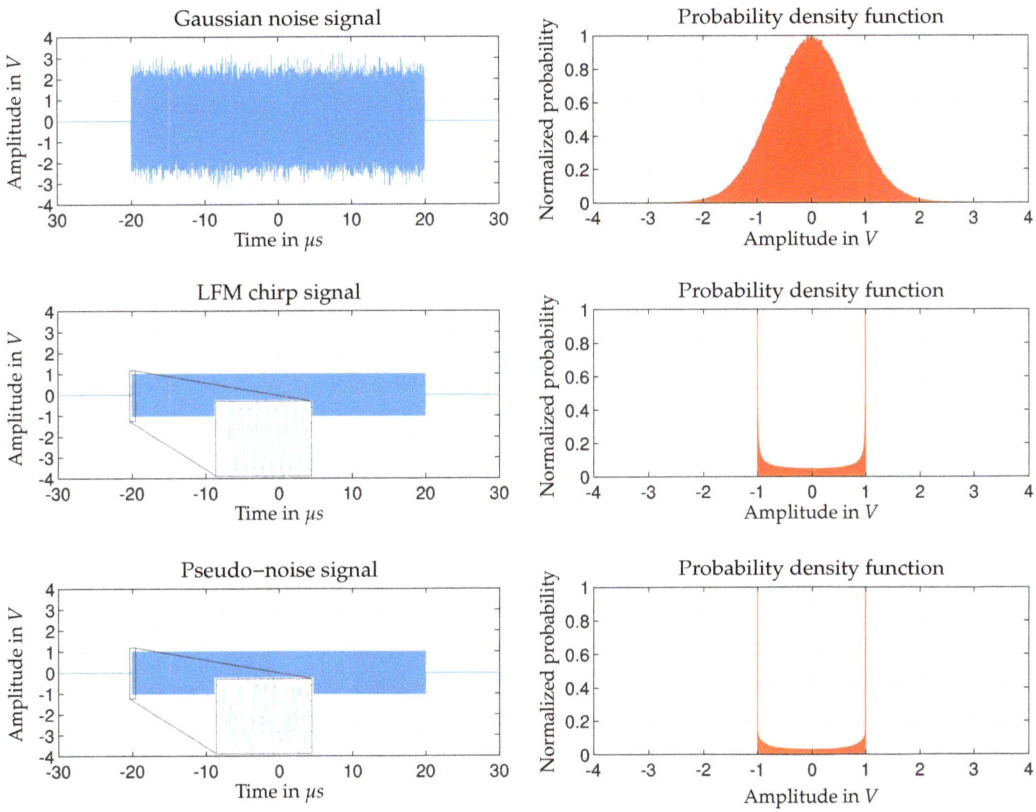

Figure 2. Three different pulse signals with the same average power of 10 dBm. A Gaussian noise signal with the variance $\sigma^2 = 0.5$ V and the corresponding probability density function (PDF) of the time domain amplitude values. A classical linear frequency modulated chirp signal and the corresponding PDF. A sharp amplitude limited pseudo-noise signal generated with the procedure described in Figure 1 and the corresponding PDF.

Further, for the normalized power of a time-continuous random signal or the ergodic process, we can write [18]:

$$P_N = \lim_{T \to \infty} \frac{1}{2T} \int_{-T}^{T} s^2(t)\, dt \tag{5}$$

In the case where the linear time average of the signal is zero, we can see by comparing (4) and (5) that $\sigma^2 \equiv P_N$ [19]. Consequently, the variance of the signal is the normalized (reference impedance 1 Ω) power of our Gaussian signal. With the intended average signal power of 10 dBm and an impedance of 50 Ω, this results in a variance of $\sigma^2 = 0.5$ V and a standard deviation of $\sigma = 0.707$ V (Figure 2 (top)). The time domain signal and the PDF show that a significant number of values up to around 3 V appear in a Gaussian noise signal with an average power of 10 dBm. Hence, the signal exhibits a high PAPR of about 34 compared to the chirp signal with a PAPR of 2. A chirp signal, however, has a sharp limited amplitude in the time domain (Figure 2 (center)). Considering an average signal power of 10 dBm, the chirp signal exhibits a peak amplitude of 1 V. This also applies to the PNS generated, as described before (Figure 2 (bottom)). Furthermore, the PDFs from chirp and PNS exhibit strikingly similar behavior. This is not immediately apparent when only the enlarged section of the two signals is compared. However, when the entire length of the signal is considered, the majority of the amplitude values fall within the range of approximately 1 or -1 volt, which gives rise to this distinctive PDF. Given that the chirp signal and the PNS signal exhibit a comparable PNS, it follows that the average power and, consequently, the PAPR of both signals are identical.

2.2. Short-Time Fourier Analysis

To analyze the spread-spectrum properties of the PNS, the results of a short-time Fourier transformation (STFT) are shown in Figure 3. The LFM chirp signal shows its typical characteristic where the signal energy is focused at a small instantaneous bandwidth. The PNS signal, however, has a characteristic very similar to that of the Gaussian noise and, as a result, it also exhibits as Gaussian noise a large instantaneous bandwidth [20], indicating a proper LPI. The spread-spectrum properties are a distinct advantage in the reduction in interference with systems that share the same frequency spectrum. Especially when very high pulse powers are used in combination with very large transmission bandwidths, appropriate interference mitigation is essential.

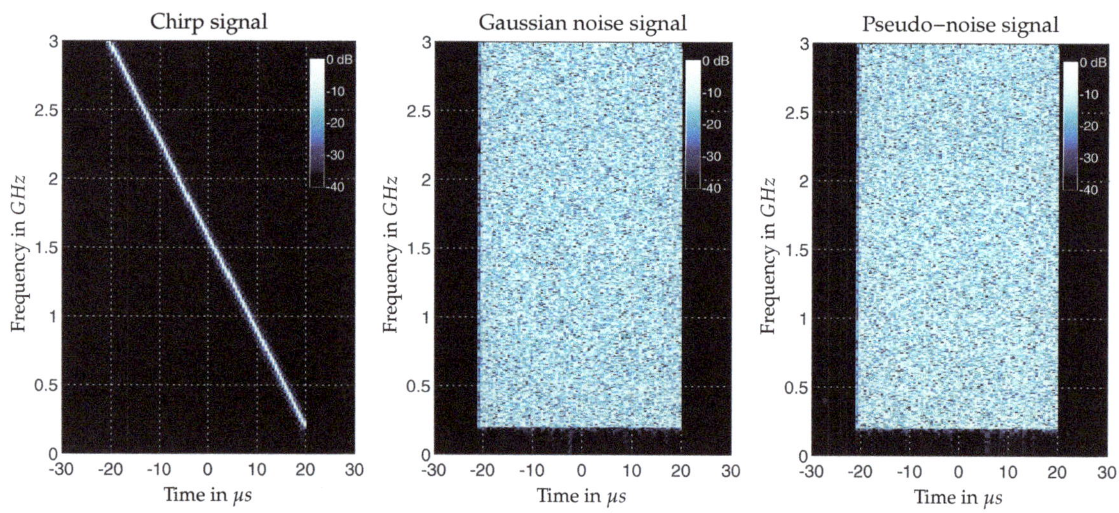

Figure 3. Short-time Fourier transform (STFT) of a chirp, Gaussian noise, and the generated pseudo-noise signal. The STFT indicates that the pseudo-noise signal exhibits very similar characteristics to a Gaussian noise signal.

2.3. Point Spread Function and Cross-Correlation Analyses

A key feature when using the PNS signal as a radar transmit signal is an appropriate autocorrelation function (ACF). The ACF defines the range profile of an ideal, frequency-independent point target, neglecting noise, attenuation, and distortion. When using a chirp signal with bandwidth B and a rectangular envelope in the frequency domain, the power representation of the range profile of a single point target ideally corresponds to a $sin(x)^2/x^2$ function, showing a typical suppression of the first side lobe level of about 13.3 dB. Further, with c the speed of light, the first zero of the sinc-function can be calculated with c/B and the half-power beam width (HPBW) corresponds to $0.89 \cdot (c/2B)$, representing the range resolution of the radar [2].

In radar systems, a matched filter approach represents the correlation between the transmit signal $s_t(t)$ and the conjugated, time-reversed version of the received signal $s_r(t)$. This convolution is often carried out in the frequency domain due to the lower computational effort. With $\mathfrak{F}(\cdot)$ the Discrete Fourier Transform (DFT) and $S_t(f) = \mathfrak{F}(s_t)$ and $S_r(f) = \mathfrak{F}(s_r)$, we obtain the following matched filter output:

$$s_{mf}(t) = \mathfrak{F}^{-1}\Big(S_t(f) \circ S_r^*(f)\Big) \tag{6}$$

where $\circ$ denotes the point-wise multiplication (Hadamard product). In the ideal case, $s_r(t)$ equals $s_t(t - \Delta t)$. If we further neglect the time shift Δt between the transmit and receive signal, the output of the matched filter represents the ACF defined by:

$$s_{ACF}(\tau) = \int_{-\infty}^{\infty} s(t)s^*(t-\tau)\,dt \qquad -\infty < \tau < \infty \tag{7}$$

Figure 4 shows the ACF of a Gaussian noise signal in comparison with the ACF of the PNS. Both signals have a bandwidth of 2.8 GHz, an initial pulse duration of 40 µs and a rectangular envelope in the frequency domain. Both show a nearly ideal characteristic down to an amplitude below −35 dB and a suppression of the first side lobe of 13.3 dB. The zeros between the side lobes have a slightly higher amplitude in the PNS case. However, in the practical case of experimental measurements, this small deviation can be neglected. The HPBW of about 54 mm corresponds to the theoretical value.

In addition, the cross-correlation function (CCF) between two different Gaussian and PNS pairs is shown in Figure 4. Again, both cross-correlated signals have a bandwidth of 2.8 GHz, a rectangular frequency envelope, a pulse duration of 40 µs, and a rectangular envelope in the frequency domain. With respect to the peak amplitude, the CCF values are mostly below −50 dBc for both signal types. The result shows that the generated PNSs are also well-suited for multiple access applications in NRT [21].

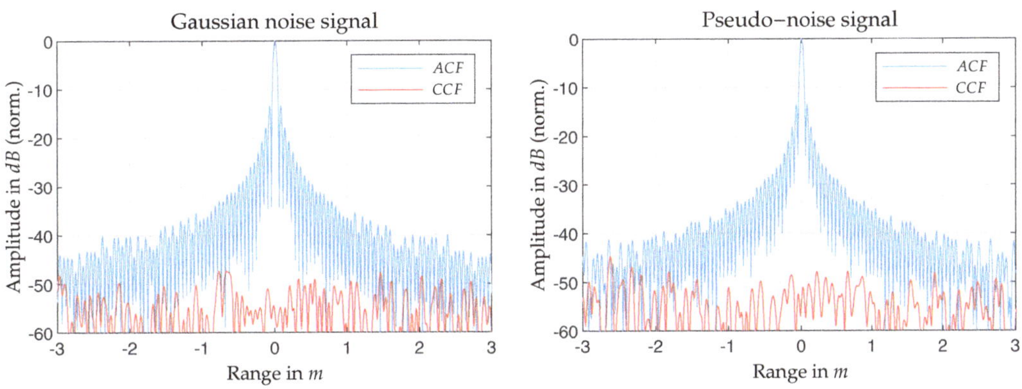

Figure 4. Simulated point spread functions using a Gaussian noise as well as the pseudo-noise signal (PNS) with a signal bandwidth of 2.8 GHz. Both exhibit a nearly ideal characteristic down to an amplitude below −35 dB. Similar to the Gaussian noise signal, a pair of PNSs exhibits very low cross-correlation (CCF) values, indicating its quasi-orthogonality.

To demonstrate the beneficial impact of a more expansive signal bandwidth, Figure 5 illustrates the ACF and CCF with half and double bandwidth compared to the characteristics shown in Figure 4. The larger bandwidth has a positive influence on the ACF as well as the CCF. In the case of a signal bandwidth of 5.6 GHz, the attenuation of the side lobes increases faster corresponding to the finer range resolution. The CCF also shows lower values when broadband signal characteristics are used.

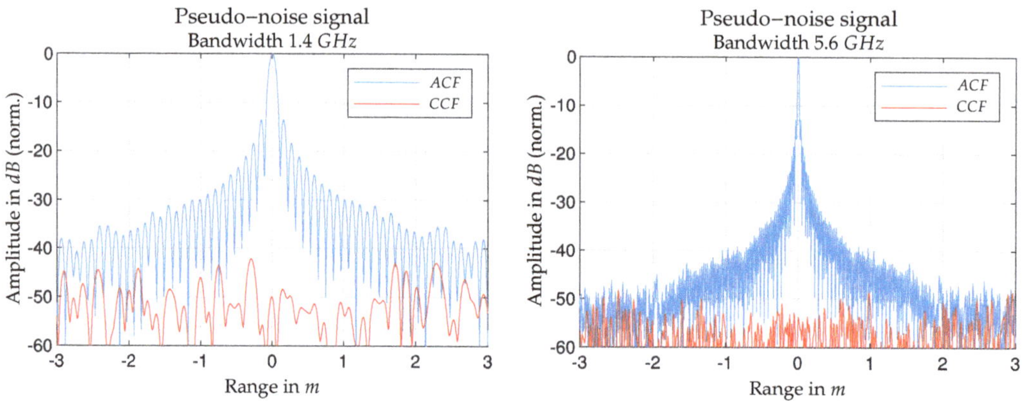

Figure 5. Simulated point spread functions using the PNS with different signal bandwidths.

2.4. Ambiguity Function

The ambiguity function illustrates the sensitivity of the PSF regarding a shift in time or frequency, respectively. In the signal processing of pulsed radar, an ambiguity function is a two-dimensional function of the propagation delay Δt and the Doppler shift frequency f_D. For a given complex base band pulse $s(t)$, the narrow band ambiguity function is given by [1,22]:

$$\chi(\Delta t, f_D) = \int_{-\infty}^{\infty} s(t)s^*(t - \Delta t)e^{i2\pi f_D t}\, dt \tag{8}$$

For broadband signals, the Doppler shift, represented in (8) by $e^{i2\pi f_D t}$, can no longer be assumed to be constant with respect to the instantaneous frequency within a broadband pulse signal. The movement of the target causes the wide-band signal to expand or compress in time. This effect can no longer be approximated by a simple frequency "shift". Therefore the wideband ambiguity function (WBA) was introduced [23–26]:

$$\chi_{wb}(\Delta t, v_r) = \sqrt{|\alpha|} \int_{-\infty}^{\infty} s(t)s^*(\alpha(t - \Delta t))\, dt \qquad with \qquad \alpha = \frac{c + v_r}{c - v_r} \tag{9}$$

Here, V is the radial target velocity with respect to the radar and c is the speed of light.

The simulated WBA of an LFM chirp signal and a generated PNS is shown in Figure 6 for two different pulse duration's τ_{pulse} and a bandwidth of 2.8 GHz. Satellites in a low Earth orbit travel with an orbital velocity of about 7.6 km/s, with the result that very high radial velocities can occur during the imaging process [27,28]. The radial velocities of $v_r = \pm 1$ km/s, used for the simulation, hence are typical values in space target imaging; however, they can be even higher at very low elevation angels of the antenna, pointing towards the space target during the pass.

In contrast to the chirp signal, stochastic signals are very sensitive to frequency shifts caused by a Doppler frequency. This also applies to the PNS signal, which can be recognized by the rapid decrease in amplitude with increasing $|v_r|$. The longer the pulse duration, the more sensitively the PNS reacts to a Doppler shift. A longer pulse duration leads to an increase in time compression or expansion of the Doppler effect, and hence, the negative effect of the radial velocity component increases significantly.

The WBA of an LFM chirp signal shows a temporal shift with increasing $|v_r|$, which can also be expressed as a shift along the range direction. At higher $|v_r|$, there is a defocusing effect which becomes more pronounced as the pulse duration becomes longer due to an increase in time compression or expansion, respectively.

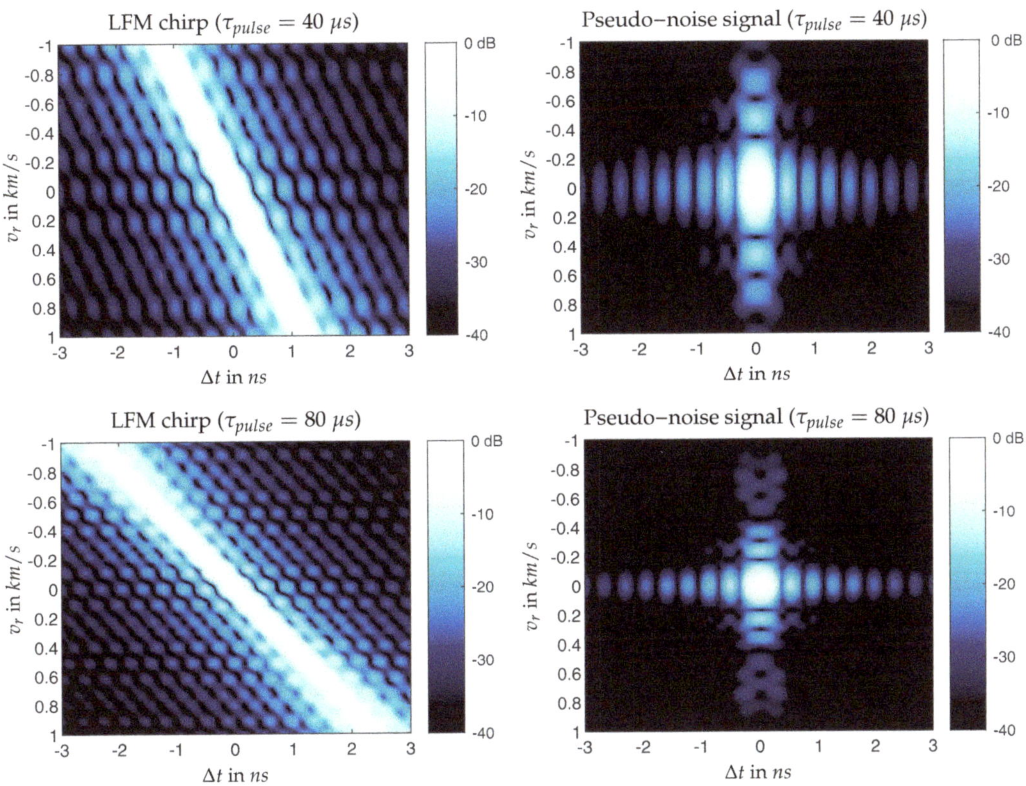

Figure 6. Wideband ambiguity function of a linear frequency modulated (LFM) chirp signal (**left**) and a generated pseudo-noise (**right**) for typical radial velocities v_r as they occur in radar-based imaging of satellites.

When considering the PNS, the Doppler sensitivity must be taken into account, and an appropriate mitigation procedure must be applied to avoid severe loss of SNR. In the case of space target imaging, v_r can be properly determined and its degradation effect mitigated because the trajectory and spatial velocity vectors of the target are well known [28].

2.5. Spectral Kurtosis

Spectral kurtosis (SK) can be used to analyze the PNS for its stationary characteristics and to quantify its LPI properties [5]. SK is a statistical tool that can indicate and locate non-stationary or non-Gaussian behavior in the frequency domain. A stationary time series has statistical properties or moments, for example, mean and variance, that do not vary in time. A small value of SK for a frequency bin indicates the existence of stationary Gaussian noise, whereas a high positive value indicates the existence of a transient or structured signal at the corresponding frequency [29,30]. The SK of a signal $S(f_k)$ can be calculated with:

$$SK(f) = \frac{\sum\limits_{k=b_1}^{b_2} (f_k - C_1)^4 S(f_k)}{(C_2)^4 \sum\limits_{k=b_1}^{b_2} S(f_k)} \tag{10}$$

with C_1 the spectral centroid:

$$C_1(f) = \frac{\sum_{k=b_1}^{b_2} f_k S(f_k)}{\sum_{k=b_1}^{b_2} S(f_k)} \tag{11}$$

and C_2 the spectral spread:

$$C_2(f) = \sqrt{\frac{\sum_{k=b_1}^{b_2} (f_k - C_1)^2 S(f_k)}{\sum_{k-b_1}^{b_2} S(f_k)}} \tag{12}$$

here b_1 and b_2 are the band edges in bins of the frequency spectrum under investigation. f_k is the frequency in Hz, and $S(f_k)$ is the signal value of the kth bin, respectively [31].

The SK of a Gaussian noise signal and the PNS are shown in Figure 7. The section of the baseband signal spectrum being investigated, ranging from 200 MHz to 3 GHz, is highlighted. Both signals exhibit low SK values and thus show their stationary characteristics. This shows that the iteratively generated PNS still contains its stationary characteristic and an LPI. By way of comparison, the SK values of a chirp signal are in the region of 100, which is indicative of its deterministic, non-stationary nature.

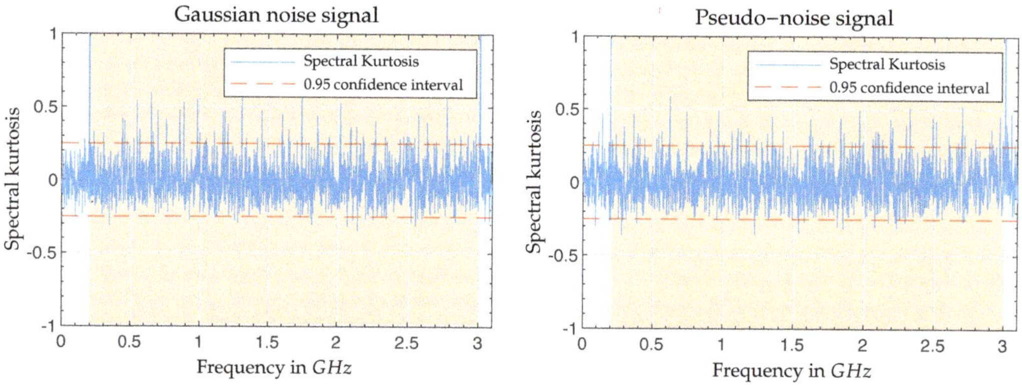

Figure 7. Comparison of the Spektral Kurtosis (SK) of Gaussian noise and the generated pseudo-noise signal. The orange highlighted section shows the frequency range of the signal from 200 MHz to 3 GHz. The low SK values of the pseudo-noise signal show its stationary signal behavior and hence its low LPI properties.

3. Experimental Verification Measurements

For the experimental verification measurements of the generated PNS, the IoSiS radar system is used [14]. The first step is to investigate whether the radar can obtain high-resolution range profiles by using the PNS. This is performed using an internal calibration path measurement. The space target used is the ISS (International Space Station), which is a well-known target and has been imaged several times with IoSiS using chirp signals [32].

3.1. IoSiS System Description

IoSiS is an experimental satellite imaging radar using pulse radar techniques and working around the X band. Figure 8 shows the hardware setup containing a multi-dish-based steerable antenna system, a TWTA (white box behind the main reflector), and a container for the main radar electronics. The main 9 m dish antenna is used as the transmitting antenna and two 1.8 m antennas are used for receiving. In the results presented in the following, only one receive antenna, i.e., one receive channel, has been used. The radar electronics enable fully digital signal generation and acquisition across the entire simultaneous radar bandwidth. In the current mode, IoSiS achieves a maximum simultaneous bandwidth of 2.8 GHz with an RF center frequency of 10.2 GHz. This corresponds to the lower sideband (LSB) considering a local oscillator (LO) frequency of 11 GHz.

Figure 9 shows the RF schematic of the transmit and one receive path. Isolators, amplifiers, and attenuators have been partially omitted for the sake of simplicity. In the transmit path, the AWG enables the digital-to-analog conversion of the generated PNS with a sample rate of 10 GS/s in the baseband. After the up-conversion into the RF band, bandpass filters work as image rejection filters, where the LSB filter is used in the measurement results presented. The TWTA at a short distance from the feed network of the Cassegrain antenna ensures the required high pulse power. A common LO source ensures coherent up- and down-conversion in the transmit and receive path, which is a prerequisite for the application of ISAR imaging techniques. In the receive path, a low-noise amplifier (LNA) ensures an appropriate receiver noise temperature. Bandpass filters in turn ensure a proper image rejection, and after down-conversion, a digital-to-analog converter samples the signal in the time domain at a sampling rate of up to 10 GS/s.

For calibration, in which the system's transfer function is corrected, a calibration path is provided that diverts an attenuated version of the transmit signal directly into the receive path. In this work, the calibration path is used to check the system behavior when using the PNS as a transmission signal before a real imaging measurement is performed.

Figure 8. A photo of the IoSiS setup is shown, featuring a 9 m transmitting antenna in a Cassegrain configuration and two 1.8 m receiving antennas. The radar electronics responsible for signal generation and acquisition are housed in a nearby container.

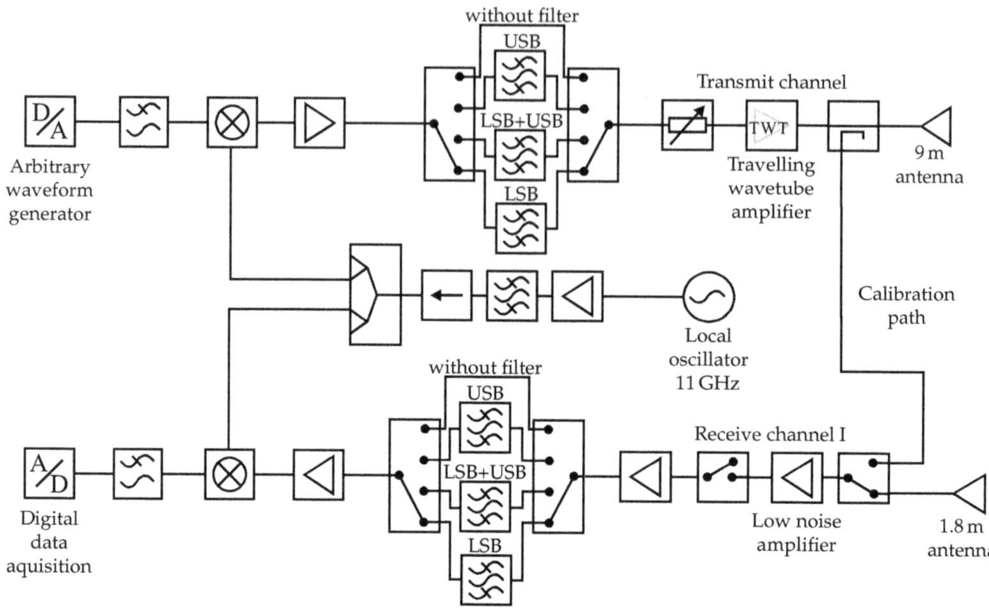

Figure 9. Simplified RF diagram of the IoSiS radar, showing the transmit channel and one of the two receive channels. Fully digital signal generation and acquisition enables the generation of arbitrary transmit signals. Filter banks in each channel enable operation in the upper (USB) and lower side band (LSB).

3.2. Point Spread Function Analysis Measurements

Essential for high-resolution ISAR imaging is a focused point spread function (PSF) together with the coherence, i.e., phase stability over time. When considering the PNS as a transmit signal, the behavior of the TWTA is of particular interest. A TWTA is operated at or even in saturation. Therefore, distortion and harmonics may be introduced into the amplified signal due to non-linearity between the input and output signals, which can degrade the PSF. The PNS, with its spread spectrum characteristics, provides the TWTA with an input signal that delivers the full system bandwidth at the same time. Therefore, in combination with the non-linear characteristic of the TWTA and in contrast to a chirp signal, which has a low instantaneous bandwidth, harmonics and intermodulation distortions can occur.

In the following, the calibration path (see Figure 9) is used to determine the PSF when using the PNS. For this, the first step is to perform a calibration of the system for the correction of a non-ideal transfer function (TF). A chirp signal is used to determine the TF and to generate a calibration signal to compensate for frequency-dependent characteristics (non-constant group delay) and amplitude variations of radar system components such as amplifiers and filters. The detailed calibration procedure can be found in [16,17].

The measured PSF using the PNS is shown in Figure 10 with a rectangular (unweighted) spectrum and a Hamming weighting. The measuremt parameters are listed in Table 1. The non-ideal transfer function of the radar system, including the dispersion characteristics and the time shift caused by the propagation delay of the radar system components, is already compensated by using the calibration signal mentioned above. In the unweighted case, the PSF represents the expected sinc-function with an HPBW of 54 mm, corresponding to the bandwidth used, and the first side lobes are around −13.3 dBc. As can already be seen in the simulation results (see Figure 4), the side lobes below −35 dBc show a slight fluctuation of the amplitude.

Also shown is the CCF of a pair of PNS signals, both measured using the calibration path. As in the simulation results, the CCF values are mostly below −50 dBc with respect to the ACF where signal power is maximized. This demonstrates that the quasi-orthogonality is not significantly affected by the system.

Coherence analysis is performed by evaluating the phase of the range profiles at the point of maximum signal power (peak). The phase values of the 2000 range profiles are displayed in Figure 11, along with the phase PDF. The phase values exhibit a Gaussian distribution with a standard deviation of merely 0.3 degrees. This extremely small range of values is not expected to have any negative effects on the experimental measurements. It confirms the phase stability of the system, along with the PNS.

The results of the PSF analysis demonstrate that high-resolution range profiles can be achieved using the PNS together with the IoSiS system. Furthermore, the coherence of the measurements is guaranteed. The following section presents the measurement results of a space target using the PNS as the transmit signal.

Table 1. Measurement parameter for the determination of the PSF using the PNS.

Parameter	Symbol	Parameter Value	Unit
Signal type	-	PNS	-
Center frequency	f_c	10.2	GHz
Bandwidth	B	2.8	GHz
Range resolution	Δr	54	mm
DAC sampling rate	f_{DAC}	10	GS/s
ADC sampling rate	f_{ADC}	10	GS/s
Pulse duration	τ_{pulse}	45	µs
PRF	prf	200	Hz
Signal type	-	PNS	-
Mode	-	LSB	-

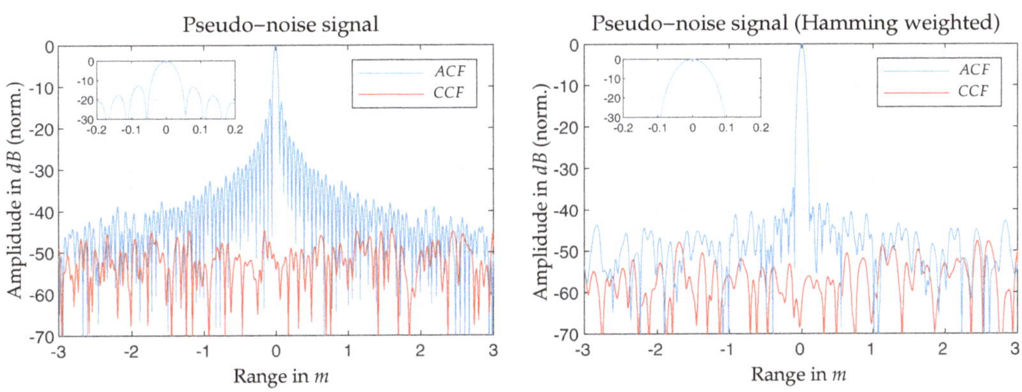

Figure 10. Measured point spread function (PSF) using the pseudo-noise signal routed through the calibration path of the radar system (**left**). Spectral weighted PSF of the pseudo-noise signal (**right**). In the non-weighted case, the half-power beam width (HPBW) corresponds to the applied pulse signal bandwidth of 2.8 GHz and equals about 54 mm. In both cases, the low cross-correlation (CCF) values show the quasi-orthogonality of a pair of pseudo-noise signals.

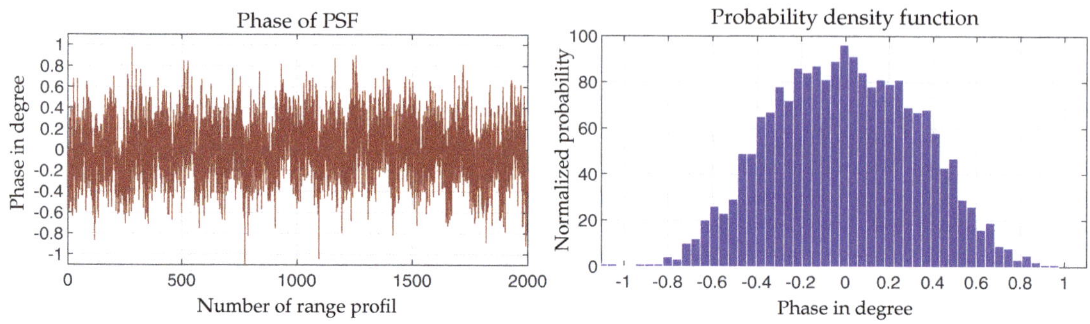

Figure 11. Phase values extracted in the maximum amplitude of the ACF point spread function (PSF) measured using the calibration path of the IoSiS radar (**left**). Corresponding phase probability function (PDF) of the measured phase values (**right**). With a phase standard deviation of $\sigma = 0.3\,°$, the phase is highly stable, which is a prerequisite for using the PNS in an ISAR system.

3.3. ISAR Space Target Measurement Results

An ISAR measurement of the ISS is performed using IoSiS to verify the PNS in a real space target measurement. The same PNS is used here for the PSF verification measurements described in the previous section. The same calibration steps were applied to achieve high spatial resolution as in measurements with a chirp signal. This includes antenna calibration and calibration of the radar system transfer function. A detailed description of the processing steps can be found in [14].

The measurement parameters and the resulting theoretical image parameters are shown in Table 2. The theoretical slant range and azimuth resolution are 54 mm and 53 mm respectively, corresponding to the signal bandwidth and azimuth integration angle [32]. As the focal plane, used for the back-projection of the range profiles, was chosen according to the spatial orientation of the ISS, we have to consider the degradation of the range resolution based on the focal plane incident angle θ_{FP}. With the slant range resolution Δr , the FP resolution is $\Delta r_{FP} = \Delta r \cdot \cos(\theta_{FP})^{-1}$. The actual spatial resolution of the FP is therefore between 67 mm and 74 mm, taking into account the changes in the angle of incidence during the measurement.

The final ISAR imaging result, after completing all processing steps is depicted in Figure 12. The image shows great detail of the entire ISS with high backscatter along the vertically aligned main truss segment as well as along the horizontally aligned various modules. The grid structures between the individual solar arrays are slightly visible, as is the end of the solar arrays with their characteristic RCS. The planar solar panels themselves are not visible. They are rotated on a continuous basis so that they are aligned with the sun. For this reason, they can only be seen if the time of day of the measurement and the imaging geometry match.

To provide an objective assessment of the spatial resolution achieved and to evaluate the PNS signal employed, two robust scatterers are analyzed in depth. The positions of these scatterers under investigation are indicated with arrows in the ISAR image. Both are located either on or in close proximity to the FP, ensuring an undisturbed PSF. These scatterers near the main truss segment have been observed in various measurements of the ISS and exhibit a wide-angle backscatter behavior that allows for comprehensive, fundamental radar verification. The distance between the scatterers is approximately 15 m, which enables examination of the focus over a wide area. However, it is important to note that these structures are not ideal isolated calibration targets as they are surrounded by other components. There are several radar calibration spheres orbiting the earth, but they cannot be used with the IoSiS system due to their high altitude together with its low RCS [33].

Table 2. Measurement parameter for the ISAR image shown in Figure 12 using the PNS .

Parameter	Symbol	Parameter Value	Unit
Transmit signal type	-	PNS	-
Center frequency	f_c	10.2	GHz
Bandwidth	B	2.8	GHz
Integration angle	ϕ_{Az}	16	degree
Slant range resolution	Δr	54	mm
Azimuth resolution	Δr_{Az}	53	mm
FP incident angle	θ_{FP}	36.2–43	degree
FP range resolution	Δr_{FP}	67–74	mm
DAC sampling rate	f_{DAC}	10	GS/s
ADC sampling rate	f_{ADC}	10	GS/s
Pulse duration	τ_{pulse}	45	µs
PRF	prf	190	Hz
Target distance	r_t	600–680	km
Processed pulses	N_{pulses}	5000	km
Processing algorithm	-	Back-projection	-
Image size	-	6000 × 9000	Pixel
Mode	-	LSB	-

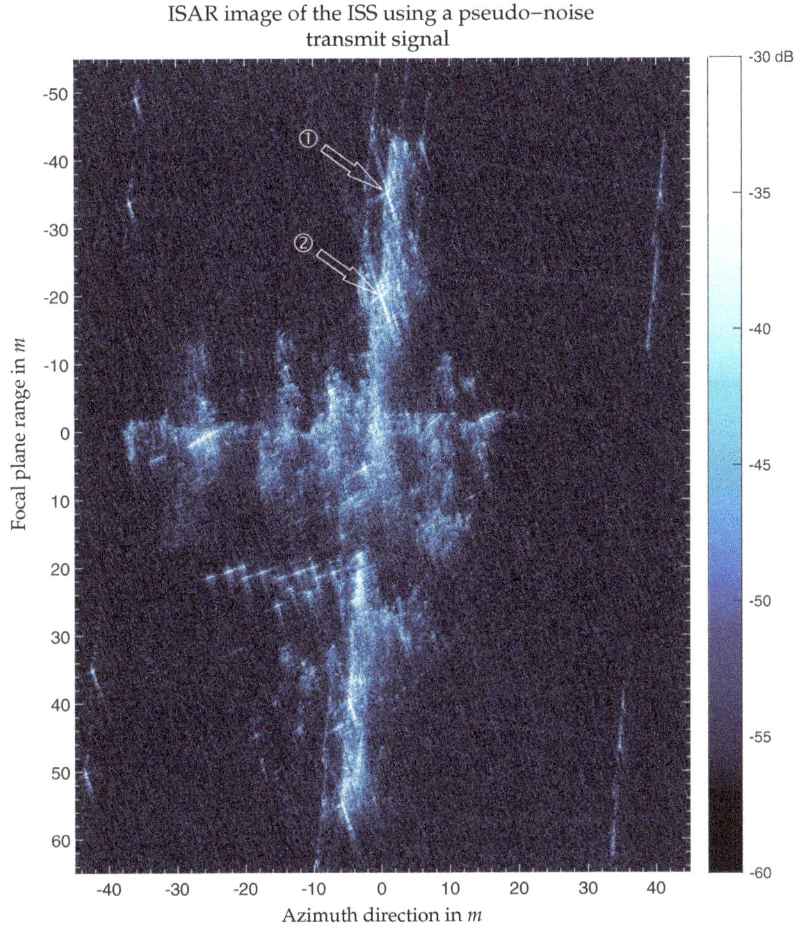

Figure 12. ISAR image of the ISS using a PNS transmit signal together with the IoSiS radar system. The achieved spatial resolution in range and azimuth direction was evaluated by analyzing two robust scatterers. The arrows indicate the positions of the scatterers under investigation.

Both sections of the ISAR image containing the robust scatterers are depicted in Figure 13. The amplitude of each image is normalized based on its own maximum to independently extract the characteristics of the PSF. The image is accompanied by the cuts in the FP range and azimuth directions for each PSF. It is obvious that there are interfering structures in the second PSF at the bottom left of the main lobe area that could affect the robustness of the analyzed target. But the cuts still show that a decent resolution could be achieved. The resolution values around 70 mm are within expectation in the FP range (refer to Table 2). In the azimuth direction, both PSFs exhibit a slightly worse resolution compared to the theoretical value. This indicates that there may be minor residual errors that have not been mitigated or non-idealities of the target.

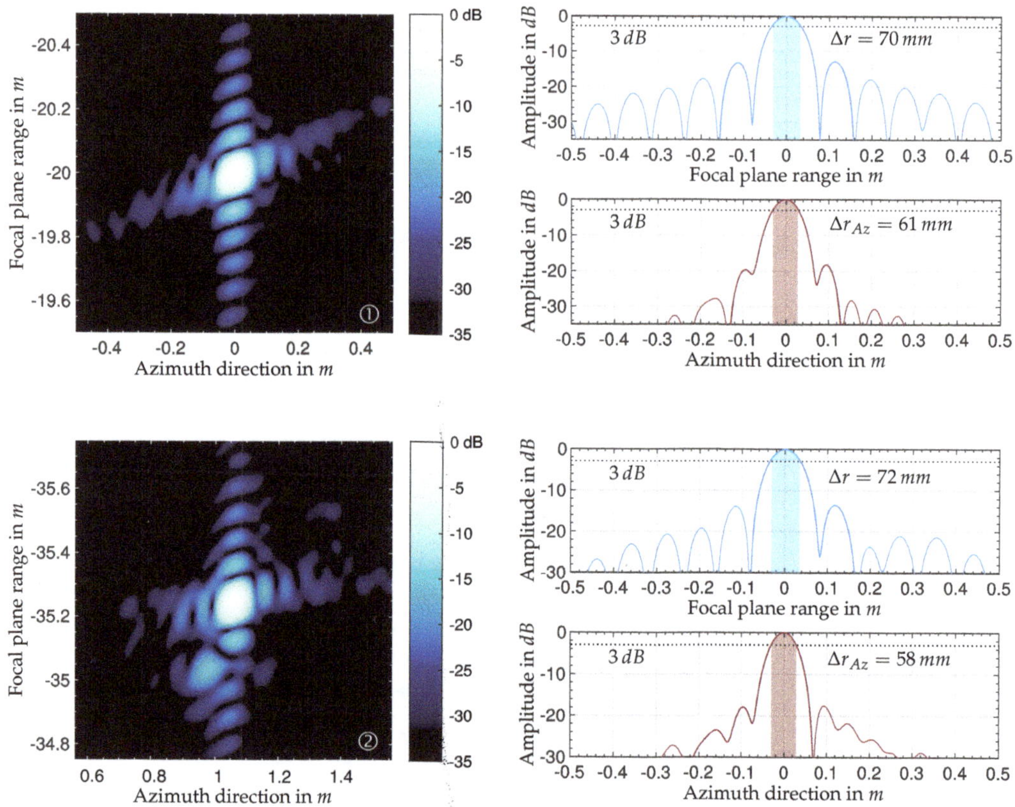

Figure 13. ISAR image of the isolated robust scatterers of the ISS, together with the cuts in the cardinal direction, to analyze the PSF and to evaluate the PNS signal used. The HPBW in the focal plane (Δr) of both are in the expected range. In the azimuth direction, both exhibit a slightly higher HPBW (Δr_{Az}) considering the integration angle.

The suppression of the range side lobes is almost ideal, although the second PSF shows some deviations, which are certainly caused by the aforementioned surrounding structure. In azimuth, the ISAR images of both PSFs clearly show that the sidelobes are not perfectly aligned with the azimuth direction. Hence, the cuts of azimuthal side lobes should be viewed with caution as the cuts do not pass through the maximum amplitude of each side lobe. The slanted side lobes are due to the imaging geometry combined with the very large integration angle and projection onto the FP.

In summary, however, the results demonstrate accurate focusing in both spatial directions of two widely separated scatterers, confirming the effectiveness of using a PNS for imaging space targets.

4. Conclusions

This paper describes the investigation of pseudo-noise signals that can be used in particular in ground-based high-resolution ISAR satellite imagery. The non-deterministic nature of noise-like transmission signals can have advantages over deterministic signals. Noise signals and their spread spectrum characteristics can, for example, reduce the mutual interference with other radar systems and communication systems that are operating in the same frequency spectrum.

The studies presented are carried out from the perspective of experimental measurements with the IoSiS satellite imaging radar. The parameters of the IoSiS system were therefore the basis for the investigations. When considering noise signals, the peak-to-average power ratio (PAPR) in particular has been a key point of optimization. A routine is introduced that iteratively generates a pseudo-noise signal (PNS) that has a chirp-like PAPR. In addition to a low PAPR, the aim was to maintain noise characteristics such as spread spectrum, low probability of intercept, and low cross-correlation values. In the presented work, simulation results show that the general properties of stationary noise are still valid for the iteratively generated PNS. Short-term Fourier transformation, ambiguity function, and spectral kurtosis of the PNS were analyzed and evaluated. The simulation results are compared to either Gaussian noise or chirp signals, demonstrating that the PNS still has all the intended characteristics of a stationary noise signal. Experiments using the IoSiS radar system are presented to verify the theoretical results. The first step of the experimental test was the verification of the applicability of the PNS with the IoSiS hardware setup via the calibration path. The analysis of the point spread function (PSF), together with the phase analysis, demonstrates that it is possible to achieve highly resolved range profiles with highly coherent phase behavior using the PNS. Furthermore, the measurement results validate the low cross-correlation values. The potential of the PNS in satellite imaging is demonstrated by performing an ISAR measurement of the ISS.

Author Contributions: S.A.: Conceptualization, data curation, formal analysis, investigation, software , validation, visualization, and writing—original draft. M.J.: Data curation, investigation, writing—original draft, and writing—review and editing. S.D.: Conceptualization, investigation, validation, writing—original draft, and writing—review and editing. M.P.: Conceptualization, funding acquisition, project administration, supervision, and writing—review and editing. All authors have read and agreed to the published version of the manuscript.

Funding: This work has been partially supported by the Federal Office of Bundeswehr Equipment, Information Technology and In-Service Support.

Data Availability Statement: The data that support the findings of this research are partly available from the corresponding author upon reasonable request. The data are not publicly available due to privacy restrictions.

Conflicts of Interest: The authors declare no conflicts of interest.

Abbreviations

The following abbreviations are used in this manuscript:

ACF	Auto correlation function
ADC	Analog-to-digital converter
AF	Ambiguity function
AWG	Arbitrary waveform generator
CCF	Cross-correlation function
DAC	Digital-to-analog converter
ELINT	Electronic intelligence systems

ESM	Electronic support measures
HPA	High power amplifier
HPBW	Half-power beam width
IoSiS	Imaging of satellites in space
ISS	International Space Station
LFM	Linear frequency modulated
LNA	Low noise amplifier
LO	Local oscillator
LPE	Low probability of exploitation
LPI	Low probability of intercept
LSB	Lower side band
NRT	Noise radar technology
PAPR	Peak-to-average power ratio
PDF	Probability density function
PNS	Pseudo-noise signal
PSF	Point spread function
RCS	Radar cross section
RMS	Root mean square
SK	Spectral Kurtosis
SNR	Signal-to-noise ration
STFT	Short-time Fourier transformation
TF	Transfer function
TWTA	Travelling wavetube amplifier
USB	Upper side band
WBA	Wideband ambiguity function

References

1. Levanon, N.; Mozeson, E. *Radar Signals*; John Wiley & Sons, Ltd.: Hoboken, NJ, USA, 2004; Chapter 4, pp. 53–73. [CrossRef]
2. Skolnik, M. *Introduction to Radar Systems*; Electrical Engineering Series; McGraw-Hill: New York, NY, USA, 2001.
3. Palo, F.D.; Galati, G.; Pavan, G.; Wasserzier, C.; Savci, K. Introduction to Noise Radar and Its Waveforms. *Sensors* **2020**, *20*, 5187. [CrossRef] [PubMed]
4. Galati, G.; Pavan, G.; De Palo, F.; Stove, A. Potential applications of noise radar technology and related waveform diversity. In Proceedings of the 17th International Radar Symposium (IRS), Krakow, Poland, 10–12 May 2016; pp. 1–5. [CrossRef]
5. Savci, K.; Galati, G.; Pavan, G. Low-PAPR Waveforms with Shaped Spectrum for Enhanced Low Probability of Intercept Noise Radars. *Remote Sens.* **2021**, *13*, 2372. [CrossRef]
6. Galati, G.; Pavan, G.; Savci, K.; Wasserzier, C. Noise Radar Technology: Waveforms Design and Field Trials. *Sensors* **2021**, *21*, 3216. [CrossRef] [PubMed]
7. De Palo, F.; Galati, G. Orthogonal waveforms for multiradar and MIMO radar using noise radar technology. In Proceedings of the Signal Processing Symposium (SPSympo), Debe, Poland, 10–12 June 2015; pp. 1–4. [CrossRef]
8. Galati, G.; Pavan, G. Noise Radar Technology as an Interference Prevention Method. *J. Electr. Comput. Eng.* **2013**, *2013*, 146986. [CrossRef]
9. De Martino, A. *Introduction to Modern EW Systems*; Artech House: London, UK, 2018.
10. Kulpa, K. *Signal Processing in Noise Waveform Radar*; Artech House: London, UK, 2013.
11. Galati, G.; Pavan, G.; Savci, K.; Wasserzier, C. Counter-Interception and Counter-Exploitation Features of Noise Radar Technology. *Remote Sens.* **2021**, *13*, 4509. [CrossRef]
12. Kulpa, K.; Lukin, K.; Miceli, W.; Thayaparan, T. Signal processing in noise radar technology. *IET Radar Sonar Navig.* **2008**, *2*, 229–323. [CrossRef]
13. Narayanan, R.M.; Xu, X. Principles and applications of coherent random noise radar technology. *Noise Devices Circuits* **2003**, *5113*, 503–514. [CrossRef]
14. Anger, S.; Jirousek, M.; Dill, S.; Kempf, T.; Peichl, M. High-resolution inverse synthetic aperture radar imaging of satellites in space. *IET Radar Sonar Navig.* **2023**, *18*, 544–563. [CrossRef]
15. Anger, S.; Jirousek, M.; Dill, S.; Peichl, M. Imaging of objects in space using arbitrary transmit signals. In Proceedings of the 14th European Conference on Synthetic Aperture Radar, Leipzig, Germany, 25–27 July 2022; pp. 1–5.
16. Jirousek, M.; Iff, S.; Anger, S.; Peichl, M. GigaRad—A multi-purpose high-resolution ground-based radar system concept, error correction strategies and performance verification. *Int. J. Microw. Wirel. Technol.* **2015**, *7*, 443–451. [CrossRef]
17. Jirousek, M.; Dill, S.; Schreiber, E.; Anger, S.; Peichl, M.; Schreiber, H. Characteristics of the high-performance highly digitized multi-purpose radar system GigaRad: System Concept, System Correction and Calibration, Applications. In Proceedings of the 2018 11th German Microwave Conference (GeMiC), Freiburg, Germany, 12–14 March 2018; pp. 395–398. [CrossRef]
18. Lindner, J. *Informationsübertragung*; Springer-Lehrbuch; Springer: Berlin/Heidelberg, Germany, 2005.

19. Ohm, J.R.; Lüke, H.D. *Signalübertragung Grundlagen der Digitalen und Analogen Nachrichtenübertragungssysteme*, 11th ed.; Springer: Berlin/Heidelberg, Germany; New York, NY, USA, 2010. [CrossRef]
20. Boashash, B. Estimating and interpreting the instantaneous frequency of a signal. I. Fundamentals. *Proc. IEEE* **1992**, *80*, 520–538. [CrossRef]
21. Ipatov, V. *Spread Spectrum and CDMA: Principles and Applications*; Wiley: West Sussex, UK, 2005.
22. Balleri, A.; Farina, A. Ambiguity function and accuracy of the hyperbolic chirp: Comparison with the linear chirp. *IET Radar Sonar Navig.* **2017**, *11*, 142–153. [CrossRef]
23. Dawood, M.; Narayanan, R. Generalised wideband ambiguity function of a coherent ultrawideband random noise radar. *IEE Proc. Radar Sonar Navig.* **2003**, *150*, 379. [CrossRef]
24. Sune, R.A. Generalized Ambiguity Functions for Ultra Wide Band Random Waveforms. In Proceedings of the International Radar Symposium, Krakow, Poland, 24–26 May 2006; pp. 1–4. [CrossRef]
25. Sibul, L.; Ziomek, L. Generalized wideband crossambiguity function. *IEEE Int. Conf. Acoust. Speech Signal Process.* **1981**, *6*, 1239–1242. [CrossRef]
26. Sen, S.; Nehorai, A. Adaptive Design of OFDM Radar Signal With Improved Wideband Ambiguity Function. *IEEE Trans. Signal Process.* **2010**, *58*, 928–933. [CrossRef]
27. Anger, S.; Jirousek, M.; Dill, S.; Schreiber, E.; Peichl, M. Imaging of satellites in space (IoSiS): Challenges in image processing of ground-based high-resolution ISAR data. In *Proceedings of the Radar Sensor Technology XXII*; Ranney, K.I., Doerry, A., Eds.; International Society for Optics and Photonics, SPIE: Bellingham, WA, USA, 2018; Volume 10633, p. 106330L. [CrossRef]
28. Montenbruck, O.; Gill, E. *Satellite Orbits: Models, Methods, and Applications*; Physics and Astronomy Online Library; Springer: Berlin/Heidelberg, Germany, 2000.
29. Antoni, J. The spectral kurtosis: A useful tool for characterising non-stationary signals. *Mech. Syst. Signal Process.* **2006**, *20*, 282–307. [CrossRef]
30. Antoni, J.; Randall, R. The spectral kurtosis: Application to the vibratory surveillance and diagnostics of rotating machines. *Mech. Syst. Signal Process.* **2006**, *20*, 308–331. [CrossRef]
31. Peeters, G. *A Large Set of Audio Features for Sound Description (Similarity and Classification) in the CUIDADO Project*; Technical Report; Icram: Paris, Paris, 2004.
32. Anger, S.; Jirousek, M.; Dill, S.; Peichl, M. ISAR imaging of space objects using large observation angles. In Proceedings of the 21st International Radar Symposium (IRS), Berlin, Germany, 21–22 June 2021.
33. McDowell, J. Navel Research Laboratory Satellites 1960–1989. *J. Br. Interplanet. Soc.* **1997**, *50*, 427–432.

remote sensing

MDPI

Article

The Radar Signal Processor of the First Romanian Space Surveillance Radar

Călin Bîră [1], Liviu Ionescu [2] and Alexandru Rusu-Casandra [1,*]

[1] Faculty of Electronics, Telecommunications and Information Technology, Politehnica University of Bucharest, 060042 Bucharest, Romania; calin.bira@upb.ro
[2] RARTEL S.A., 011041 Bucharest, Romania; liviu.ionescu@rartel.ro
* Correspondence: alexandru.rusu@upb.ro

Abstract: This paper describes the work for the radar signal processor, the core of the Cheia space surveillance radar. It presents the basic operation, the requirements and the achieved processing performance together with a description of optimizations both in terms of signal-to-noise ratio and in terms of software processing. The Cheia Radar is financed by the European Space Agency and will be used for the tracking of low Earth orbit objects and for refining the international catalogs of space objects.

Keywords: radar; space debris; space situation awareness; low Earth orbit; digital signal processing

1. Introduction

Founded according to a decision in 2014 by the European Parliament and Council [1], the Space Surveillance and Tracking (SST) Consortium is currently composed of seven Member States of the European Union (EU), including Romania. As defined in [1], the SST system is a network of ground-based and space-based sensors capable of surveying and tracking space objects, together with processing capabilities focusing on providing information and services on space objects that orbit around our planet.

In order to extend and improve Romanian contribution to the EU SST system, studies have been undertaken by the Romanian company Rartel, funded by the European Space Agency (ESA), for the evaluation of the existing infrastructure that could be used for this scope. The purpose of a new Romanian SST Radar sensor is to detect and track objects and space debris in the low Earth orbit (LEO), in order to improve the orbital information of such objects and contribute to a European database for space debris.

It was decided that a promising option is to retrofit the two existing C-band 32 m dish antennas, built and originally commissioned in 1976 and 1979, available in the National Space Communication Center in Cheia [2], situated 120 km north of Bucharest. A subsequent project [3] has expanded the study into assessing the status of the antennas, considered at a system level, and the main changes and improvements which should be implemented to design and build a ground-based SST radar, by using a quasi-monostatic configuration, where one antenna is transmitting and the other is receiving.

This paper describes the processes of design, implementation, and testing of the software and hardware blocks that make up the signal processor (SP) of the Cheia SST Radar. Section 1 presents similar signal processing systems from other radars. Section 2 contains the description of the Cheia Radar system and the purpose of its signal processor. Section 3 shows the results of the signal processor benchmarking. The paper ends with the Conclusions.

State of the Art

Modern radar systems are complex signal processing systems, built of subsystems and functional components which interact with each other via well-defined and often standardized interfaces. The signal processor is one of the most important components, as it is responsible for generating the output information used to detect targets. The signal processors of several space radars are detailed in the following paragraphs.

The Italian Bistatic Radar for LEO Tracking (BIRALET) uses a radio frequency (RF) transmitter located close to Calgari, Sardinia, with a fully steerable 7 m diameter parabolic antenna and up to 10 kW of output power [4,5]. The receiver is located 20 km away, close to San Basilio, Sardinia. This bistatic P-band (420 MHz) radar receives the target-scattered signal with a 64 m parabolic antenna and feeds it to an RF receiver block, which is intended for increasing the signal-to-noise ratio, changing the polarization from linear to circular, noise calibration, and selecting a 20 MHz bandwidth of the received signal with a bandpass filter. The next subsystem of the BIRALET radar is the new back-end that is capable of measuring also the slant range of the target, together with the Doppler shift. The new back-end uses a commercial field-programmable-gate-array (FPGA) software-defined-radio (SDR) ETTUS/NI USRP 2954R board. A second one is installed also in the transmitter subsystem, with the purpose of generating the digitally modulated signal to be transmitted. Both are synchronized with Global Positioning System (GPS) timing signals.

The received signal outputted by the RF receiver passes through an antenna focus selector and a power combiner, and then is fed to the USRP, where it is conditioned, down-converted to a frequency of 10 MHz, and digitized using 14 bits. The resulting in-phase and quadrature data are stored and processed on a workstation. This is using the ETUUS drivers, a LabView radar processing application, and a graphical user interface (GUI). The processing program generates the slant range and Doppler shift measurements of the detected target. A matched filter is used for correlating the well-known transmitted signal, with the unknown received signal, in order to detect the presence of the transmitted signal in the unknown signal. The matched filter is an optimal linear filter for maximizing the SNR in the presence of additive stochastic noise [4]. The transmitted waveform is a mixed signal composed of a frequency-modulated continuous wave (FMCW) signal (chirp), plus a CW tone, with an overall bandwidth of 5 MHz. The pulse repetition interval and pulse repetition frequency of the chirp assure the detection of targets at distances up to 3000 km. Because of the mixed transmitted signal, the system acts as a pulsed radar and also as a CW radar. The GUI contains the control panel, the status display, the measurements extractor, a spectrogram, and plots for range and Doppler for the received signal.

Another European sensor is ATLAS, a tracking radar system for space object detection situated in Portugal, operating in the C-band (5.56 GHz). It is a monostatic pulse radar with a maximum transmission power of 5 kW for tracking debris objects in LEO. The receiver stage is completely coherent, with completely digital detection and processing, with a bandwidth of 50 MHz, and with the capability to detect Doppler speeds up to 10.79 km/s [6]. The receiver includes an analog block that has the role of amplifying the input signals to the necessary digital input levels of the analog-to-digital conversion ADC board. The receiver begins with a low-noise amplifier. Next, there is a mixer to convert the input signal down to a 400 MHz intermediate frequency. The down-converted signal then passes through a bandpass filter and a digital variable gain amplifier, which is commanded by the controller board.

At last, the in-phase/quadrature detector demodulates the input to the baseband and outputs the I and Q parts of the echo signal. The data acquisition contains the acquisition board and a single-board computer. The acquisition is commanded by the controller board and is in charge of converting the I/Q signal parts to the digital domain, by using two 16-bit ADCs at a maximum sample frequency of 100 MS/s. The I/Q signal in the digital domain is fed to the computer (an MIO 2360 from Advantech headquarted in Taipei, Taiwan), which performs the required digital signal processing [6]. Communication between the acquisition board and the single-board computer is conducted via universal serial bus USB 3.0 (100 MS/s). A Fast/Slow mode buffer is available, acting as an alternative between increased sample resolution and increased sample rate.

The Spanish Space Surveillance and Tracking Radar (S3TSR) is an L-band (1400 MHz) radar, that works in a close monostatic configuration. Its architecture is scalable and is using electronically scanning arrays as transmitter and receiver antennas. The receiver uses direct Radio frequency undersampling technology. Each reception module, which is connected to four antenna array elements, employs four RF blocks for low noise amplification, filtering, and equalizing the signal, as well as a divider and switches for inserting a test signal in the four channels from a common input. The test signal is necessary for synchronization and calibration uses. The four augmented RF signals are fed to ADCs and the digital samples are used by a high-performance FPGA for digital down-conversion and beamforming. The receiver (Rx) array connects with the Processing and Control system inside the Rx Building. It includes the subsystem for the generation of all signals (transmitter Tx, test, and reference oscillator) required by the Rx antenna and the up to three Tx antennas. It also contains power amplifiers for the reference oscillator and test signal used by the Rx antenna, as well as passive splitting networks for both signals [7].

Signal and data processing capacities are offered by commercial high-end computers. They are integrated into the Processing and Control system. The signal processor is working on a single computer fitted with a PCIe board with fiber optic communication and a high-end FPGA. This board connects with the Rx array controlling and receiving the beamforming channel samples from the Reception Modules. It is responsible for the final stage of beamforming, concerning the integration of samples from the different antenna array rings. Samples for all Rx antenna patterns are fed to the processing software running on host processors through PCIe interfaces. The results of the signal processor are target detections that are fed to the Data Processor (DP). Initially, a correlation operation is applied by the DP for generating a sum plot of all detections for the same object at the same moment, obtained from processing the different Rx beams of the group assigned to the same Tx signal. The DP is responsible for the object tracking process, which starts with the results of the correlation operation. The tracking involves the identification of the group of plots that are generated by the passing of a space object through the field of view and the elimination of the false alarm plots generated by noise. For this scope, a recursive Least Square Error minimization algorithm is used, with an improvement of the track state vector in each iteration. After a trace is obtained, the set of related plot data, consisting of measured positions and radar cross-section estimations, is organized and sent to the connected users. The DP performs also a continuous recording and data dissemination task. Users connected to the radar through local or remote connections can ask for raw plots and track data from the recorded period. The signal and data processing algorithms are implemented in standard C/C++ language and executed on real-time Linux [7].

Another European space debris radar system is the German Tracking and Imaging Radar (TIRA)-Effelsberg bistatic system. It contains a 34 m fully computer-controlled parabolic antenna with a Cassegrain reflector, installed in Wachtberg, Germany. It can operate both in L-band (1.33 GHz) as a narrow-band monopulse fully coherent tracking radar and Ku-band (16.7 GHz) as a high-range resolution imaging radar [8]. The L-band transmitter of TIRA can transmit high-frequency pulses, usually from 1 to 2 MW peak power, in Right-Handed Circular Polarization (RHCP), while the Ku-band transmitter has a peak power of 13 kW. Working as a monostatic radar, the TIRA can detect objects with a diameter of 2 cm at a range of 1000 km. Moreover, when connected with the Effelsberg radio telescope of the Max-Planck-Institute of Radio Astronomy (MPIfR), comprising a parabolic dish with a diameter of 100 m located in Bad Münstereifel (Germany), used as a receiver, the size of detectable objects reaches 0.9 cm in diameter. The TIRA antenna can be turned at a radial speed of 24° per second and is protected inside a radome [9].

GESTRA (German Experimental Space Surveillance and Tracking Radar) is a quasi-monostatic scientific radar. This L-Band phased array radar is designed to monitor the low Earth orbit. It supports the German Space Situational Awareness Center to generate a catalog of orbital data for objects at altitudes between 300 and 3000 km. It was developed at the Fraunhofer Institute for high-frequency physics and radar techniques, based in

Wachtberg, Germany [10]. The radar uses different antennas and shelters for transmitting and receiving. They are set up at a distance of about 100 m. Both antenna apertures consist of 256 active cavity-backed stacked patch antennas in a circular arrangement, whereby the transmitting radiators are excited with linear polarization, and the receiver's radiator elements are designed with double polarization. The corresponding transmitter modules at the rear of the transmitting antenna are water-cooled. The receiving antenna uses the principle of digital beamforming. In addition, the antenna array can be rotated in any direction using a mechanical steering system [11].

GRAVES (Grand réseau adapté à la veille spatiale) is a bistatic radar system operated by the French Air Force in the VHF band at 143 MHz for orbit determination of artificial earth satellites at an altitude up to 1000 km. It is a space surveillance system developed by the French Aerospace Research Establishment (ONERA). It is a full electronic scanning bistatic continuous wave (CW) sensor, operating in very high frequency (VHF), using Doppler detection and digital beam forming reception technique. It transmits in up to three envelopes, at different elevations, for maximizing the number of detected targets. The transmission is implemented using a phased array and the reception is by a technique of digital beam forming [12]. The reception array comprises 100 antennas distributed over a metallic disc forming a ground-level surface. Each antenna is linked to an individual receiver whose signals are then digitized. The addition, in phase, of all the signals issuing from the reception antennas results in the creation of a vertical reception beam equivalent to that which would be created by a single antenna of the same size as the disc carrying the antennas.

With the signals issuing from each antenna being digitized, it is possible to assign to the signal from each antenna a specific phase shift. The choice of an adapted phase set allows the creation of a beam oriented in any direction, not simply the vertical. This technique of digital beam forming allows, given a real-time calculator of sufficient power, simultaneous observation in all directions potentially illuminated by the transmission. The detection of the satellites is then made possible with a Doppler technique by applying a Fourier transform to the sum signal. The transmitter site is located at Broye-Aubigney-Montseugny, and the receiver site is 364 km south of it at Revest-du-Bion. The radar was put into operation in 2005.

The ABISS project (Antenna BIstatic for Space Surveillance) is a ground-based surveillance radar demonstrator for European Space Agency. This European Space Surveillance radar breadboard is a bistatic system that is located on two separate sites in the northern part of France around Paris. This radar operates in L-band for the detection of space objects over a steerable Field of Regard (FoR) defined as 30° in azimuth by 25° in elevation oriented preferably to the south. One of the advantages of such a system is its operation with a continuous wave (CW) signal, which is easier for spectral management as it is strictly limited to the carrier frequency. The demonstrator has been designed on the basis of simulations of performances, as a radar which should have a reference detection range at 500 km for a reference target of 0 dBm^2 as radar cross section (RCS) for a revisit time of 10 s, meaning that the beam of the radar should revisit the same azimuth and elevation position within 10 s. These are the reference values that have been used in the power budget computation for the design of the system. The principle of the detection for such radar is only based on Doppler measurement; there is no evaluation of the distance of the target.

The transmitting system (Tx) is constituted of one antenna array of surface area 1.5 m^2 with 49 radiating elements, and 49 CW solid state power amplifiers, controlled in both phase and gain, associated with each radiating element. All electronic equipment for transmitting and control is in a shelter. The antenna array is set up under a radome on the roof oriented to the south. The positioner which hosts the antenna array can be steered through $+/-90°$ in azimuth around a central position oriented in the south direction, and 0° to 90° in elevation. The receiving site (Rx) is within the ONERA center in Palaiseau. The antenna panel, with a surface area of 10.5 m^2 and 64 receiving antennas, is installed on the roof of an office building. The antenna array can be steered through $+/-90°$ in azimuth around a central position oriented in the south direction, and it can be inclined from 0° to 90° in elevation. The receiving system is constituted of 64 analog front-end stages in one cabinet, four digital receivers with sixteen channels, and a cluster of ten computer servers to process the data.

The Field of Regard is completely covered with an average gain greater than 22 dB. The signal processing implemented in the radar consists of Doppler/acceleration measurements (i.e., speed estimation), detection, and monopulse compression for precise direction finding. As the Field of Regard is covered by 10 positions of the transmitting beam in 10 s, one second of signal is analyzed at each position. Consequently, the coherent integration time is 1 s. Thus, detections of objects that pass through the Field of Regard are characterized by Doppler traces versus time. The passage of one space object is signaled by a series of closely spaced radar plots. Each extracted plot contains all measurement parameters, namely azimuth, elevation, Doppler (speed), radial acceleration, and time of detection. Tracking of a previously detected target is realized with consecutive plots using the Track-While-Scan (TWS) method, which combines both search and track functions [13].

CASTR (Chilbolton Advanced Satellite Tracking Radar) is an S-band radar mounted on the 25 m steerable antenna at the Chilbolton Observatory in southern England. The dish is a fully steerable elevation over azimuth mount and is capable of tracking most satellites in LEO. A pulse repetition frequency (PRF) of 71.428 Hz, corresponding to a pulse repetition interval (PRI) of 14.0 ms and a maximum unambiguous range of 2100 km, was chosen so as to achieve alias-free range measurements for targets in low Earth orbit (LEO). Under these conditions, although the peak power is some 700 kW, the average transmitted power is only 25 W. The system's polarizer was configured to transmit pulses of fixed, horizontal polarization, while the radar's receivers simultaneously recorded both co-polar (horizontal, H) and cross-polar (vertical, V) target returns [14].

2. Materials and Methods

2.1. Radar System Description and Purpose of Radar Signal Processor (SP)

The Cheia Radar set-up has been finalized and has begun functional tests at the end of 2022. The ultimate set-up of the space radar installation is shown in Figure 1, containing: a transmitter (Tx), receivers (Rx), monitoring and control (M&C) server, and other subsystems.

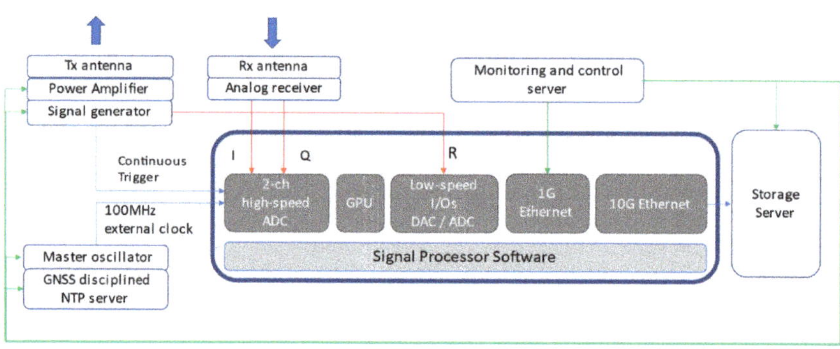

Figure 1. System schematic of the radar. In the center—the signal processor (blue rectangle). The antennas transmit the signal and receive the echo from the space object, the master oscillator provides a very reliable, very precise external clock for the SP's ADCs, the monitoring and control (M&C) server offers the user interface and manages the system, and the storage server stores the acquired data. Red arrows denote analogue signals, green arrows denote digital signals.

The master oscillator (MO), the signal generator and the power amplifier (PA) are subsystems of the Tx. The transmitted signal and trigger pulses are generated by the low-noise signal generator, which is controlled by the M&C server and uses a reference signal from the MO. The generated signals are increased to the necessary level by the linear PA and then fed to the Tx antenna. A limited fraction of the transmitted signal is withdrawn in order to be employed as a local oscillator signal in the receiver [15].

The input signals are received by the Rx antenna, filtered, frequency translated/ demodulated, and amplified into the Rx subsystem, to the level required for the best functioning of the analog-to-digital converter (ADC), that belongs to the signal processor (SP) [16].

The RF receiver (Rx) and the signal processor (SP) are the components of the Cheia radar's receiver. The SP transfers the input signal into the frequency domain, time stamps, and analyzes it, using a constant false alarm rate method. The output is the space object's range, relative speed and signal-to-noise ratio (SNR), which is used to evaluate the target's radar cross-section. These output data are transferred to the M&C server. The M&C server is a computational system that controls the radar's functioning through software process-based commands, implemented through the M&C software [17]. The hardware specifications and the functional parameters of the radar can be found in [18].

The purpose of the SP is to acquire analog data from the receiver, transform it into digital data, process it, detect space objects and describe their trajectory (distance to the object in time, speed of the object in time). A full system schematic is drawn in Figure 1 below where one can see top-level data exchange between the components of the radar.

The radar's signal processor is a 4U Xeon server fitted with high-speed high-resolution ADC cards used in acquiring analog signals from the receiver and also fitted with a high-performance GPU used to accelerate digital data processing. The SP runs a Linux-based OS and on top of it a full suite of drivers topped with the FSM (finite state machine) of the SP, the top-most level software running on it.

2.2. The Hardware of the Signal Processor (SP)

In Table 1 below, one may see the main hardware components of the signal processor. The SP is a 4U server, with redundant power supply units and enough PCI-e slots to fit all components.

Table 1. Components and characteristics of the signal processor.

Component Type	Model	Purpose
CPU	Intel(R) Xeon(R) W-3223 8C/16T @ 3.50 GHz	Runs the OS. Moves data to/from GPU, builds logs, assures command and control
GPU	Tesla V100S-32 GB/5 k cores	SIMD accelerator: parallel processing for digital wave data
ADCs	AlazarTech ATS9462, 2 × 16-bit data @ 180 MSa/s	Acquires analog data from the receiver and converts it to digital data
NIC	10 Gbps	Interface used to transfer acquired data to storage server
NIC	1 Gbps	Interface used to receive command and control from M&C system
SSDs	1TB PCI-x4 NVMe, Samsung SM981/PM891/PM983	Short-term data storage before transfer to storage server
RAM	3 × 32 GB DDR4 2933 MT/s	Working memory

2.3. The Software of the Signal Processor (SP)

The SP runs a Linux-based OS on which a kill-safe multithreaded application runs (SPapp). The application handles the GPU using OpenCL API, the ADCs, the general-purpose IO board which also provides time tags (Ts) for timestamping the received signals, the TCP server which allows SP to be controlled remotely, the WDT, which restarts the

SPapp if it becomes non-responsive and last, but not least, a service to reliably move acquired data towards the storage server.

In Figure 2, one may see the layered hardware (gray), and low-mid-high level software (up to down).

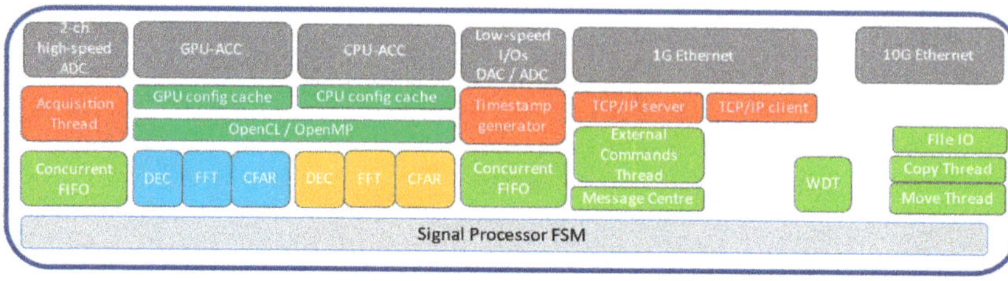

Figure 2. Signal processor inner schematic. Top is hardware and towards the bottom, software of increasing abstractness and complexity. The dark-grey components are hardware, the red components are low-latency threads, the blue components are GPU-enabled parallel processing components, the yellow components are the same parallel-processing components, but optimized for CPU. The dark green components are the middleware used to improve the performance of the GPU tasks, and the light green are the glue-logic or additional components without specific low-latency or high-performance requirements. The light-grey component is the finite state machine that commands and controls every other software subsystem in the Signal Processor.

2.4. The Performance Requirements for the Signal Processor

The SP should assure real-time data processing of incoming data. Analog data are sampled at 100 MSa/s using 4 channels (two ADCs) of 16-bit; this results in a data rate of 100 M $\times$ 4 $\times$ 16 = 6.4 Gbps

As seen in Table 2 below, multiple scales for distances between 400 and 4000 km were used (and accordingly multiple settings for RF modulation and digital processing); this is due to the following reasons:

- It allows to keep a constant, relatively small RF bandwidth (2 MHz) which allows the receiver chain to keep noise under control (the noise is a product of bandwidth with spectral density). Constant bandwidth allows simpler receiver design, and one would prefer to modify the parameters of the receiving pipeline, in software (digital domain) in the SP, as opposed to temper with the analog part in the receiver. The complex receiver means complex tuning across different frequencies (we use frequency modulation).
- The power scattered by the target (the echo signal) decreases with a power of 4 of distance, so the further away the target is, the much worse would be for the SNR. This requires a much better noise cancellation algorithm, which requires a larger FFT, which requires a longer signal in time (continuous emission over larger distances requires a longer time-duration of the signal due to longer round-trip time).

There is only one performance requirement, and it is for the processing of an acquired wave, depending on the processing time of this required acquisition, as can be seen in Table 2:

Table 2. Radar scales for different ranges.

Scale Number	Description (Range)	Range (km)	Trigger Period (ms)	Acquired Wave Size (Samples)	Maximum Processing Time (ms)
1	Close	<400	40	256 k	40
2	Mid-close	<900	100	512 k	100
3	Mid-far	<1800	250	1 M	186
4	Far	<3900	500	2 M	290

2.5. The FSM of the Main Signal Processor

The signal processor acquires data from the receiver, processes it, and logs data and results. In addition to this functionality, it also allows for fine-tuning of its 40+ parameters. Its finite state machine diagram is visible in Figure 3.

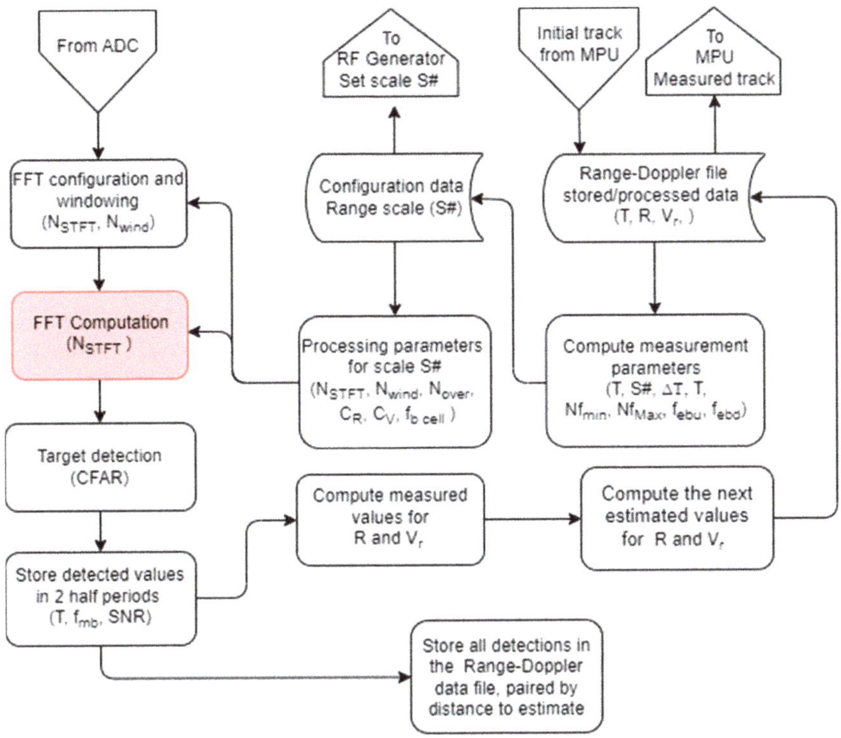

Figure 3. The FSM of the signal processor. Digital data flows inside from the ADCs, and they are processed by decimation and windowing, FFT, absolute value, and CFAR. The FFT processing is highlighted in red, as this has a very large size, quite unusual in a radar system.

2.6. The Modifiable Parameters of the Signal Processor

The behavior of the signal processor may be altered by changing the values of several parameters. A list of such parameters is presented in Table 3.

Table 3. Working parameters of the signal processor (selection).

Param Index	Param Name	Param Type	Param Description
1	Thr_ADC_AM	Int32	(1) ADC level—threshold value for MSP
3	N_Thresh_AM	Int32	Percent of detections over threshold @ 10 k
5	N_dec	Int32	Decimation factor
6	N_S	Int32	Number of range scales (4, 5 or 6)
7	N_STFT	Int32array	Size for Fourier transforms
8	W_Type	Int32	Windowing type
9	N_cfarSP	Int32	CFAR Number of samples used for avg
10	N_guardSP	Int32	CFAR Number of guard samples
11	K_detCA	Int32	CA-CFAR Detection factor
12	K_detGO	Int32	GO-CFAR Detection factor
13	K_detLO	Int32	LO-CFAR Detection factor

Table 3. *Cont.*

Param Index	Param Name	Param Type	Param Description
14	CFAR_Type	Text	CFAR algorithm type (CA, GO, LO)
15	KW_det	Fp64	Detection window coefficient for the beat frequencies
16	SFW_det	Fp64	Scale factor for computing the CFAR windows on all scales
17	SC_det	Int32	MSP detections selection criteria
18	KSC_det	Fp64	MSP multiple detection selection coefficient
19	S_Max	Int32array	Maximum range for all scales
20	T_trig	Int32array	Trigger period (UP-DOWN-UP) for all scales
21	Nf_min	Int32array	Min freq bin cell to be processed in all scales
22	Nf_max	Int32array	Max freq bin cell to be processed in all scales
23	C_R1	Int32array	Range computation factor 1 for fbu in all scales
24	C_R2	Int32array	Range computation factor 2 for fbu in all scales
25	C_V	Int32array	Doppler computation factor in all scales
26	Fb_cell	Int32array	STFT frequency cell on all scales
27	R_alpha	Fp64	Range ABG filter coefficient—alpha
28	R_beta	Fp64	Range ABG filter coefficient—beta
29	R_gamma	Fp64	Range ABG filter coefficient—gamma
30	V_alpha	Fp64	Doppler ABG filter coefficient—alpha
31	V_beta	Fp64	Doppler ABG filter coefficient—beta
32	V_gamma	Fp64	Doppler ABG filter coefficient—gamma

2.7. Other Considerations Regarding the System Reliability and Performance

A few techniques were used to improve the quality of the signal processor:

- Creating and using a simulator to play back the data without the need for a real space target;
- Using references instead of memory pointers;
- Using the software Watchdog Timer;
- Intercepting all OS signals sent to the application, to graciously close used resources in case of system errors/crashes;
- Auto-restart the app with the last auto-saved settings and commands received from the M&C system;
- All data processing operations were tested against Octave implementation to check for precision loss;
- All computation was performed in double precision (fp64).

2.8. Software Inside the Signal Processor

2.8.1. Software Class Diagram

The software for the SP (illustrated in the diagram from Figure 4) is split into various components:

- Components that assure communication with the outside systems (range setter, M&C thread, Message Center, Message);
- Components that assure communication between SP's components (SWave, ConcurrentQueue);
- Components that assure data processing (CFAR, ABfilter, FFT, Windowing);
- Components that assure data acquisition (Data Generator, Config Sink);
- Components that enable data output (File Move Thread, File Copy Thread, BB*file);
- Components that enable testing;
- Components that assure reliable behavior in case of errors (WatchDog Thread, CPU-usage Thread);
- Top-level component of MSP (Main Signal Processor which uses all the other components).

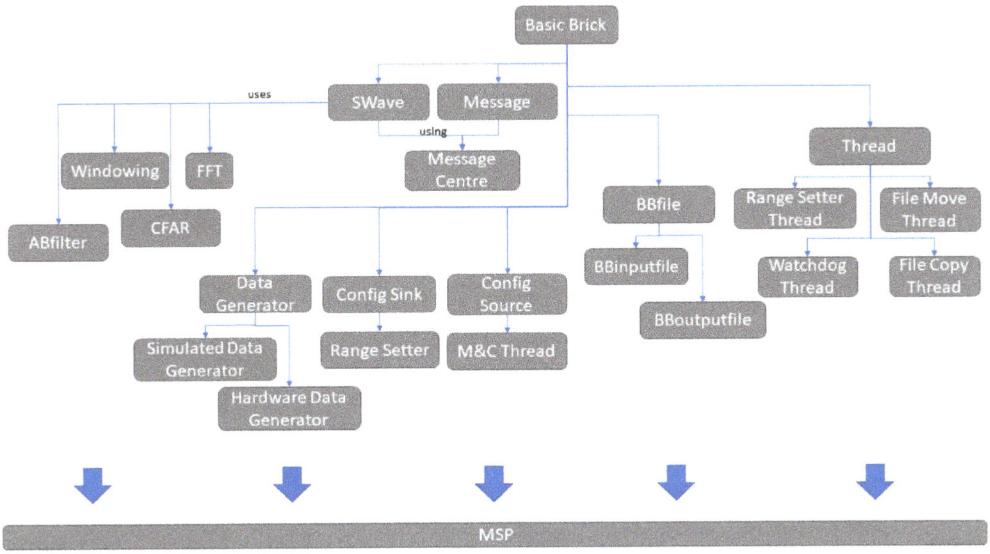

Figure 4. The class diagram for the SP software. MSP is assembled by composition using objects of all the classes above (denoted with blue arrows).

Regarding the data flow, the SP was fitted with an inheritance class hierarchy so that data may come and go from either hardware or software (simulator) source/sink.

2.8.2. Digital Processing Operations

The intensive digital processing operations performed by the SP are:

- DEC (decimation using a custom factor 5);
- FFT (1D complex-to-complex conversion of time-domain to frequency-domain data);
- ABS (absolute value of complex numbers);
- CFAR (applying a constant false alarm rate detection algorithm);
- ABF (alpha-beta filter, for predictive range and radial speed).

DEC (Decimation)

Two versions of decimation with factor F were tested:

- Decimation by sampling: output sample is one of the F consecutive samples;
- Decimation by averaging: output sample is the average of F consecutive samples.

The DEC computation since it was in O(N), was more efficient to be performed in parallel in the CPU using OpenMP.

FFT (Fast Fourier Transform)

The FFT is a faster way (as opposed to DFT) of creating a frequency representation of a periodic signal. Initially discovered by Gauss around the same period of time as Fourier discovered the Fourier transform (1800s), it was rediscovered in the 1960s in the context of Soviet Union-US meetings for differentiating underground nuclear tests from the earthquakes, with the higher purpose of the worldwide partial nuclear test ban. In 1963, it effectively reduced the time needed to process an incoming seismograph data stream of an analyzed event, from 3 years to 30 min (that is the difference between O(NlogN) DFT's complexity of $O(N^2)$). FFT takes advantage of similarities found in the sinewaves that have the frequencies multiple of the fundamental frequency, which allowed for far fewer multiplications than in DFT. In both cases, FFT and DFT, the more samples the time-signal contains, the more frequency bins we obtain; this is why we thought of using a larger FFT transform to obtain a better determination of frequency components in the radar's received signal (it effectively reduces noise in that bin and allows for weaker signals to be detected with a frequency-based CFAR algorithm)

We used an OpenCL implementation of O (5*N*log(N)) similar to the Cooley–Tuckey version. In addition, an extra layer of cache was developed (the first run takes more, subsequent ones take less time); in this software cache, we saved the plans (mainly the fiddle factors) and GPU memory allocation buffers and just-in-time compiled programs, for common sizes, so a warm-up phase is executed at application start-up to improved performance on subsequent runs.

ABS (Absolute Value)

The ABS is a straightforward absolute value computation of complex numbers. This is also an O(N) complexity so unless other advantages are drawn running in the GPU, it will be run in the CPU. Since the log is applied to extract SNR, at some point, the sqrt may be skipped (let us note that operation, as absm) by using a factor of 2 afterward ($\log a^b = b \log a$).

Assuming $x = a + bi$, a complex number, the $abs(x) = \sqrt{a^2 + b^2}$, and therefore the $absm(x) = (a^2 + b^2)$.

CFAR (Constant False-Alarm Rate Algorithm)

The CFAR algorithm uses a moving window to detect peaks in a wave. It can be used in both time-domain and frequency-domain, to detect local peaks. Local peaks are selected using an average of left training cells and right training cells on one side and the other of the peak-candidate value (cell under test). To avoid false peak detections near an actual peak, usually one uses "guard" samples between the candidate value and the left/right samples, as depicted in Figures 5 and 6.

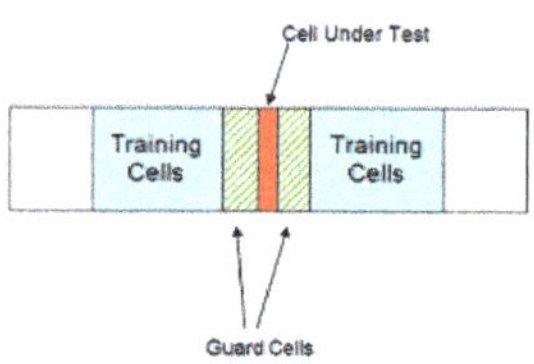

Figure 5. The CFAR moving window—choice of cells. Reprinted/adapted with permission from Ref. [19]. © 1994–2023 The MathWorks, Inc.

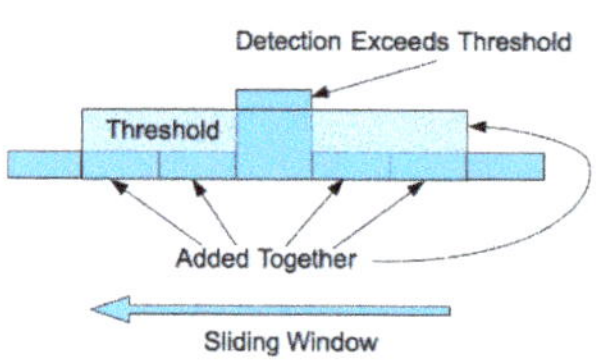

Figure 6. The CFAR moving window—choice of peak [20]. Usually, the threshold is multiplied with a factor, to obtain the actual threshold that will be utilized in peak detection. Credit: Nanoatzin License: CC BY-SA 3.0.

An avg of left-side training cells is computed AVGL = sum (left_samples)/ num_left_samples.

An avg of right-side training cells is computed AVGR = sum (right_samples)/ num_left_samples.

There are three versions of threshold calculation, usually used to compute the threshold, depending on how much weight the two halves (left and right) have:

- CA-CFAR (Cell Averaging), Threshold = (AVGL + AVGR)/2;

- GO-CFAR (Greatest Of), Threshold = max(AVGL, AVGR);
- LO-CFAR (Least Of), Threshold = min(AVGL, AVGR).

ABF (Alpha-Beta filter)

The alpha-beta filter was fitted with multiple simulated trajectories to see its behavior when trying to predict the next range or radial speed. An example of the trajectory that was tested may be found in Section 3.1.2 below.

2.9. Graphical User Interface of Signal Processor (SP gui)

While the SP may be used without a graphical user interface (gui), during test phases (where various hardware or satellites to be tracked may be unavailable) one needs to be able to simulate the SP's activity. For this purpose, a simulator module was developed (which is based on real acquisitions made previously). The graphical interface can connect itself either to the real SP or to the simulated SP and provides an insight into the data acquisition and processing (configuration, status, and data) for either a real-time run ("real-time mode") or a previous run ("analysis mode"). In addition, the gui allows issuing commands as if they were coming from the M&C system, to allow SP's testing even when M&C was in development or upgrade.

In real-time mode, all runtime parameters, configuration, and limited data of the SP are monitored while the acquisition is performed.

In analysis mode, the previously captured data (and acquisition settings), may be visually inspected and played back frame by frame to see if the SP works as expected. If the captured data need to be detailed, an Octave script will be invoked from the gui, to parse the large debugging data. Figure 7 shows a caption from the SP gui application, while in real-time mode with real-time data disabled, Figure 8 shows the SP gui application in analysis mode, and Figure 9 shows the Octave script's figure display for an acquisition that produced valid data.

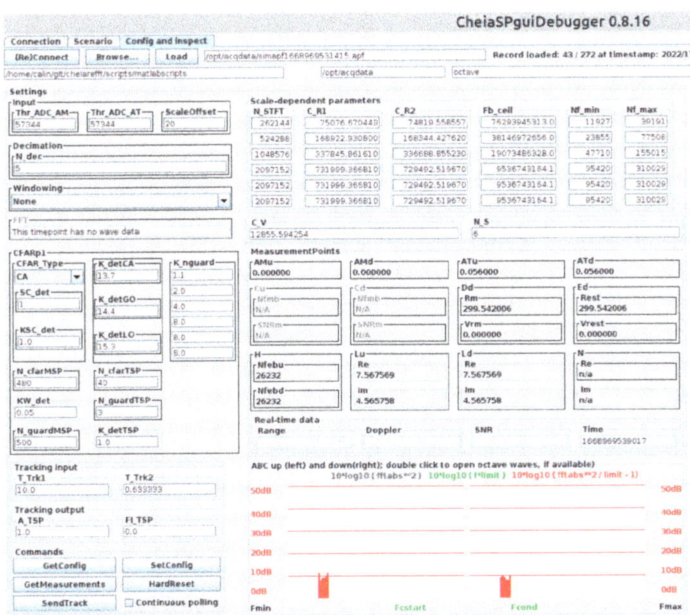

Figure 7. SP gui application when in real-time mode. The top-left part of the GUI is populated with configuration data; the bottom and middle part contains real-time data (during acquisition); "**"expresses pow function of 2 (squared).

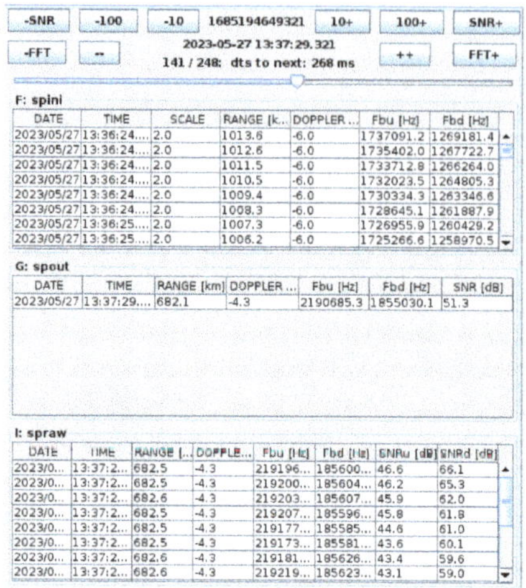

Figure 8. Extract of SP gui application while in analysis mode, with test data. The tables summarize the produced data (input track, raw data, filtered data, data seen and parsed by the CFAR engine, etc.). The top part contains a navigation menu, used to go through the records of the previously acquired data: the navigation may be made 1-by-1, 10-by-10, 100-by-100, or from/to the next/previous frame that contains a full FFT output wave or SNR data over a predetermined value.

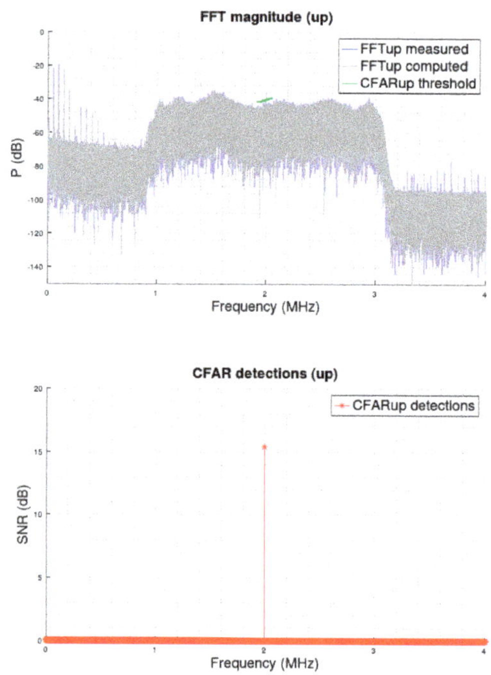

Figure 9. Octave script output. The SP GUI was running in analysis mode, and the user requested a detailed view of the monitored frequency spectrum. The top part contains the 0–4 MHz monitored spectrum: with blue, it is the output of the FFT as computed by SP, with gray is the output of the FFT computed by Octave, based on the raw data from the SP, and with green is the threshold of the CFAR algorithm. The bottom part contains the same 0–4 MHz spectrum but only with the peaks that exceeded the CFAR threshold.

3. Results of the Signal Processor Benchmarking

The results of SP processing may be split in two categories:

- quality of computation
- performance of computation

The performance was evaluated on a Xeon W-3223 CPU @ 3.50 GHz (8C/16T) with 3 × 32 GB of DDR4 and an Nvidia GPU Tesla V100s-32 GB with CUDA 11.0 and Driver Version: 450.80.02 on Linux 5.4.0-89.

3.1. Measuring the Quality of Digital Processing

3.1.1. Measuring the Quality of the FFT

We tested our CPU and GPU-accelerated version against the Octave implementation, then we calculated the sum of squared differences for various FFT sizes. The results are summarized in Table 4.

Table 4. FFT computation error CPU or GPU implementation when compared with baseline Octave implementation for a rectangular periodic signal. CPU-only OpenCL driver crashed when testing with sizes larger than 512 k (hence, n/a).

FFT Size	CPU rms Error	GPU rms Error	Allowed Error
512	2.99×10^{-15}	3.01×10^{-15}	1×10^{-9}
128 k	8.48×10^{-14}	8.27×10^{-14}	1×10^{-9}
256 k	9.47×10^{-14}	1.00×10^{-13}	1×10^{-9}
512 k	1.28×10^{-13}	1.28×10^{-13}	1×10^{-9}
1024 k	n/a	1.93×10^{-13}	1×10^{-9}
2048 k	n/a	2.91×10^{-13}	1×10^{-9}
4096 k	n/a	4.18×10^{-9}	1×10^{-9}

3.1.2. Measuring the Quality of the ABF

We first generated simulated range and Doppler tracks (22 scenarios, generated by experts in space trajectories), added noise then filtered them to check how well the alpha-beta filter works. Then we swept all values from 0.01 to 0.99 step 0.01 for both alpha and beta (in total 10,000 combinations) for all range or Doppler scenarios. The best (least RMS between filtered and initial values) was selected out of them and transformed into drawings to allow visual checks. Such a scenario is depicted in Figures 10–13 below, where alpha is set to 0.26 and beta to 0.03. We plotted for each case, the filtered–initial and the noisy–initial so we can compare them visually. Figures 10 and 11 are for range, Figures 12 and 13 for radial speed (but with alpha of 0.18 and beta of 0.03).

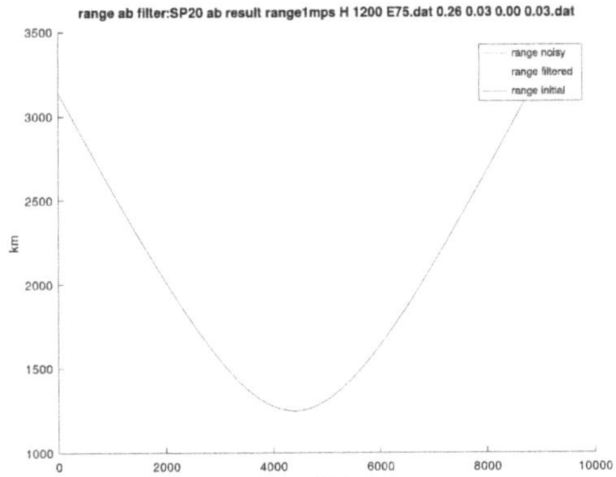

Figure 10. Example with a scenario for the alpha–beta filtering. The range (distance to the object, in km) decreases until it is as close as possible to the radar, then it goes further away.

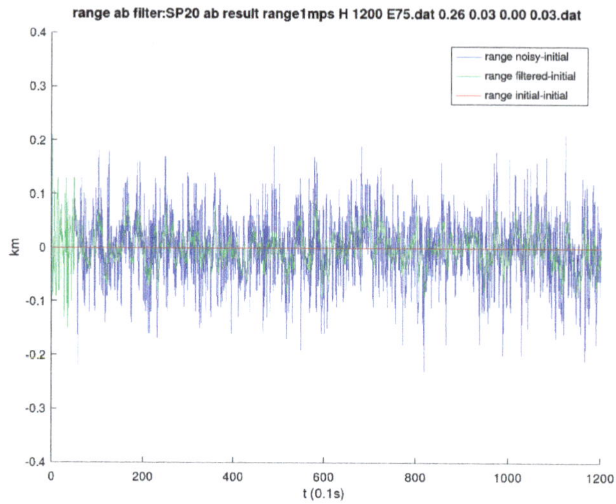

Figure 11. Example with a scenario for the alpha–beta filtering. The differences in filtered–initial values (green) are compared to the differences noisy–initial values, in time. The result is as expected since the green wave has a smaller spread than the blue one.

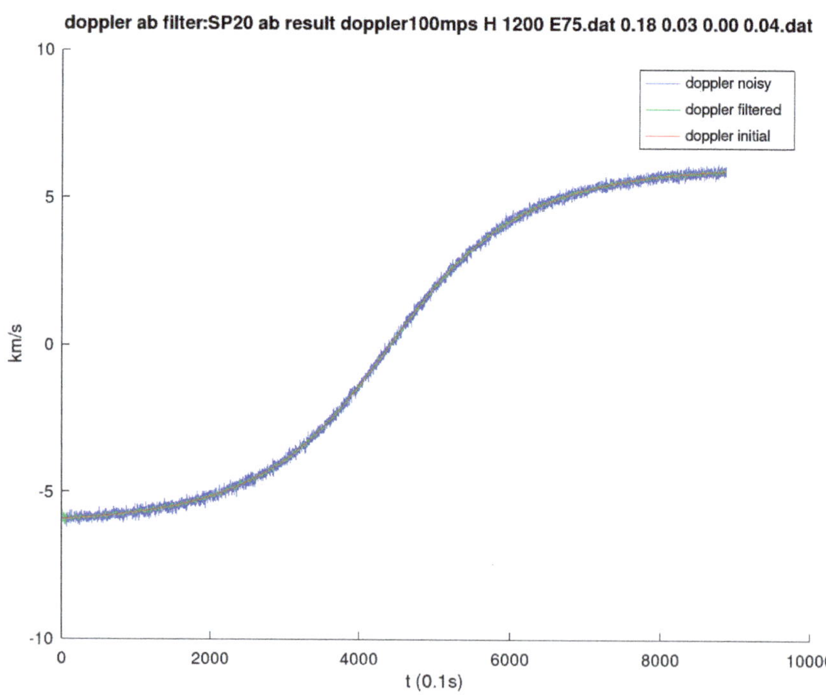

Figure 12. Example with a scenario for the alpha–beta filtering of radial speed. The Doppler (radial speed in fact) shows that the object approaches the radar up until some point in time, and then it goes further away.

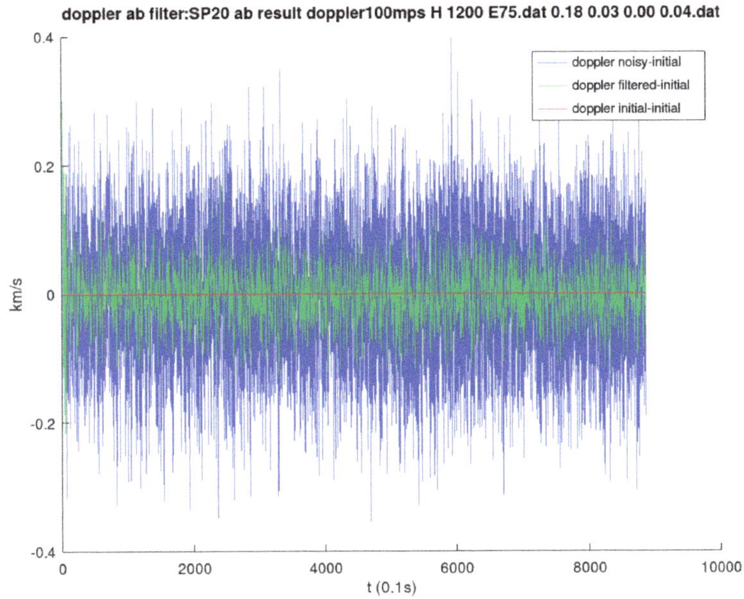

Figure 13. Example with a scenario for the alpha–beta filtering of radial speed. The differences in filtered–initial values (green) are compared to the differences noisy–initial values, in time. The result is as expected since the green wave has a smaller spread than the blue one.

3.1.3. Checking the Quality of CFAR Algorithms

We generated scenarios with sinewaves of 1 MHz and 2.9 MHz (expected minimum and maximum values) and we swept various ranges of guard cells, averaging window size, and CFAR factor. An example of such a scenario is in Figure 14 below. The algorithms were thus validated.

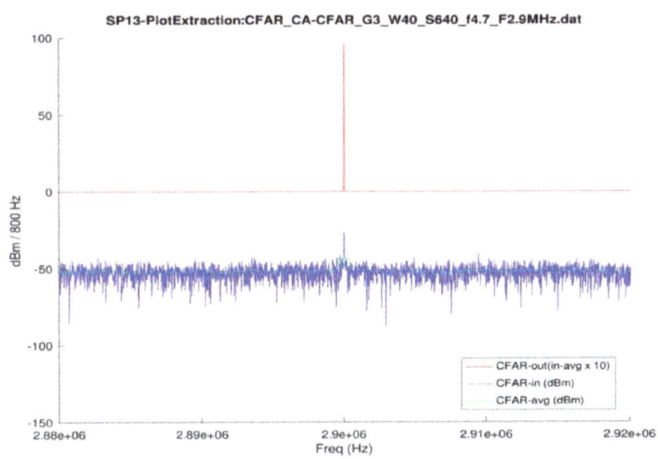

Figure 14. Example with a CA-CFAR scenario with 3 guard cells, 20 left-cells, 20 right-cells, and a factor of 4.7; the input of CFAR is plotted with blue; the cell averaging is plotted with green and the output is plotted with red. The actual threshold is computed by multiplying the cell averaging with a factor.

3.2. Measuring the Performance of Digital Processing

The main performance-intensive operations for the longest data acquisition are listed in Table 5 below. The operation of WIN_FFT_ABS_CFAR does all these ops in a chained manner (without transfers to/from GPU memory while between the ops). The difference between chained operations and non-chained operations is quite large. For example, the WIN_FFT_ABS_CFAR if chained, can be performed in 16 ms whereas when WIN + FFT + ABS + CFAR are performed independently, one will only finish the operations in 61 ms; the advantage of having the operations independently is the ability to save intermediate results for later investigations. This was mitigated in the end, by creating a buffer in the GPU memory where only one such intermediate result may be saved, per each WIN_FFT_ABS_CFAR operation (usually it was used for the absolute value output).

Table 5. Performance results for intensive data processing tasks for the largest scale (4), using 10 M elements before decimation and 2M FFT bins. GPUa timings are measured with warm-up. GPU timings are measured without warm-up. Green values are the best values, data copy, and decimation will therefore always be run on the CPU.

Operations (ops)	GPUa (ms)	GPU (ms)	CPU (ms)
DEC [1]	60	198	16
COPY	15	105	1
FFT	15	126	5
ABS	15	116	5
CFAR	16	n/a	13
WIN	15	119	28
FFT_ABS	16	174	40
FFT_ABS_CFAR	17	134	25
WIN_FFT_ABS_CFAR	16	133	30

[1] Decimation has 5 times more data, as it defaults to decimation by 5, using average. COPY is the operation of moving data from CPU-RAM to CPU-RAM; FFT performs Fast Fourier Transform using O(5*N*logN) Cooley-Tuckey variant. ABS computes absolute value of complex numbers. FFTABS does the FFT and then ABS without moving data to/from GPU between those operations. CFAR is the operation of a constant false alarm rate, and WINDOW applies a time-window to the data. GPUa version has warm-up performed, as opposed to the GPU version.

3.3. Measuring the Network Performance

Multiple software commands are available for using the signal processor (SP). The SP has two hardware network interfaces:

- A NIC with 1 Gbps is used for command and control (an SP server listens on a TCP port);
- A NIC with 10 Gbps is used for large data transfer/long-term storage (over Samba/CIFS protocol).

3.3.1. Measuring the Network Performance (Commands)

To test network latency, we chose the get_status command, which returns a JSON-based string over a TCP connection with data about the state of the SP. Usually, the command contains 20 bytes whereas the answer usually contains 166 Bytes. We benchmarked the command in the local scenario (client and server are on the same machine, the signal processor) and remote scenarios (the server is on the signal processor machine and the client is around 111 km afar). The turnaround time is measured from the moment the client sends the command until the client receives the answer from the SP's server. The results are listed in Table 6. The spread (histogram) of the turnaround time is plotted in Figure 15 below. It shows that 8 ms is the most probable turnaround time.

Table 6. Network latency using the get_status command.

Characteristic Name	Local	Remote1 [1]	Remote2 [2]	Units
Turnaround time (min)	0	6	6	milliseconds
Turnaround time (avg)	0.22	8.52	8.70	milliseconds
Turnaround time (max)	5	32	234	milliseconds

[1] Client's location is somewhat central to Bucharest and the Internet provider is Telekom; the client is about 111 km afar from the Cheia SP's server. [2] Client's location is somewhat more peripheric in Bucharest and the Internet provider is Digi Communications; the client is about 112 km afar from Cheia SP's server.

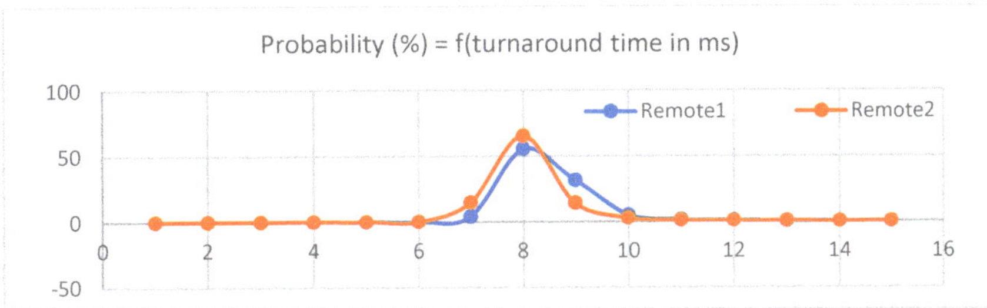

Figure 15. Spread of network turnaround time (in ms) when measured on 10 k command loops, when using the get_status command. Remote1 and Remote2 clients, receive the answer in around 8 ms in 55% and 65% cases, respectively.

3.3.2. Measuring the Network Performance (Large Data)

As previously stated, the 10 Gbps network interface card is used to transfer large data from the SP to the storage server, for future analysis. The SP functions as follows:

- SP is powered-up and its main software starts;
- SP runs self-tests (speed and accuracy tests) to ensure correct operation;
- SP warms-up (builds a data cache with most common operations and compiles all the OpenCL programs to be run on GPU, to save time when executing a real-life measurement);
- The SP receives a command from the Monitor & Control (M&C) system, where the expected track during expected timestamps (beginning (T), ending and a few timestamps in the middle) are given;
- The SP arms itself, interpolates the track (with position and time data) at 100 ms;
- At about T-4 s a short system check is performed to make sure SP module work as expected; in case of problems various misbehaving subsystems are restarted;
- At about T-1 s, data acquisition starts (using a parallel software thread), but data are dropped until T time occurs;
- Data are acquired every 100 ms and are pushed into data queues. A separate software thread starts processing it. The SP switches from debug mode (acquiring and storing large data wave from ADC output, decimated output, FFT output, and CFAR output) to fast mode (only one debug wave is saved, usually the FFT output) dynamically depending on the time budget;
- When the acquisition end occurs, SP stops data acquisition and starts transferring output and debug data into the storage server; simultaneously the M&C system polls it to check the status and retrieve the results.

The data generated for debugging purposes (or later analysis) may be quite large, so the faster 10 Gbps link was used to transfer it to the storage server. In Table 7, we present the results of such measurements of data transfer (the data contained about 6 GB of waves: DEC output, ABS output, CFAR output, CFAR threshold, and CFAR output; all waves were 8 MB long and were acquired at a rate of 2× per every 100 ms, therefore would amount for about 8 s of processed data).

Table 7. Data rate measurement for large data transfers (via 10 Gbe network interface).

Characteristic Name	CHEIA	Units
Data rate (min)	392	MB/s
Data rate (avg)	450	MB/s
Data rate (max)	525	MB/s

The results are somewhat expectable since the data storage is using SSD storage memory for source and large HDDs in RAID5 configuration, as destination.

3.4. Performance Impact of Hardware and Digital Processing

The ABISS radar uses two phased array antennas and low transmitted power but is intended only for re-entry (altitude < 200 km) and objects of size higher than 2 m.

The Cheia Radar is designed for LEO tracking (200–2000 km altitude), having a potential detection capability of 5 cm at 1000 km. The Cheia Radar makes use of 2 decommissioned Cassegrain antennas, previously used for satellite communication, and exhibits detection capabilities similar to TIRA but, due to a special signal processing needs to transmit only a fraction of TIRA's transmitting power.

Although the Main Signal Processor is not the only contributor to these results, together with a good choice of frequency generator and a custom implementation of the analog receiver, they enable the radar to perform side by side with the BIRALET system, as seen in Table 8.

Table 8. Comparison with the BIRALET Radar [4,5,8,18,21].

Characteristic Name	BIRALET	Cheia	Units
Frequency stability	5×10^{-9} *	3×10^{-12}	Hz
Phase noise	??	−130	dBc/Hz
Receiver noise coef.	5–7 *	1.74	dB
Rx ADC resolution	14 *	16	bits
Max receiver gain	37.5 *	63	dB
Receiver bandwidth	5 ^	2 [1]	MHz

[1] Instantaneous receiver bandwidth is further decreased to 10 Hz for far-away targets using a large FFT, which is highly unusual in radars. The large FFT of N size assures full bandwidth coverage with a number of N receivers used simultaneously, where the band of each receiver is dictated by the duration of the processed signal. ?? depicts undeclared/unavailable data. * These data are extracted from [21] Copyright © 2023 National Instruments Corp. ^ These data are extracted from [4].

4. Discussion and Conclusions

In radars, there is always a tradeoff between the transmitted power and the minimum RCS of the detectable target, for a given range or between the transmitted power and the maximum range, for a given RCS of the target.

The limit is always determined by the target's signal-to-noise ratio (SNR) at the input of the target's extractor. The article shows that this limit can be drastically reduced by signal processing. For example, assuming a Swerling 5 radar case model, to obtain a detection probability of 0.9 and a false alarm probability of 10^{-6}, an SNR of 13.2 dB is required at the input of the target detector. The noise power is proportional to the receiver bandwidth, which is in the ranges of MHz to tens of MHz, for a regular time processing radar. A frequency processing radar, using a 1 M sample FFT would then generate a bank of 1 million filters with a bandwidth of Hz to tens of Hz each. In each of those filters, the noise power will be 60 dB less, due to the 60 dB reduction of the bandwidth, while the signal power will be contained by a single or a reduced number of filters. If the total power of the received signal is contained within one filter bandwidth, the SNR at the input of a target detector analyzing the output of each filter is reduced by 60 dB, thus requiring an SNR of −47 dB (60 dB less) at the output of the receiver. Some additional digital processing losses of 2 dB should be also considered.

Reducing the required SNR at the output of the receiver translates in significantly greater sensitivity and/or reduced transmitted power.

Reduced transmitted power translates into smaller costs of operation, reduced power requirements, and consequently a reduced ecological fingerprint.

This reduction comes with the cost of increased complexity of the signal processing as well as with some restrictions that have to be well understood. For the space radars, operating at very large ranges, this kind of processing is particularly suitable.

The advances in signal processing-related devices, especially high dynamic range ADCs and high-speed GPUs or FPGAs will make this kind of processing more and more common once its advantages are understood and mastered.

Future research directions include further reducing the required SNR by increasing the dynamic range of the ADC and designing new types of probing signals and processing algorithms to overcome the received signal length limitations.

The improvements in increasing the dynamic range of the ADC need to be studied and tested as soon as ADCs having more than 16 bits will be available in the video domain.

The design of new types of probing signals and processing algorithms have to be studied for implementation in shorter-range radars that are limited by the short time length of the received signal and by the reduced GPU and FPGA processing speed.

In conclusion, this paper presents in detail, the design, implementation, and testing of the software and hardware elements that compose the signal processor used by the Cheia Space radar for the acquisition and tracking of the targets. The first part of this work contains a review of the state of the art, with similar signal processing systems from other radars. Next, the architecture of the signal processor is explained, together with its state diagram and parameters. Also, a comparison is shown between the outputs of the signal processor and the results obtained with benchmark processing algorithms.

The novel contributions of the presented architecture come from the uncommonly large FFT length (up to 4096 k points) that is employed, allowing the tracking of targets with low RCS, in the context of Tx power lower than what other space radars are emitting [18]. The usage of frequency processing, using large-sized FFTs, allows a significant reduction of transmission power requirements and consequently a reduced ecological fingerprint in radars, to a degree similar to the implementation of the chirp pulse instead of the unmodulated pulse. The strong points of the Cheia receiver are also illustrated by the comparison with the BIRALET bistatic radar, which shows the higher optimization and sensitivity of the Romanian design, due mostly to the superior signal processor.

As of the first quarter of 2023, the Cheia Radar is completely implemented on-site and is undergoing calibration, functional testing, and validation procedures according to EUSST requirements.

Author Contributions: Conceptualization, C.B. and L.I.; data curation, C.B. and L.I.; formal analysis, L.I. and C.B.; investigation, L.I., A.R.-C. and C.B.; methodology, L.I. and A.R.-C.; project administration, L.I. and A.R.-C.; software, C.B. and L.I.; supervision, L.I. and A.R.-C.; validation, L.I.; visualization, L.I. and C.B.; writing—original draft, C.B. and A.R.-C.; writing—review and editing, C.B., A.R.-C. and L.I. All authors have read and agreed to the published version of the manuscript.

Funding: The research and development activities described in this paper were carried out within a series of three contracts financed by the European Space Agency (ESA) and awarded to a consortium led by RARTEL S.A., Bucharest, Romania: SSA for Romania, Contract No. 4000108671/13/D/MRP; SSA P3-SST-IV-CHEIA PHASE 1; SSA P3-SST-V-CHEIA PHASE 2.

Data Availability Statement: The data presented in this study are available on request from the corresponding author. The data are not publicly available due to the clauses of the previously-mentioned contracts.

Acknowledgments: We would like to thank RARTEL's, Silicon Acuity SRL's, and AdHoc Telecom Solution's employees for their hard work in making the project, and therefore, this paper, possible. In addition, we would like to thank the Politehnica University of Bucharest for supporting us by funding various taxes for conferences and journals, therefore making our scientific journey, possible.

Conflicts of Interest: The authors declare no conflict of interest. The funders had no role in the writing of the manuscript or in the decision to publish the results.

References

1. EUSST Consortium. ESA Safeguarding European Space Infrastructure. Available online: https://www.eusst.eu/about-us/ (accessed on 2 July 2022).
2. Societatea Naţională de Radiocomunicaţii S.A. (RADIOCOM), Centrul de Comunicaţii Prin Satelit CHEIA. Available online: https://www.radiocom.ro/Materiale/Cheia/Centrul%20de%20Comunicatii%20prin%20Satelit%20CHEIA.pdf (accessed on 2 February 2023).
3. Ionescu, I.; Scagnoli, R.; Istriteanu, D.; Turcu, V. Cheia Antennas Retrofit to a Space Tracking Radar. In Proceedings of the European Space Agency 1st NEO and Debris Detection Conference NEOSST1, Darmstadt, Germany, 22–24 January 2019; Available online: https://conference.sdo.esoc.esa.int/proceedings/neosst1/paper/415/NEOSST1-paper415.pdf (accessed on 20 February 2023).
4. Schirru, L.; Pisanu, T.; Podda, A. The Ad Hoc Back-End of the BIRALET Radar to Measure Slant-Range and Doppler Shift of Resident Space Objects. *Electronics* **2021**, *10*, 577. [CrossRef]
5. Muntoni, G.; Schirru, L.; Montisci, G.; Pisanu, T.; Valente, G.; Ortu, P.; Concu, R.; Melis, A.; Urru, E.; Saba, A.; et al. A Space Debris-Dedicated Channel for the P-Band Receiver of the Sardinia Radio Telescope: A Detailed Description and Characterization. *IEEE Antennas Propag. Mag.* **2020**, *62*, 45–57. [CrossRef]
6. Pandeirada, J.; Bergano, M.; Neves, J.; Marques, P.; Barbosa, D.; Coelho, B.; Ribeiro, V. Development of the First Portuguese Radar Tracking Sensor for Space Debris. *Signals* **2021**, *2*, 122–137. [CrossRef]
7. Casado Gómez, R.; Martínez-Villa Salmerón, J.; Besso, P.; Alessandrini, M.; Pinna, G.; Prada, M. Initial operations of the breakthrough Spanish Space Surveillance and Tracking Radar (S3TSR) in the European context. In Proceedings of the 1st NEO and Debris Detection Conference, Darmstadt, Germany, 22–24 January 2019; Available online: https://conference.sdo.esoc.esa.int/proceedings/neosst1/pap'er/479 (accessed on 20 February 2023).
8. Muntoni, G.; Montisci, G.; Pisanu, T.; Andronico, P.; Valente, G. Crowded Space: A Review on Radar Measurements for Space Debris Monitoring and Tracking. *Appl. Sci.* **2021**, *11*, 1364. [CrossRef]
9. Space Observation Radar TIRA [Online]. Available online: https://www.fhr.fraunhofer.de/en/the-institute/technical-equipment/Space-observation-radar-TIRA.html (accessed on 27 May 2023).
10. GESTRA Space Radar Passes Its First Test, [Online]. Available online: https://www.dlr.de/en/latest/news/2019/04/20191129_latest-radar-technology (accessed on 27 May 2023).
11. GESTRA [Online]. Available online: https://www.radartutorial.eu/19.kartei/02.surv/karte075.en.html (accessed on 27 May 2023).
12. Michal, T.; Eglizeaud, J.P.; Bouchard, J. GRAVES: The new French System for Space Surveillance. In Proceedings of the Fourth European Conference on Space Debris, Darmstadt, Germany, 18–20 April 2005.
13. Saillant, S.; Flécheux, M.; Mourot, Y. Detection of ExoMars launcher during its passage over Europe with Space Surveillance radar breadboard. *Adv. Sci. Technol. Eng. Syst. J.* **2018**, *3*, 99–105. [CrossRef]
14. Ladd, D.; Reeves, R.; Rumi, E.; Trethewey, M.; Fortescue, M.; Appleby, G.; Wilkinson, M.; Sherwood, R.; Ash, A.; Cooper, C.; et al. Technical Description of a Novel Sensor Network Architecture and Results of Radar and Optical Sensors contributing to a UK Cueing Experiment. In Proceedings of the Advanced Maui Optical and Space Surveillance (AMOS) Technologies Conference, Maui, HI, USA, 19–22 September 2017.
15. Tatomirescu, A.; Rusu-Casandra, A. C-Band FMCW RADAR Receiver Architecture for Space Debris Observation. In Proceedings of the 2022 14th International Conference on Communications (COMM), Bucharest, Romania, 16–18 June 2022; pp. 1–5. [CrossRef]
16. Bira, C.; Rusu-Casandra, A. Dataset of Spatial Objects Acquired Using Romania's First Ground-Based Space Tracking Radar. In Proceedings of the 14th International Conference on Communications COMM 2022, Bucharest, Romania, 16–18 June 2022. [CrossRef]
17. Tramandan, I.; Rusu-Casandra, A. A Real-Time Monitoring and Command System for the Cheia Space Surveillance and Tracking Radar. In Proceedings of the 2022 International Symposium on Electronics and Telecommunications (ISETC), Timisoara, Romania, 10–11 November 2022; pp. 1–5. [CrossRef]
18. Ionescu, L.; Rusu-Casandra, A.; Bira, C.; Tatomirescu, A.; Tramandan, I.; Scagnoli, R.; Istriteanu, D.; Popa, A.E. Development of the Romanian Radar Sensor for Space Surveillance and Tracking Activities. *Sensors* **2022**, *22*, 3546. [CrossRef] [PubMed]
19. MathWorks Help Center, Constant False Alarm Rate (CFAR) Detection [Online]. Available online: https://nl.mathworks.com/help/phased/ug/constant-false-alarm-rate-cfar-detection.html (accessed on 20 February 2023).
20. Constant False Alarm Rate (CFAR) [Online]. Available online: https://www.wikiwand.com/en/Constant_false_alarm_rate (accessed on 20 February 2023).
21. USRP-2954 Specifications [Online]. Available online: https://www.ni.com/docs/en-US/bundle/usrp-2954-specs/page/specs.html (accessed on 27 May 2023).

remote sensing

MDPI

Technical Note

Micro-Doppler Signature Analysis for Space Domain Awareness Using VHF Radar

Emma Heading [1,*], Si Tran Nguyen [1], David Holdsworth [1,2] and Iain M. Reid [2,3]

[1] Sensors and Effectors Division, DST Group, Edinburgh, SA 5111, Australia
[2] School of Physical Sciences, University of Adelaide, Adelaide, SA 5000, Australia
[3] ATRAD, 154 Ashley Street, Underdale, SA 5032, Australia
* Correspondence: emma.heading@defence.gov.au

Abstract: The large quantity of resident space objects orbiting Earth poses a threat to safety and efficient operations in space. Radar sensors are well suited to detecting objects in space including decommissioned satellites and debris, whereas the more commonly used optical sensors are limited by daylight and weather conditions. Observations of three non-operational satellites using a VHF radar system are presented in this paper in the form of micro Doppler signatures associated with rotational motion. Micro Doppler signatures are particularly useful for characterising resident space objects at VHF given the limited bandwidth resulting in poor range resolution. Electromagnetic simulations of the micro Doppler signatures of the defunct satellites are also presented using simple computer-aided design (CAD) models to assist with interpretation of the radar observations. The simulated micro Doppler results are verified using the VHF radar data and provide insight into the attitude and spin axis of the three resident space objects. As future work, this approach will be extended to a larger number of resident space objects which requires a automated processing.

Keywords: space domain awareness; micro-Doppler signature; VHF radar; satellite rotations

1. Introduction

Space domain awareness (SDA) has been a topic of global concern due to the large number of resident space objects orbiting Earth [1–7]. Resident space objects refer to both active and inactive satellites, spacecraft, debris or any human-made object in space. The number of satellites being launched into orbit is rapidly increasing due to mega constellations developed by companies including SpaceX, OneWeb, Amazon, Telesat and GW [8]. Space debris in particular poses a significant risk to the operation of global positioning systems, telecommunications and weather forecasting because the debris can deorbit, explode or cause collisions creating even more debris [9].

Space operations requiring SDA are supported with networks of space surveillance sensors including ground-based radar and optical sensors in addition to space-based optical sensors [1]. In Australia, there are various radar systems that perform SDA functions. The C band Space Surveillance Radar located in Exmouth, Western Australia, is capable of tracking several hundreds of space objects per day and forms a node in the US global Space Surveillance Network [5]. Two S band radar systems, known as the Western Australia Space Radar (WASR) located in Bunbury, Western Australia, have been operational since 2023 and form part of the the LeoLabs global network of phased array radars. Existing radar systems have been considered to contribute to the SDA effort in Australia including an experimental HF line-of-sight system [6,7], the Murchison Widefield Array used in a passive radar configuration [3,4] and the Australian Telescope Compact Array used in a bistatic configuration [10]. In South Australia, a VHF radar system at Buckland Park typically performs lower atmosphere wind profiling measurements, and further research dedicated to SDA has resulted in reliable detections of up to 1000 space objects per day [1,11].

Rotational motion of resident space objects has been observed using photometric light curves from optical sensors and basic polyhedron shapes have been determined when using multiple sites [12–14]. The rotational motion can provide information about the stability of the resident space object in orbit. In [13], the attitude and spin axis of a non-operational satellite TOPEX/Poseidon (NORAD ID: 22076) indicated a great possibility of overturn at precession angles greater than 55 degrees. In comparison to optical sensors, radar sensors have the advantage of operating in poor visibility and can also provide wide-area surveillance for the detection and tracking of resident space objects by utilising broad transmit beamwidths, and multiple receive beams from antenna arrays. The Buckland Park VHF radar in South Australia demonstrated that the defunct satellite Telkom 3 (NORAD ID: 38744) increased in rotation rate prior to decaying to Earth, indicating the importance of monitoring these features [11]. Earlier work with the S band Chibolton radar also observed a periodic radar cross section (RCS) from the defunct GEO-IK-2 satellite (NORAD ID: 37362) at a rate of 0.47 Hz from the object's rotational motion [15]. Further investigations were conducted into the effect of the attitude of the satellite RCS which introduced a second-order "stability index" [16]. More recently, a long baseline bistatic radar system has observed the tumbling motion of the Delta 4 rocket body (NORAD ID: 40535) using a network of radio telescopes operating at L band [17]. The variation in RCS from the tumbling motion of the rocket body was used to estimate the rotation period and tomographic images were formed to estimate the approximate size and shape of the resident space object.

Historically, radar imaging has been used to identify object features in two dimensions and Inverse Synthetic Aperture Radar (ISAR) imaging has been explored based on the rotational motion of resident space objects [2]. However, ISAR imaging is not applicable for radar systems operating at low frequencies with limited bandwidth. Direct analysis of time-frequency transforms has been found to be suitable for feature extraction of rotating resident space objects [11,18] and is well suited to observations at VHF. Computational electromagnetic simulations offer the flexibility of easily generating data for different attitude and spin axes in order to characterise a given resident space object. Large amounts of simulated data can also be produced and used as training data in supervised learning algorithms to automatically classify different shaped objects [19,20]. However, a deep understanding of the features in the data and the scattering effects in the surrounding environment is needed for reliable and meaningful solutions.

This paper presents the analysis of micro Doppler signatures induced by the rotational motion of three non-operational satellites, which were chosen to be representative of typical observations provided by the Buckland Park VHF Radar in South Australia. The Buckland Park VHF radar is described in further detail in Section 2 followed by the translational motion compensation processing in Section 3. Section 4 illustrates the CAD models for the three resident space objects and describes the electromagnetic simulations. Further analysis of the simulated and measured micro Doppler signatures is presented in Section 5 to gain insights into the attitude, spin axis and object features. The limitations and approximations used in the simulations are discussed in further detail in Section 6, and the complexity of extending this approach to a larger number of resident space objects is discussed as future work.

2. Buckland Park VHF Radar

The Buckland Park Stratosphere–Troposphere (BPST) VHF radar shown in Figure 1 ($-34°37'36.03''$, $138°28'3.91''$) is located 35 km north of Adelaide in South Australia. The radar is a pulsed monostatic system operating in the VHF band at a frequency of 55 MHz ($\lambda = 5.45$ m), with a peak transmit power of 40 kW and maximum duty cycle of 10%. The BPST radar was designed and manufactured by ATRAD and is operated by the University of Adelaide and ATRAD. The radar is designed primarily for the measurement of tropospheric and stratospheric winds (0.5–20 km) [21]. The transmit array (hereafter "main array") consists of a 12×12 array of gamma-matched linearly-polarised Yagi antennas. The inter-antenna spacing is 0.644λ, giving an aperture size of 38.6 m by 38.6 m. The array allows 5-beam directions to be used: vertical and 15° off zenith in cardinal directions (north, east, south, west). For the observations described in this paper, reception is performed using a single receiver connected to the main array. The combined transmit/receive main array beam width is 6° (half-power beamwidth). The use of the BPST radar for resident space objects (RSO) observations is described in [1], revealing detection rates of 200 RSO's per day. Subsequent signal processing and scheduling improvements now allow detection rates up to 1000 RSO's per day. The typical parameters used for the majority of observations presented in this paper are shown in Table 1, which illustrates the limited range resolution and the need for Doppler processing to obtain high resolution radar signatures. Coherent averaging was performed for data compression as outlined in [1].

Figure 1. The location of the Buckland Park Stratosphere–Troposphere VHF radar in South Australia shown approximately 35 km north of Adelaide (**left**) and the 12 × 12 antenna array [22] (**right**).

Table 1. Typical radar parameters used for the Buckland Park Radar for satellite observations.

PRF	2 kHz
No. Coherent Averages	2
Temporal Sampling Frequency	1 kHz
Waveform	13-bit Barker code
Baud Length	0.5 km
Pulse Length	6.5 km
Receive Filter Width	64 kHz
Range Resolution	0.25 km

The analysis employed in this paper is a variation on the "catalogue maintenance mode" (CMM) described in [1], referred to as the "quick catalogue maintenance mode" (QCMM). This uses "Keystone-like" processing using the predicted RSO state-vectors (generated through the Simplified General Perturbations Model (SGP4) forward propagated two-line elements (TLEs) or using Special Perturbations (SP) data) to transform the range–time data into range difference and Doppler difference data. The TLE and SP data are made available from the US Space Surveillance Network. This processing is computationally less intensive than CMM as the second step in the Keystone-like processing also removes the effects of radial acceleration. This obviates the need to incorporate a bank of acceleration hypotheses into the processing, thereby reducing the analysis dimension from 3-dim (range, time, acceleration) to 2-dim (range difference, Doppler difference). However, like CMM, the QCMM processing assumes that the RSO state-vectors are accurately predicted, and it is not applicable to non-catalogued RSOs for which state-vector predictions are unavailable. In the following section, a detailed description of the analysis is discussed.

3. Automatic Translational Motion Compensation

Generally, the translational motion is dominant compared to the rotational motion of the resident space object. For example, the translational velocity of a satellite in low earth orbit is approximately 7–8 km/s, while the spin rate of RHESSI (NORAD ID: 27370) has been reported to have a nominal spin rate of 15 rpm. In order to observe the micro Doppler signature due to the potential spinning motion of the space object of interest, the effects of translational motion will first need to be compensated in the radar data. It is desirable to automate the process of obtaining the translational motion compensated signal when there are large datasets.

The method used to automate the translational motion compensation is based on the Keystone transform. The Keystone-like processing is applied to the range–time radar returns $x(r,t)$ in two steps. The range and phase is corrected using the propagated range $R(t)$, i.e., the range from the radar to the object over time obtained from the propagated state vector, as demonstrated in Figure 2:

1. Range migration correction: The propagated range is used to convert the range–time data $x(r,t)$ into "range-corrected" time data $x_{r'}(r',t)$. This is achieved by interpolating $x(r,t)$ in the range dimension similar to the technique used in the Keystone transform [23]. Alternatively, a circular shift can be performed on the data in the range dimension by $\lceil (R(t)/\Delta R) \rceil$ samples, where $\lceil . \rceil$ represents the rounding up operator and ΔR is the range resolution. The latter option is feasible for the data presented in this paper as $x(r,t)$ is oversampled by at least a factor of two.
2. Phase correction: The "range-corrected" time data $x_{r'}(r',t)$ are then multiplied by a phase term based on the propagated range in (1) to compensate for phase differences introduced in step one.

$$x_k(r',t) = x_{r'}(r',t)e^{\left(\frac{4\pi i R(t)}{\lambda}\right)} \quad (1)$$

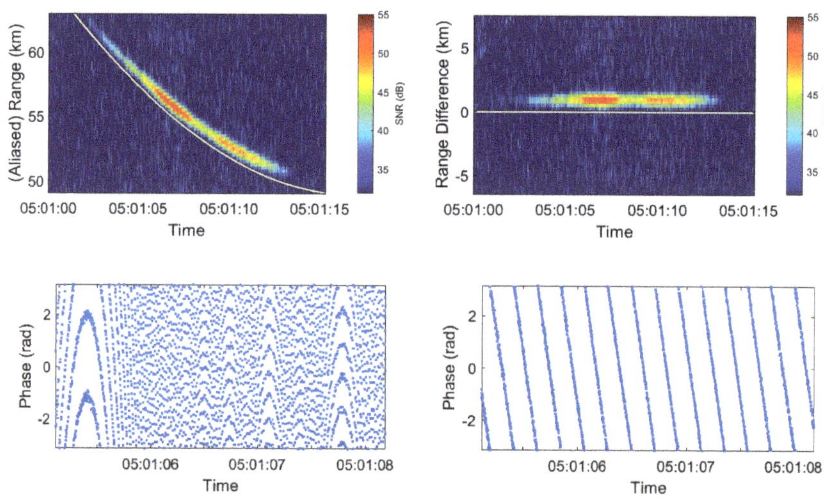

Figure 2. Example of the application of "Keystone-like" processing. Top left: range–time signal $x(r,t)$ representing the transit of the International Space Station (ISS) where the white line indicates the satellite propagated range. Top right: Associated range-corrected time plot $x_{r'}(r',t)$. Bottom left: phase difference obtained at the corrected range where signal power is maximised after applying the first step of the Keystone-like processing. Bottom right: phase difference variation after applying the final step of the Keystone–like processing.

The satellite data collected using the BPST data are subject to the effects of ionospheric group retardation which is evident by the measured range exceeding the predicted range shown in the top left plot in Figure 2. The top right plot in Figure 2 shows that the corrected range exceeds zero, which typically varies from 0.5 km (night) to 1.5 km (day). Following the Keystone-like processing in which translational motion has been accounted for, a spectrogram is produced using data at the appropriate range difference cell containing the target of interest. If the propagated state vectors are sufficiently accurate, the resulting phase difference will be zero and the corresponding spectrogram will be centred at zero Doppler. In typical circumstances where no rotational/spin motion exists, the spectrogram features a single peak centred around zero Hz, as shown for the ALOS-2 satellite (NORAD ID 39766) at the top of Figure 3. The backscattered power is relatively constant during the observation time. However, in some circumstances, this is not the case, as is illustrated for RHESSI (NORAD ID 27370) in the middle plot of Figure 3. This spectrogram shows significant Doppler broadening compared to ALOS-2, and a one-second periodicity. This periodicity occurs because RHESSI is designed to spin at 15 revolutions per minute [24] and has four perpendicular banks of solar panels, such that reflections from the solar panels occur every second. The bottom of Figure 3 shows the Telkom 3 satellite (NORAD ID 38744) with identifying features of the solar panels as the satellite rotates in time.

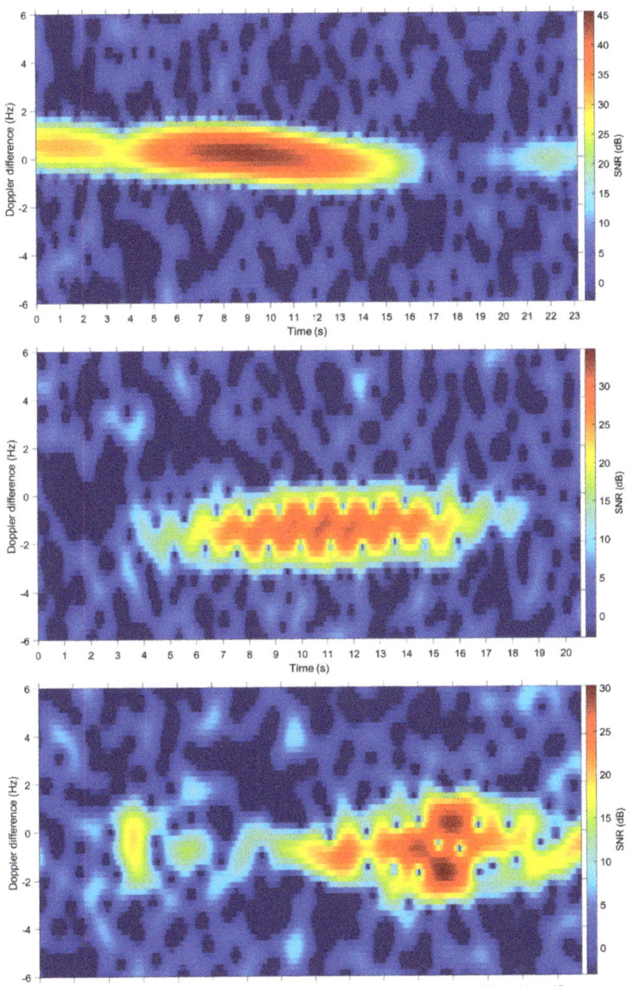

Figure 3. After Keystone-like processing. Top: spectrogram of ALOS-2 showing a single peak; middle: spectrogram of RHESSI showing Doppler broadening and periodic RCS pattern; bottom: spectrogram of Telkom 3 showing micro Doppler features.

4. Simulations

Simulated data provide a means of understanding the scattering characteristics of resident space objects of interest. For ease of interpretation, purely rotational motion was considered in isolation from noise and ionospheric effects. A simple computer-aided design (CAD) model provided the basis for electromagnetic simulation, where fine details were excluded and the material properties of the CAD model were assumed to be perfectly electrically conducting due to the relatively large wavelength at VHF. Three space objects of different electrical sizes are described in the following sections with the associated CAD model used in the electromagnetic simulations in the CST Microwave Studio. The system scenario for the simulations is described in Figure 4 defining the plane of rotation, spin axis and elevation angle relative to the BPST VHF radar.

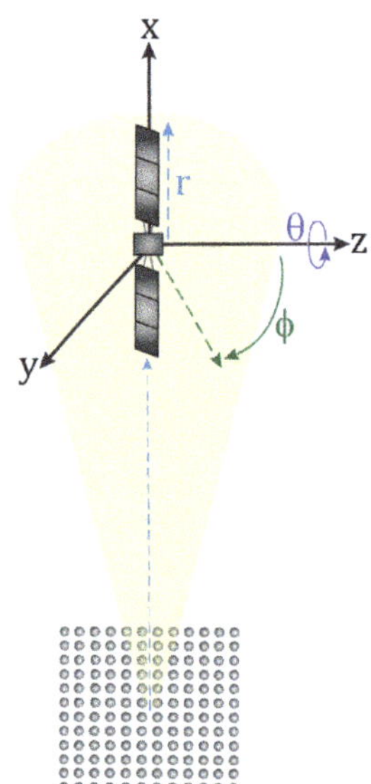

Figure 4. Radar system scenario with the object's azimuth angle θ defined as a rotation about the z axis (spin axis) and elevation angle ϕ defined as angle from the spin axis to the x–y plane (plane of rotation), which defines the object's attitude.

4.1. Satellite Information

The NASA Ramaty High Energy Solar Spectroscopic Imager (RHESSI) was launched into low earth orbit in 2002 to perform high resolution imaging of solar flares. During operations, the satellite was reported to rotate on an axis at a rate of 15 rpm with the imager directed at the sun [25]. RHESSI was deactivated in 2018 and remained in low earth orbit until April 2023 [26]. An image of RHESSI is depicted in Figure 5 with the CAD model designed in CST Microwave Studio. RHESSI is 2.16 m by 5.76 m in size which is comparable to the BPST VHF radar wavelength of 5.45 m [27].

The TOPEX/Poseidon satellite was a joint development between the United States and French space agencies to measure the ocean topography of the Earth from space. The satellite was launched in 1992 and decommissioned in 2006 after a stabilisation system malfunction resulting in the satellite being unable to reside in its intended orbital orientation. Photometric observations of TOPEX/Poseidon have shown that that the satellite is undergoing spinning motion at a rotation rate of about 10.6 s [13]. An image of TOPEX/Poseidon is shown in Figure 6 with the corresponding CAD design in CST Studio Suite. The main bus structure is 5.5 m which is comparable to the radar wavelength with a single solar panel of dimensions 8.9 m by 3.3 m [28]. The electrical size is approximately twice the wavelength of the BPST VHF radar.

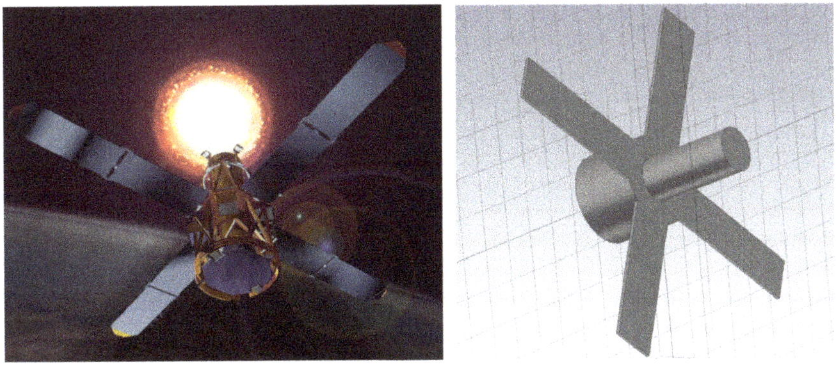

Figure 5. Image of RHESSI (**left**) and simplified CAD model designed in CST Studio Suite (**right**).

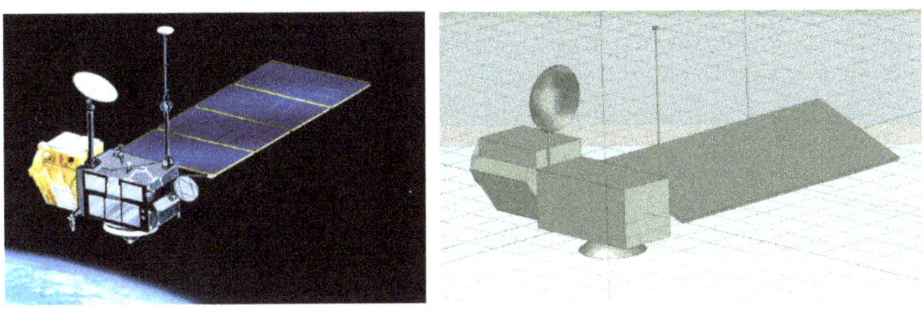

Figure 6. Image of TOPEX/Poseidon (**left**) and simplified CAD model designed in CST Studio Suite (**right**).

Telkom 3 was an Indonesian telecommunications satellite built by ISS Reshetnev and was launched in February in 2012. An image of Telkom 3 is shown in Figure 7 with the associated CAD model designed in CST Studio Suite. The satellite was intended for geostationary orbit but was stranded in an unusable orbit due to an upper stage failure of the Proton-M/Breeze-M launch vehicle [29]. The satellite decayed to Earth in February 2021. There is limited information publicly available on the physical dimensions of the satellite; however, the main bus structure was based on the Ekspress 1000N platform produced by Reshetnev. An estimation of the bus size for Telkom 3 from [30] was approximately 3 m × 2 m × 2.6 m. The solar panels were assumed to be the same dimensions as the sides of the bus structure, with a length multiplied by three when extended from the stowed position. The maximum dimension of the satellite is 23 m, providing an electrical size of approximately four wavelengths of the BPST VHF radar which is expected to be a good candidate for representing micro Doppler features.

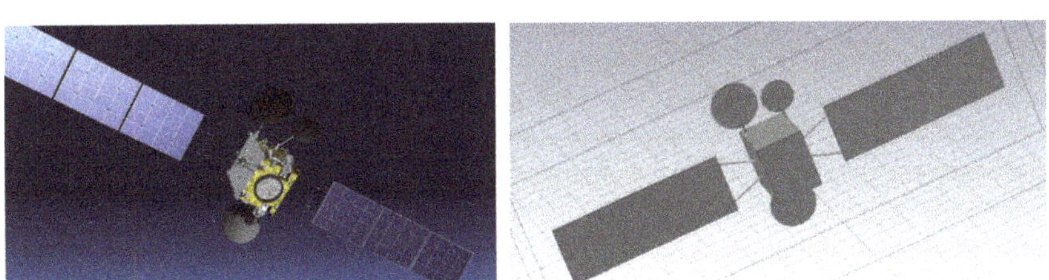

Figure 7. Image of Telkom 3 (**left**) and simplified CAD model designed in CST Studio Suite (**right**) [31].

4.2. Simulated Results

CST Studio Suite was used to simulate the electric field scattered from the satellite CAD models. Due to the electrical size being comparable to the radar wavelength, a full wave solver was required to capture the appropriate scattering characteristics. The electric field was calculated spanning 360° around the CAD model with a 0.005° step size in the plane of rotation at a defined elevation angle. This effectively provides the scattered signal as though the object was rotating uniformly in a plane.

Typically, the radar cross section is extracted from electromagnetic simulations; however, Doppler information can be provided if a common phase centre is defined across the angular samples for the electric field calculations. Doppler information can be retrieved from angular samples equivalently to time samples [32], which effectively creates an angular sampling rate conversion of $PRF_a = PRF/\omega$ in samples/radian [33]. In practice, the angular sampling rate is defined as the total number of angular samples, divided by the maximum angular extent. The angular Doppler frequency changing in time due to the effective object rotation can be observed using a short–time Fourier transform (STFT) by means of a spectrogram. This kind of signal representation is collectively known as a micro Doppler signature. The extent of the micro Doppler frequency is given by (2) where r_{max} is the maximum radial distance of the object from the origin of rotation and ω is the rotation rate. From inspection of a spectrogram, the micro Doppler frequency can be converted back to a measurement of distance or size with the simple multiplication factor of $\lambda/(2\omega)$.

$$f_d = \left(\frac{2\omega r_{max}}{\lambda}\right) \tag{2}$$

The resolution of the spectrogram is constrained by three factors based on the Doppler ambiguity condition, the linear approximation of the Fourier transform and the Doppler migration condition [33]. The Doppler migration condition $\Delta\Theta_{DM}$ is inversely proportional to the electrical size $(2r/\lambda)$ of the resident space object and determines the angular bin size used in the spectrogram. The results provided in Section 5 use a Doppler migration constraint of 28° to generate sufficient resolution to observe features in the spectrogram, which may include some non-linear effects.

$$\Delta\Theta_{DM} \leq \left(\frac{\lambda}{2r_{max}}\right)^{0.5} \tag{3}$$

5. Micro-Doppler Signature Analysis

The micro-Doppler signature is presented for three defunct satellites using both measured data from the BPST VHF radar owned and operated by The University of Adelaide and ATRAD in South Australia and the simulated electric field from CST Studio Suite manufactured by Dassault Systemes in France.

5.1. RHESSI

A typical observation of RHESSI from the BPST VHF radar is presented in Figure 8, highlighting the Doppler broadening as the satellite rotates with a periodic signature across time. The Doppler spread is indicative of the rotational motion of the solar panels. From inspection of the data, the time between the solar panels is approximately 0.9 s resulting in a rotation period of 4.275 s and rotation rate of 1.3288 radians per second. The TLE information used for the translational motion compensation provides a good estimate of the radial velocity and range over time resulting in the centring of the signal around zero Doppler.

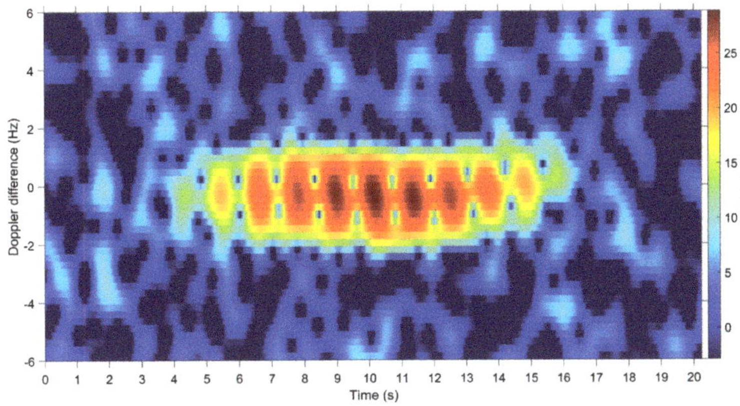

Figure 8. Spectrogram obtained using observations of the RHESSI satellite from the BPST VHF radar.

The corresponding spectrogram using simulated data from the RHESSI CAD model shown in Figure 5 is presented in Figure 9 for a single rotation. The RHESSI satellite is known to spin on its axis with the solar panels rotating around the axis of the imager. The scenario for maximum Doppler return occurs when the satellite is rotating in the same plane as the radar beam with the solar panels oriented in the same plane. The features of the solar panels are most prominent with vertical polarisation and similarly present a spread in Doppler frequency with a repeated pattern as shown previously with the measured data in Figure 8.

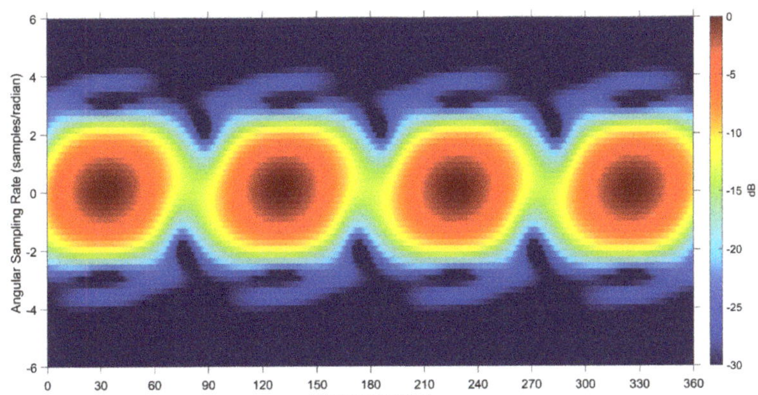

Figure 9. Spectrogram using simulated data using the RHESSI CAD model at an elevation of 40° with vertical polarisation at a frequency of 55 MHz.

5.2. *TOPEX/Poseidon*

The measured spectrogram for Topex/Poseidon from the BPST VHF radar is provided in Figure 10, indicating a general sinusoidal pattern with Doppler spread indicative of a solar panel. From inspection of the data, the time between the solar panels is approximately 5.5 s resulting in a rotation period of 11 s and rotation rate of 0.5712 radians per second.

Using the CAD model of TOPEX/Poseidon in Figure 6, the simulated result is provided in Figure 11. The scenario for the maximum Doppler frequency occurs when the solar panel is rotating in the same plane as the radar line of sight. A sinusoidal pattern is shown with intense specular flashes at 90° and 270° similar to the measured data in Figure 10. The main bus structure of TOPEX/Poseidon is oriented asymmetrically from the location of the solar panel which causes the asymmetrical Doppler signature in Figure 11.

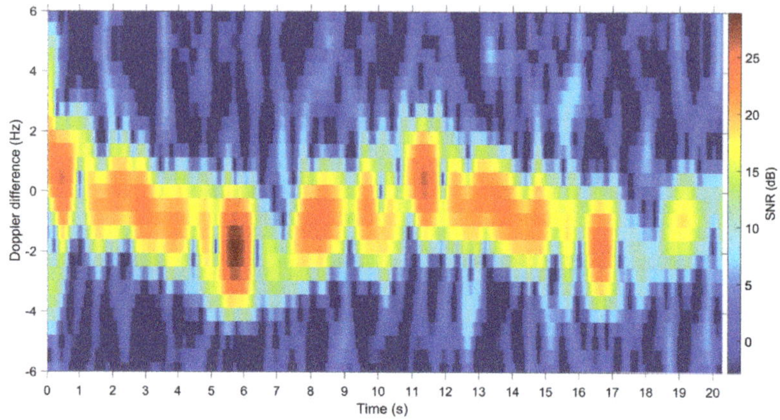

Figure 10. Spectrogram obtained using observations of TOPEX/Poseidon satellite from the BPST VHF radar.

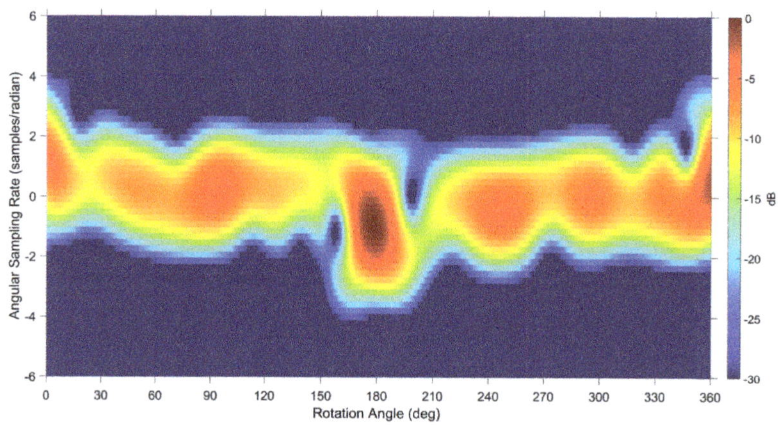

Figure 11. Spectrogram using simulated data with TOPEX/Poseidon CAD model at an elevation of 40° with vertical polarisation at a frequency of 55 MHz.

5.3. *Telkom 3*

The spectrogram from the BPST VHF radar for Telkom 3 is presented in Figure 12. The measured data shows specular reflections corresponding to the two solar panels and adjacent returns close to zero Doppler from the antennas. From inspection of the data, the time between the solar panels is approximately 7.75 s resulting in a rotation period of 15.5 s and rotation rate of 0.4054 radians per second.

The resulting spectrogram using the Telkom 3 CAD model from Figure 7 simulated at 55 MHz with linear vertical polarisation is presented in Figure 13 for 0° elevation angle. The specular reflections from the solar panels are present across positive and negative Doppler centred at a rotation angle of 180° with adjacent reflections close to zero Doppler corresponding to the antenna dishes.

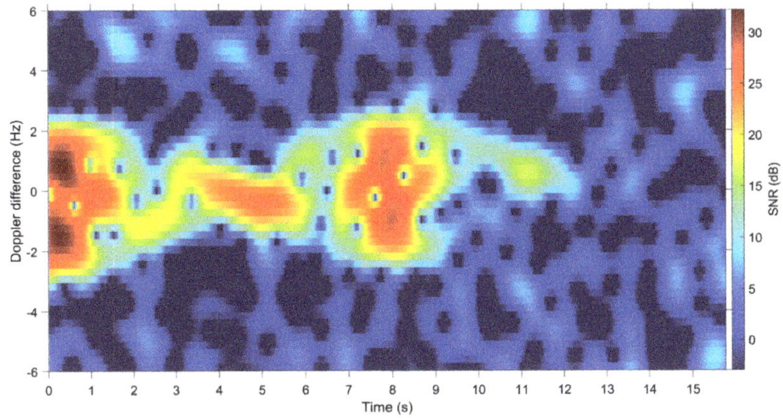

Figure 12. Spectrogram obtained using observations of Telkom 3 satellite from the BPST VHF radar.

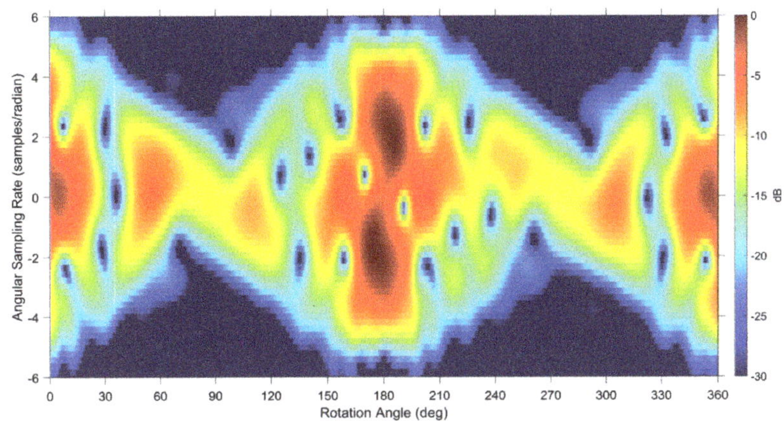

Figure 13. Spectrogram using simulated data with Telkom 3 CAD model at an elevation of 0° with vertical polarisation at a frequency of 55 MHz.

The corresponding spectrogram at an elevation angle of 40° is shown in Figure 14, illustrating a different micro Doppler signature than is shown in Figure 13, highlighting the dependency of the received radar returns on the satellite orientation.

Finer details are not present in the spectrogram when using a lower number of samples in the Doppler or angular sampling rate dimension to process the micro Doppler information. For example, the same data were used to generate the results in Figure 15 as used in Figure 13 but with half the number of samples. The bottom plot in Figure 3 shows the measured data with a similar micro Doppler pattern as Figure 15 illustrating that care needs to be taken in choosing an appropriate number of samples as outlined in Section 4.2.

If the rotation rate ω is known or estimated, then the angular sampling rate PRF_a can be converted to the traditional time-based sampling rate PRF. Measured radar observations can be used to estimate ω using the time taken between the solar panel specular reflections in a rotation; however, each observation of the satellite may present variations in the rotation rate.

As an example, the simulated RCS pattern of Telkom 3 is provided in Figure 16 which indicates maximum peaks occurring every 180° corresponding to the broadside orientation of the satellite. The RCS pattern could be used in a similar way to monitor the spinning motion as observed by the Chibolton radar in [15].

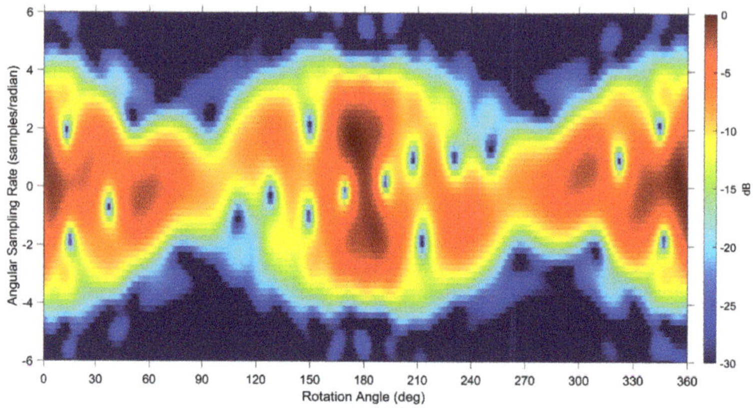

Figure 14. Spectrogram using simulated data with Telkom 3 CAD model at an elevation of 40° with vertical polarisation at a frequency of 55 MHz.

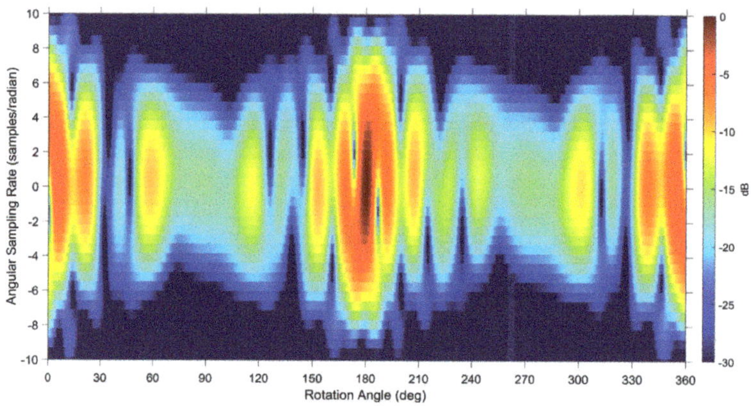

Figure 15. Spectrogram using the Telkom 3 CAD model at an elevation of 0° with the angular sampling rate downsampled by a factor of 2 with vertical polarisation at a frequency of 55 MHz.

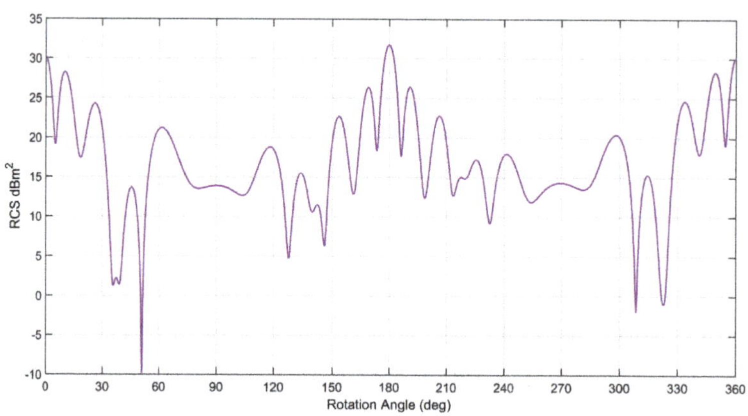

Figure 16. Simulated RCS of Telkom 3 at an elevation of 0° with vertical polarisation at a frequency of 55 MHz.

6. Discussion

The observed micro-Doppler signatures for all three defunct satellites using the BPST VHF radar in Section 5 indicate that there are features changing in time, indicative of rotational motion. The simulated data assisted with interpreting the resident space object's spin axis as well as identifying returns from the antennas and solar panels. It was observed that the largest micro-Doppler extent corresponded to the length and number of solar panels. The symmetry of the micro-Doppler pattern enables the number of solar panels to be extracted from the data. In Figure 11, the asymmetrical returns around zero Doppler indicate an odd number of solar panels only occurring at 0° and 180° indicative of a single panel for TOPEX/Poseidon. The simulated micro-Doppler features for RHESSI are symmetrical in Figure 9 indicating an even number of solar panels, where the returns occur every 90 degrees indicating that there are four panels, whereas the results for Telkom 3 in Figure 13 show symmetrical returns from the panels that occur every 180° indicative of two panels. When the simulated data are compared to the measured data, the number of solar panels can be verified and assessed as to whether a solar panel is damaged or missing.

A range of spin axes were examined in the simulated data in order to find the best agreement with the measured data. Figure 17 shows the resulting attitude for each resident space object including RHESSI with the spin axis along the x direction, TOPEX/Poseidon with spin axis in the y direction and Telkom 3 with the spin axis in the z direction assuming the radar is looking in the negative y direction. Different elevation angles were also investigated when comparing the simulated and measured data, where the elevation angle ϕ is defined as the angle from the spin axis to the rotation plane. The results in Figure 9 show that the elevation angle of 40° degrees has the best agreement with the measurements of RHESSI and TOPEX/Poseidon. However, the zero degree elevation angle in Figure 13 showed the best agreement for Telkom 3.

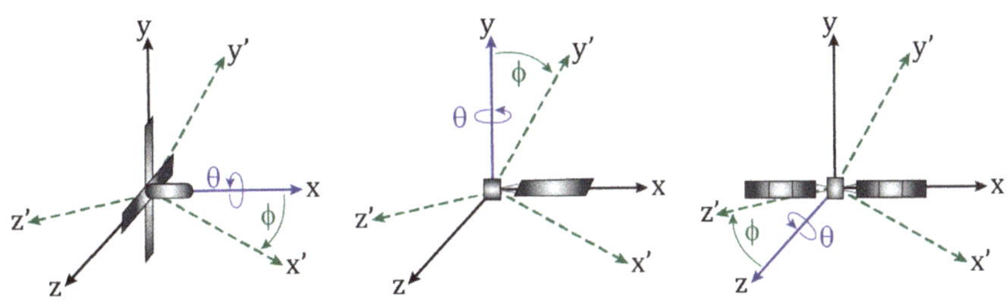

Figure 17. The rotation axis identified for RHESSI (**left**), TOPEX/Poseidon (**middle**) and Telkom3 (**right**).

A summary of the insights gained from the analysis of the simulated micro-Doppler signatures is provided in Table 2. The spin axis, elevation angle and polarisation estimations were able to be determined from the comparison of the micro-Doppler pattern from the simulated data and the BPST VHF radar observations.

Table 2. Summary of object features extracted from the simulated micro-Doppler results.

Object Feature	RHESSI	TOPEX/Poseidon	Telkom 3
No. Solar Panels	4	1	2
Spin Axis	x	y	z
Elevation Angle	0°	40°	40°
Polarisation	V	V	V

The simulated micro-Doppler pattern for a single rotation assisted with interpretation of the measured results. Once the number of solar panels were known from comparing the simulated and measured results, the estimated Doppler extent, rotation period, rotation rate and total object length were extracted from inspection of the measured BPST VHF radar data. The summary of the features extracted from the measured data is presented in Table 3.

Table 3. Summary of object features extracted from the measured micro-Doppler results.

Object Feature	RHESSI	TOPEX/Poseidon	Telkom 3
Estimated Doppler Extent	2.5 Hz	3 Hz	4 Hz
Estimated Rotation Period	4.3 s	11 s	15.5 s
Estimated Rotation Rate	1.3228 rad/s	0.5712 rad/s	0.4054 rad/s
Estimated Length	5.2 m	7.2 m	21.6 m
True Length	5.76 m	5.5 m	23 m

Several assumptions were made in order to conduct the computational electromagnetic simulations to synthesise the BPST VHF radar setup and environmental effects. Although the BPST VHF radar transmitted a linear vertically polarised signal, the received signal may be subjected to Faraday polarisation rotation caused by the interaction of the signal with the ionosphere [34]. As a result, the polarisation of the received signal could be rotated by an unknown amount in the polarisation plane. This study was confined to simulating the horizontal (aligned wit the z axis) and vertical (aligned with the x axis) polarisation to reduce the complexity of the problem. The results give an indication of whether the received signal was closer to horizontal or vertical polarisation rather than determining the exact rotation angle in the polarisation plane. This assumption means that the simulated micro-Doppler features will highlight the dominantly horizontal or vertical components of the structure rather than match the exact polarisation of the measured radar data.

There were also constraints placed on the simulated motion of the resident space object including a fixed elevation angle and constant rotation rate, whereas in reality, this may not be the case. Large variations in the motion of the resident space object could impact the micro-Doppler signature, but it is expected that the variations are minimal for non-operational satellites within the small observation times shown in this paper. All material properties of the CAD model used in the simulations were assumed to be perfectly electrically conducting which may slightly overestimate the intensity of the micro-Doppler signature. The CAD models were also rotated to examine different spin axes to find the best fit with the simulated and measured results. Consequently, the simulated micro-Doppler signatures are an approximate indication of polarisation, spin axis and elevation angle. However, the simulated data show good agreement with the measured BPST VHF radar data and validate the CAD models to show that the object structure is largely intact, and micro Doppler features were found to be consistent with the solar panels and antennas.

This study focuses on characterising a representative set of different resident space objects that are typical of what the BPST VHF radar may observe to gain insight into the scattering characteristics and extract object features. However, there are a vast number of resident space objects the BPST VHF radar is capable of observing which would require an automated process and management of a large simulated dataset incorporating a range of spin axes, elevation angles and polarisation rotation angles. This problem is well suited to machine learning to automatically extract target features and classify objects for situational awareness and monitoring of abnormal behaviour of resident space objects. The ease of producing simulated data is beneficial for generating training data for supervised learning algorithms. The results in Section 5 have shown that low–medium fidelity CAD models simulated with the full-wave solver agree with the measured BPST radar data to a good approximation. The automated process of extracting information from a larger number of resident space objects is a logical next step to progress this research in future work.

7. Conclusions

The micro Doppler signatures for three defunct satellites were studied using measured radar observations from the BPST VHF radar in South Australia. Computational electromagnetic simulations assisted with interpreting the measured radar data in order to characterise the three resident space objects. Several insights were gained from the comparison of the simulated and measured micro-Doppler signatures including an estimation of the spin axis, attitude, number of solar panels, rotation rate and approximate size. In future work, this initial research will be extended to a larger number of resident space objects requiring automated processing with particular consideration of the complexities involved in generating the simulated data.

Author Contributions: Conceptualisation, D.H., S.T.N. and E.H.; methodology, D.H., S.T.N. and E.H.; software, D.H., S.T.N. and E.H.; validation, D.H. and E.H.; formal analysis, D.H., E.H. and S.T.N.; investigation, D.H. and E.H.; resources, D.H. and I.M.R.; data curation, D.H. and I.M.R.; writing E.H. and D.H.; visualisation, D.H., S.T.N. and E.H.; supervision, I.M.R.; project administration, D.H. and E.H. All authors have read and agreed to the published version of the manuscript.

Funding: This research received no external funding.

Institutional Review Board Statement: Not applicable.

Data Availability Statement: The data used in this research are not publicly available.

Acknowledgments: The authors would like to thank the University of Adelaide and ATRAD for use of the BPST VHF radar to perform the measurements and Defence Science and Technology Group for supporting this research. We would also like to thank Bobby Yau and Stephen McClanachan for sharing their expertise on computational electromagnetic simulations, and we acknowledge discussions with Brendan Hennessy, Kruger White, An Phan and Hai-Tan Tran from Defence Science and Technology Group.

Conflicts of Interest: The authors declare no conflicts of interest. The funders had no role in the design of the study; in the collection, analyses, or interpretation of data; in the writing of the manuscript; or in the decision to publish the results.

References

1. Holdsworth, D.; Spargo, A.; Reid, I.; Adami, C. Low Earth Orbit object observations using the Buckland Park VHF radar. *Radio Sci.* **2020**, *55*, e2019RS006873 . [CrossRef]
2. Anger, S.; Anglberger, H.; Jirousek, M.; Dill, S.; Peichl, M. ISAR Imaging of Satellites in Space–Simulations and Measurements. In Proceedings of the 2019 20th International Radar Symposium (IRS), Ulm, Germany, 26–28 June 2019; pp. 1–6. [CrossRef]
3. Hennessy, B.; Tingay, S.; Hancock, P.; Young, R.; Tremblay, S.; Wayth, R.B.; Morgan, J.; McSweeney, S.; Crosse, B.; Johnston-Hollitt, M.; et al. Improved Techniques for the Surveillance of the Near Earth Space Environment with the Murchison Widefield Array. In Proceedings of the IEEE Radar Conference, Boston, MA, USA, 22–26 April 2019.
4. Palmer, J.E.; Hennessy, B.; Rutten, M.; Merrett, D.; Tingay, S.; Kaplan, D.; Ord, S.T.S.M.; Morgan, J.; Wayth, R.B. Surveillance of Space using passive radar and the Murchison Widefield Array. In Proceedings of the IEEE Radar Conference, Seattle, WA, USA, 8–12 May 2017.
5. Smith, C.H.; Greene, B.; Bold, M.; Drury, R. Development of a new SSA facility at Learmonth Australia. In Proceedings of the Advanced Maui Optical and Space Surveillance Technologies Conference, Maui, HI, USA, 11–14 September 2018.
6. Frazer, G.J.; Meehan, D.H.; Warne, G.M. Decametric measurements of the ISS using an experimental HF line-of-sight radar. In Proceedings of the 2013 International Conference on Radar, Adelaide, Australia, 9–12 September 2013.
7. Frazer, G.J.; Rutten, M.; Cheung, B.; Cervera, M.A. Orbit determination using a decametric line-of-site radar. In Proceedings of the Advanced Maui Optical and Space Surveillance Technologies Conference (AMOS), Maui, HI, USA, 9–12 September 2014.
8. Boley, A.; Byers, M. Satellite mega-constellations create risks in Low Earth Orbit, the atmosphere and on Earth. *Sci. Rep.* **2021**, *11*, 10642. [CrossRef] [PubMed]
9. NASA Office of Inspector General Office of Audits. *NASA's Efforts to Mitigate the Risks Exposed by Orbital Debris*; Technical Report IG-21-011; NASA: Washington, DA, USA , 2021.
10. Nosrati, H.; Smith, S.; Hayman, D.; Horiuchi, S.; Hellicar, A. Bi-static Radar Interferometric Localization of MEO and GEO Space Debris using Australia Telescope Compact Array. In Proceedings of the Advanced Maui Optical and Space Technologies Conference, Maui, HI, USA, 27–30 September 2022.
11. Heading, E.; Nguyen, S.; Holdsworth, D.; Field, D.; Reid, I. Analysis of RF Signatures for Space Domain Awareness using VHF radar. In Proceedings of the 2022 IEEE Radar Conference (RadarConf22), New York, NY, USA, 21–25 March 2022; pp. 1–6. [CrossRef]
12. Benson, C.; Scheeres, D.; Moskovitz, N. Light-curves of Retired Geosynchronous Satellites. In Proceedings of the 76th European Conference on Space Debris, Darmstadt, Germany, 17–21 April 2017; Volume 7.

13. Kudak, V.I.; Epishev, C.P.; Perig, V.M. Determining the orientation and spin period of TOPEX/Poseidon satellite by a photometric method. *Astrophys. Bull.* **2017**, *72*, 340–348. [CrossRef]
14. Fulcoly, D.O.; Kalamaroff, K.I.; Chun, F.K. Determining Basic Satellite Shape from Photometric Light Curves. *J. Spacecr. Rocket.* **2012**, *49*, 76–82 . [CrossRef]
15. Eastment, J.D.; Ladd, D.N.; Walden, C.J.; Trethewey, M.L. Satellite observations using the Chibolton radar during the initial ESA 'CO-Vi' tracking campaign. In Proceedings of the European Space Surveillance Conference, Madrid, Spain, 7–9 June 2011.
16. Stevenson, M.; Nicolls, M.; Rosner, C. Space Object Attitude Stability Determined from Radar Cross-Section Statistics. In Proceedings of the Advanced Maui Optical and Space Surveillance Technologies Conference, Maui, HI, USA, 17–20 September 2019.
17. Uysal, F.; van Dorp, P.; Serrano, A.; Kobsa, A.; Ghio, S.; Kintz, A.; Bassa, C.; Garrington, S.; Caro Cuenca, M.; Otten, M.; et al. Large Baseline Bistatic Radar Imaging for Space Domain Awareness. In Proceedings of the Radar 2023, Berlin, Germany, 20–22 September 2023.
18. Ghio, S.; Martorella, M. Estimation of rotating RSO parameters using radar data and joint time-frequency transforms. In Proceedings of the 7th European Conference on Space Debris, Darmstadt, Germany, 18–21 April 2017.
19. Jung, K.; Lee, J.I.; Kim, N.; Oh, S.; Seo, D.W. Classification of Space Objects by using Deep Learning with Micro Doppler Signature Images. *Sensors* **2021**, *21*, 4365. [CrossRef] [PubMed]
20. Cammenga, Z.A.; Baker, C.J.; Smith, G.E.; Ewing, R. Micro-Doppler target scattering. In Proceedings of the 2014 IEEE Radar Conference, Cincinnati, OH, USA, 9–23 May 2014; pp. 1451–1455. [CrossRef]
21. Dolman, B.; Reid, I.; Tingwell, C. Stratospheric tropospheric wind profiling radars in the Australian network. *Earth Planets Space* **2018**, *70*, 170. [CrossRef]
22. of Adelaide, T.U. Buckland Park VHF ST Radar. 2024. Available online: http://www.physics.adelaide.edu.au/atmospheric/vhf.html (accessed on 1 February 2024).
23. Richards, M. The keystone transformation for correcting range migration in range-doppler processing. *Pulse* **2014**, *1000, 13–14* .
24. Lin, R.P.; Dennis, B.R.; Hurford, G.J.; Smith, D.M.; Zehnder, A.; Harvey, P.R.; Conway, A. *The Reuven Ramaty High-Energy Solar Spectroscopic Imager (RHESSI)*; Springer: Dordrecht, The Netherlands, 2003.
25. Krebs, G.D. RHESSI (RHESSI, Reuven Ramaty, SMEX 6, Explorer 81). 2023. Available online: https://space.skyrocket.de/doc_sdat/explorer_hessi.htm (accessed on 1 February 2024).
26. Hill, D. NASA Retired Solar Energy Imager Spacecraft Reenters Atmosphere. 2023. Available online: https://www.nasa.gov/feature/nasa-retired-solar-energy-imager-spacecraft-reenters-atmosphere (accessed on 1 February 2024)
27. Dennis, B. RHESSI MISSION FACTS. 2023. Available online: https://hesperia.gsfc.nasa.gov/rhessi2/mission/mission-facts/index.html (accessed on 1 February 2024).
28. NASA. TOPEX/Poseidon. 2021. Available online: https://sealevel.jpl.nasa.gov/system/documents/files/1674_tp-fact-sheet.pdf (accessed on 1 February 2024).
29. Page, G.S. Telkom 3. 2021. Available online: https://space.skyrocket.de/doc$_$sdat/telkom-3.htm (accessed on 1 February 2024).
30. Technology, A. Telkom-3 Communications Satellite, Indonesia. 2021. Available online: https://www.aerospace-technology.com/projects/telkom3-communication-satellite-indonesia (accessed on 1 February 2024).
31. Reshetnev, J.A.M. Information Satellite Systems. 2021. Available online: http://www.iss-reshetnev.com/spacecraft/spacecraft-communications/telkom-3 (accessed on 1 February 2022).
32. Munson, D.; O'Brien, J.; Jenkins, W. A tomographic formulation of spotlight mode synthetic aperture radar. *Proc. IEEE* **1983**, *71*, 917–925. [CrossRef]
33. Tran, H.T.; Heading, E.; Ng, B.H. On the Slow-Time *k*-Space and its Augmentation in Doppler Radar Tomography. *Sensors* **2020**, *20*, 513. [CrossRef] [PubMed]
34. Klobuchar, J. *Ionospheric Effects on Earth-Space Propagation;* Technical Report AFGL-TR-84-0004; Ionospheric Physics Division: Air Force Geophysics Laboratory: Bedford, MA, USA, 1983.

remote sensing MDPI

Article

Range-Doppler-Time Tensor Processing for Deep-Space Satellite Characterization Using Narrowband Radar †

Alexander Serrano *, Jack Capper, Robert L. Morrison, Jr. and Mohamed D. Abouzahra

MIT Lincoln Laboratory, Lexington, MA 02421, USA; jack.capper@ll.mit.edu (J.C.); rmorrison@ll.mit.edu (R.L.M.J.); abouza@ll.mit.edu (M.D.A.)
* Correspondence: alexander.serrano@ll.mit.edu
† DISTRIBUTION STATEMENT A. Approved for public release. Distribution is unlimited. This material is based upon work supported under Air Force Contract No. FA8702-15-D-0001. Any opinions, findings, conclusions or recommendations expressed in this material are those of the author(s) and do not necessarily reflect the views of the U.S. Air Force.

Abstract: There is growing demand for the high-fidelity characterization of satellites in Geosynchronous Earth Orbit (GEO) to support Space Domain Awareness (SDA). This is particularly true for newly launched satellites, where it is necessary for satellite providers to ascertain whether components have deployed properly. Conventional wideband radar systems are capable of imaging satellites provided that (i) they have sufficient power aperture and bandwidth, and (ii) they observe enough target aspect change to generate a resolved image. While wideband radars are used routinely for characterizing satellites in Low-Earth Orbit (LEO), powerful radars with sensitivity sufficient for large GEO ranges (36,000 km or greater) are lacking. Thus, researchers often rely on more widely available high-power narrowband tracking radars for GEO characterization. In this paper, we present a novel range-Doppler-time (RDT) tensor processing technique for GEO characterization with narrowband radar. This technique encapsulates the strengths of previously proposed methods for narrowband-radar characterization at GEO, providing a generalized approach that can be applied in a variety of settings. The technique generates fully resolved 2D images of rotating GEO satellites in low-bandwidth scenarios. In cases where aspect change is limited, the technique provides detailed Doppler information for enhanced satellite status monitoring. This work presents a comprehensive quantitative analysis of the technique that considers the impact of key parameters on characterization performance. Simulated radar data, and radar data collected in a compact range on a scaled satellite model, are used to evaluate the technique.

Keywords: range-Doppler-time tensor (RDT); range superpulses; Doppler superpulses; Doppler tomography; bandwidth enhanced non-coherent imaging (BENI); power-sum image reconstruction; GEO characterization

1. Introduction

There is a need for detailed characterization of satellites in Geosynchronous Earth Orbit (GEO) to support Space Domain Awareness (SDA). SDA objectives such as satellite stability assessment and target-size estimation are of great interest to the space surveillance community. In addition, satellite providers need to ascertain whether components such as solar panels have deployed properly after launch. Although a number of concepts have been proposed for space–object characterization, ground-based optics and radars are the dominant sensor approaches utilized for this awareness at GEO. Earth-based radars are particularly well matched to satellite characterization, as they can provide high-resolution imagery of space objects at all times of the day and in a variety of weather conditions [1]. Additionally, in contrast to optical sensors, radar image resolution depends only on the transmitted bandwidth and the total aspect change observed, and thus, is independent of the range to the target [2,3]. Ground-based radars can image all satellites visible to the sensor in Low Earth Orbit (LEO) and Medium Earth Orbit (MEO) provided they have sufficient power aperture to generate a detectable return, and thus, they are routinely used for this purpose. However, in GEO, radar imaging is more challenging because, depending on the location of the target, the bandwidth and aspect change needed for high-resolution imaging may not be available. This is due to two factors. First, there are a small number of wideband radars with power sufficient to reach GEO ranges (36,000 km or greater), thus limiting the region of the GEO belt where high range resolution can be achieved. Second, most satellites in GEO are Earth stabilized and reside in low-inclination orbits. Thus, the satellites present very little aspect change to radars on the Earth, resulting in the target being unresolved in cross range.

Related work has explored deep-space satellite characterization through many different approaches, such as light-curve polarimetry for shape and material estimation [4,5], pattern recognition applied to unresolved images for object identification [6], as well as traditional techniques such as frame differencing [7]. Recent efforts have achieved success in applying Artificial Intelligence (AI) algorithms to the problem of SDA, as AI algorithms have proven highly capable of object detection [8]. A recently published paper explored the performance of AI-based algorithms for SDA when compared to legacy methods [9], while other papers have seen success in utilizing AI on light-curve data [10–13], orbital-observation data [14], or optical imagery [15] for characterization. However, these works do not focus on complex-valued radar data or the capabilities that narrowband imaging offers.

Despite the aforementioned limitations, there are a number of cases where either partial bandwidth or partial aspect change is available, and thus, can be exploited for characterization. There are a large number of deep-space narrowband tracking radars that provide coverage of most of the GEO belt. Thus, researchers often rely on these more widely available high-power radars for GEO characterization. The bandwidth provided by the narrowband radars, while insufficient for traditional high-resolution radar imaging, can be coupled with the Doppler information naturally provided by these sensors to generate images. In addition, satellites present opportunities for imaging when operating in modes where the body rotates, or when the satellite reaches end of life and enters into a tumble motion post-deactivation (after insertion into super-GEO graveyard orbit). Given rotation, Doppler techniques can be used to form resolved images, even in cases where the bandwidth is reduced to the point where the target is unresolved in range. Examples of Doppler techniques applicable to GEO include Doppler–Time Intensity (DTI) analysis [16–20], Doppler tomography imaging [16–19,21–23], coherent narrowband imaging [22,24–27], and the recently proposed Bandwidth Enhanced Non-coherent Imaging (BENI) algorithm [28].

Doppler–Time Intensity plots, which show the Doppler trajectories of individual scatterers on an object, have long been used to characterize spaceborne targets [1]. More recently, the use of tomography has been explored as a way to generate images of tumbling satellites in the absence of bandwidth [16,28]. Here, each column of the DTI, corresponding to a fixed time instant, provides a one-dimensional projection of the target's reflectivity. Thus, an inverse Radon transform can be applied to one tumble period of data to form an image of the satellite. As studied in [28], the resulting Doppler tomography images provide a reduced-quality representation of the target in comparison to conventional wideband imaging due to the lack of resolving capability in range from insufficient bandwidth. Noting that many narrowband tracking radars have some bandwidth available, the BENI algorithm was proposed to improve upon Doppler tomography by incorporating range information [28]. For cases where the target has complex rotational dynamics (e.g., spin-precession motion), imaging becomes more challenging. In such cases, the DTI plots by themselves provide valuable information that can be used to make assessments about satellite structure, status, and behavior. Motivated by this, an enhanced DTI product known as Doppler superpulses was developed [16]. Doppler superpulses use information from multiple DTIs (one from each range bin) to form a sensitivity-enhanced composite, which can be used as an input for tomography to produce a superior reconstruction. The utility of the Doppler techniques above was demonstrated for GEO characterization in a recent paper, using multistatic radar data collected on several tumbling space objects. Here, the Millstone Hill radar in Westford, MA and the TIRA radar in Wachtberg, Germany were used as narrowband illuminators with radio telescopes in the United Kingdom and the Netherlands as bistatic receivers [29,30]. These works highlight both the feasibility and successful demonstration of narrowband radar techniques for rotational motion determination and size estimation using real, on-orbit tumbling rocket bodies.

While a number of disparate narrowband-radar techniques have been proposed for GEO characterization, each with strengths in specific scenarios, a generalized processing framework applicable across a variety of settings is lacking. Doppler tomography can be used to image a rotating satellite when the object is unresolved in range (i.e., no or low radar bandwidth). BENI can produce improved images of the same object, but only when additional bandwidth is available. If aspect change is limited, e.g., when only partial rotation of the satellite is observed, one can obtain information through Doppler–time analysis. DTI plots are useful in inferring target type, in determining whether satellite components have deployed properly after launch, and in assessing satellite stability and rotational dynamics. Moreover, DTI plots and tomographic images, in concert, provide complimentary information that can be used to enhance satellite assessment; DTIs capture localized behavior well (e.g., specular reflections, rotational maneuvers), while tomographic images highlight the structural features of targets. Thus, the complete characterization of GEO satellites requires the application of multiple techniques, followed by the interpretation and integration of the data-processing outputs. For robust and efficient characterization at GEO, an integrated processing approach is needed.

In this paper, we present a novel range-Doppler-time (RDT) tensor processing technique for GEO characterization with narrowband radar. This technique encapsulates the strengths of previously proposed methods for narrowband-radar characterization at GEO, providing a generalized approach that can be applied in a variety of settings. Our method expresses the radar signal as a 3D tensor varying in range, Doppler, and time. By applying simple operations to the tensor, we demonstrate that a variety of characterization products can be generated. These include fully resolved 2D power-sum images of rotating GEO satellites, as well as range and Doppler superpulse plots providing detailed, time-localized characterization of space objects. In addition, the 3D framework lends itself well to new extensions of the technique for sensitivity enhancement that exploit the Higher-Order Singular Value Decomposition (HOSVD) of the multi-dimensional tensor. The RDT tensor representation makes it easy to show the relationship between previous processing techniques (e.g., Doppler tomography and BENI) and to understand the impact of key processing parameters.

The goals of this paper are (i) to develop a unified processing technique based on the 3D RDT tensor representation of the radar signal providing enhanced GEO characterization, and (ii) to characterize the performance of the technique as a function of radar and target parameters. Throughout our treatment, we will emphasize challenges encountered when deviating from ideal image models and collection geometries. The key contributions of this work are summarized as follows:

- Development of a novel RDT tensor processing technique for deep-space satellite characterization
- Quantitative performance assessment of the RDT tensor processing technique considering the impact of key parameters on characterization performance
- Demonstration of extensions to the RDT tensor framework, including tensor denoising using the Higher-Order Singular Value Decomposition (HOSVD) to produce enhanced sensitivity images

The organization of this paper is as follows. Section 2 presents the target radar-return signal model and provides a review of standard radar-characterization techniques. Section 3 develops the novel 3D RDT tensor processing technique, presenting results using both simulated and compact-range radar data. Previous imaging algorithms, Doppler tomography and BENI, are presented as related cases of the power-sum imaging method proposed herein. A quantitative performance analysis of the approach is introduced in Section 4. Section 5 discusses extensions of the 3D tensor technique. Conclusions are presented in Section 6.

2. Background on Radar Characterization of Space Objects

2.1. Radar Imaging Geometry

The derivations for the monostatic radar-return signal models and imaging geometry in our work are based on the following assumptions. We first assume the target is under pure tumbling motion (angular velocity $\vec{\omega}$ is constant during the entire satellite pass). Previous work on estimating tumble periods of inactive GEO satellites has shown this to be a valid assumption over multiple full rotations of the object [31]. For real on-orbit targets, translational motion compensation is needed to ensure the center of rotation of the object is aligned at zero Doppler [30]. This form of Doppler alignment results in well focused radar-image reconstructions, and thus best serves to maximize image contrast. The uniform rotation of different point scatterers on the target can be represented using a turntable geometry. The target shown in Figure 1a has an angular velocity $\vec{\omega}$, which is aligned with the Z-axis of the depicted coordinate system fixed to the target's frame, as shown in Figure 1b.

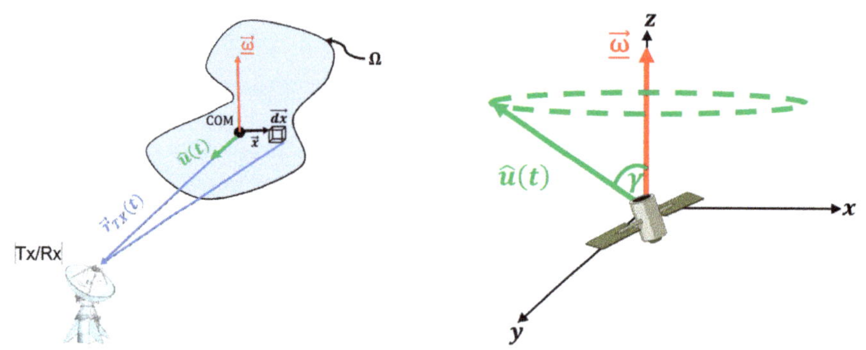

(**a**) Monostatic radar-imaging geometry (**b**) Target reference frame

Figure 1. Turntable imaging with the target's center of mass at the origin.

Using the origin of the target's frame, we define the radar's Line-Of-Sight (LOS) as the following time-varying unit vector $\hat{\mathbf{u}}(\mathbf{t})$:

$$\hat{\mathbf{u}}(t) = \begin{bmatrix} \sin(\dot{\theta}t)\sin\gamma \\ \cos(\dot{\theta}t)\sin\gamma \\ \cos\gamma \end{bmatrix} \tag{1}$$

where $\hat{\mathbf{u}}(\mathbf{t})$ points from the target center of mass to the radar, and

$$\dot{\theta} = \frac{2\pi}{T_p}$$

denotes the rotation rate of the object tumbling with period T_p, and γ is the tumble angle between the radar LOS and the target's rotation axis $\vec{\omega}$, as shown in Figure 1b. The position of the radar relative to the target's center of mass is defined as $\mathbf{r_{TX}}(\mathbf{t}) = r_{TX}(t)\hat{\mathbf{u}}(\mathbf{t})$. The target can be represented using a complex-valued radar scattering reflectivity density, $\sigma(\mathbf{x})$, defined over a spatial volume $\Omega \subset \mathbb{R}^3$, where $\mathbf{x} = [x, y, z]^T \in \Omega$ is the position of a scattering element (or scatterer) on the target. Note that both $\hat{\mathbf{u}}(t)$ and $\mathbf{x}$ are expressed in a body-frame coordinate system that is fixed to the target. The target and the coordinate system rotate in a counter-clockwise fashion, which is equivalent to the radar LOS rotating clockwise in the body frame and tracing out a cone.

The absolute monostatic range $\tilde{R}(t;\mathbf{x})$ associated with a scatterer at $\mathbf{x} \in \Omega$ can be computed and approximated as follows:

$$\begin{aligned}\tilde{R}(t;\mathbf{x}) &= \|\mathbf{x} - r_{TX}(t)\hat{\mathbf{u}}(\mathbf{t})\| \\ &= r_{TX}(t)\sqrt{1 + \frac{\|\mathbf{x}\|^2}{r_{TX}^2(t)} - \frac{2\mathbf{x}^T\hat{\mathbf{u}}(\mathbf{t})}{r_{TX}(t)}} \\ &\approx r_{TX}(t)\left[1 - \frac{\mathbf{x}^T\hat{\mathbf{r}}_{\mathbf{TX}}(\mathbf{t})}{r_{TX}(t)}\right] \\ &= r_{TX}(t) - \underbrace{\mathbf{x}^T\hat{\mathbf{u}}(\mathbf{t})}_{R(t;\mathbf{x})},\end{aligned} \tag{2}$$

$$R(t;\mathbf{x}) = \mathbf{x}^T\hat{\mathbf{u}}(\mathbf{t}), \tag{3}$$

where $R(t;\mathbf{x})$ denotes the relative range, or range to the scattering element at $\mathbf{x}$ relative to the center of mass of the object.

The expression for the scatterer relative-range trajectory in (3) makes several simplifying assumptions that are often applicable to GEO and near-GEO objects. First, the radar and satellite center of mass are fixed relative to each other. This assumption is appropriate for an Earth-based radar observing a satellite in Geosynchronous orbit (i.e., orbit with zero-inclination at GEO altitude). Second, the satellite is in a pure tumble, rotating with a fixed angular velocity. The two assumptions hold reasonably well for slowly drifting objects subject to natural forces over the timescales of interest for imaging. Later in the paper, we will discuss the impact of deviations from the assumptions.

Using (1) and (3), we have

$$R(t;\mathbf{x}) = x\sin\gamma\sin(\dot{\theta}t) + y\sin\gamma\cos(\dot{\theta}t) + z\cos\gamma. \tag{4}$$

To gather more intuition about the form of (4), we express x and y in polar coordinates: $x = \rho\cos\varphi$ and $y = \rho\sin\varphi$. This results in

$$R(t;\mathbf{x}) = \rho\sin\gamma\sin(\dot{\theta}t + \varphi) + z\cos\gamma.$$

Thus, the trajectory takes the form of a sinusoid with amplitude $\rho\sin\gamma$, phase angle φ, and offset $z\cos\gamma$. Note that the sinusoid attains maximum amplitude when $\gamma = \pi/2$ (radar LOS orthogonal to the rotation axis), and has amplitude zero for $\gamma = 0$ (LOS parallel to the rotation axis).

2.2. Radar-Return Signal Model

The radar-return signal model for an object in a pure tumble at GEO is [2]

$$G(f,t) = \int_\Omega \sigma(\mathbf{x})\exp\left(-j\frac{4\pi}{c}fR(t;\mathbf{x})\right)d\mathbf{x}, \tag{5}$$

where

$$f_c - \frac{B}{2} \le f \le f_c + \frac{B}{2}$$

is frequency, $t \in [0, T_{max}]$ is slow-time, B is the bandwidth transmitted by the radar at center frequency f_c, and $\sigma(\mathbf{x})$ is the radar reflectivity density at position $\mathbf{x} = [x,y,z]^T \in \Omega$. We refer to the radar signal in (5), which varies with frequency and time, as the phase-history data.

2.3. Compact-Range Satellite Model Case Study

To validate our methods in this paper, we tested using real compact-range radar data obtained from a scale model of a satellite. The data were collected at the Submillimeter-Wave Technology Laboratory at University of Lowell in their 160 GHz center frequency compact range [32], utilizing the turntable geometry shown in Figure 2a [33]. The model scale factor and center frequency were selected such that the ratio of the center frequency to the scale model is comparable to the full-size object. After collection, the data were scaled to X-band (10 GHz center frequency), resulting in a bandwidth of $B = 2$ GHz. The satellite scale model, displayed in Figure 2b, is structurally characterized by its hexagonal bus and segmented solar panels, with etched cells and machined hinges along the frame.

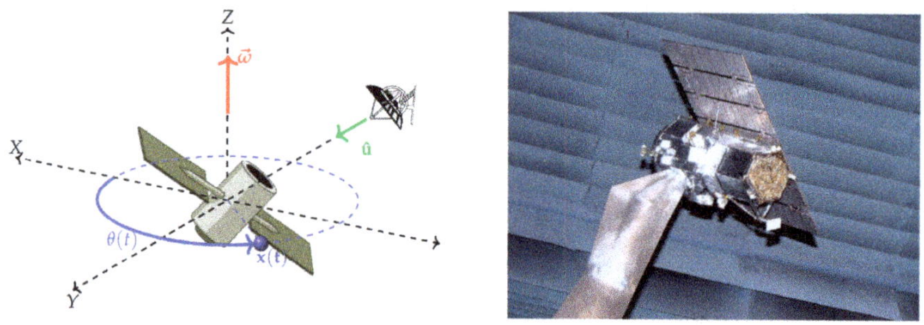

(**a**) Turntable collection geometry (**b**) Satellite model

Figure 2. Uniformly rotating compact-range satellite model [33].

2.4. Standard Radar-Characterization Processing

In this section, we review standard radar-characterization processing techniques. This overview provides necessary background before motivating the generalized 3D RDT technique.

2.4.1. Range–Time Intensity (RTI)

The Range–Time Intensity (RTI) data captures that evolution of the wideband target signature along the radar LOS as a function of time. The RTI data $\hat{g}(r,t)$ are generated by applying an inverse Fourier transform in frequency to the phase-history data in (5):

$$\hat{g}(r,t) = \int_{-B/2}^{B/2} G(f,t) \exp\left(j\frac{4\pi}{c} f_B r\right) df_B, \tag{6}$$

where f_B represents baseband frequency such that $f = f_c + f_B$. Applying (5) to (6) yields

$$\hat{g}(r,t) = B \int_{\Omega} \sigma(\mathbf{x}) \exp\left(-j\frac{4\pi}{c} f_c R(t;\mathbf{x})\right) \operatorname{sinc}\left(\frac{r - R(t;\mathbf{x})}{c/2B}\right) d\mathbf{x}. \tag{7}$$

Expression (7) shows that each scattering element contributes a sinc point response in range weighted by the reflectivity and a phase term that is proportional the relative range. The scatterers follow sinusoidal trajectories as a function of time.

Note that the RTI data in (7) are complex-valued. Because of this, the log power of (7), $20\log_{10}(|\hat{g}(r,t)|)$, is typically displayed for visual interpretation. Figure 3 shows the RTI data for the compact-range satellite model introduced in Section 2.3 over a single full rotation. The bandwidth is large enough such that individual point scatterers are well-resolved in range and trace out sinusoidal trajectories over slow-time. Bright specular flashes occur at aspect angle views $\theta = 90°$ and $270°$.

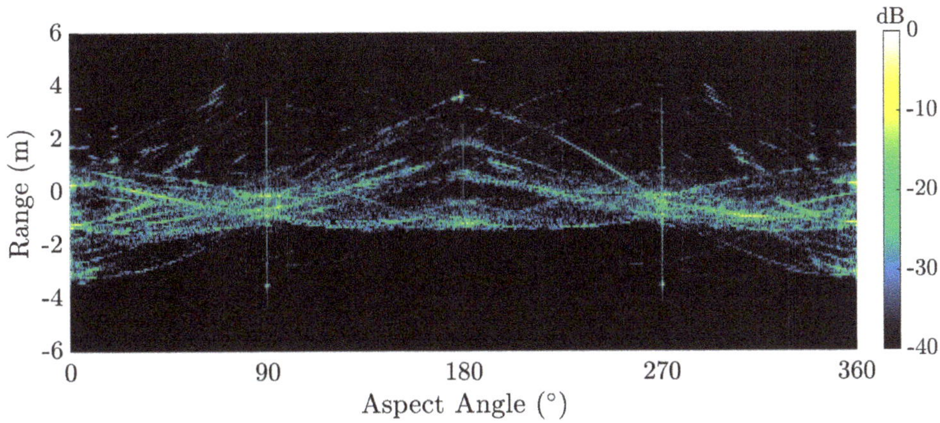

Figure 3. RTI for compact-range satellite model observed over a full rotation.

2.4.2. Range–Doppler Map (RDM)

Range–Doppler Maps (RDMs) provide two-dimensional snapshots of rotating objects in range and range rate. Here, Doppler refers to range rate derived from the signal phase. The RDM $g(r,\dot{r})$ is generated by collecting the phase-history data over a window of duration T and applying an inverse 2D Fourier transform:

$$g(r,\dot{r}) = \int_{-T/2}^{T/2}\int_{-B/2}^{B/2} G(f,t)\exp\left(j\frac{4\pi}{c}f_B r\right)\exp\left(j\frac{4\pi}{c}f_c\bar{t}\dot{r}\right)df_B d\bar{t}, \tag{8}$$

where $\bar{t} = t - t_i$ is time relative to the window midpoint t_i. Let

$$\Phi(f,t) = -\frac{4\pi}{c}fR(t;\mathbf{x})$$

denote the argument of the phase in (5). Taking a first-order Taylor expansion about $f = f_c$ and $t = t_i$ generates an approximation where the phase is linear in t and f, and is a separable function of the two variables:

$$\Phi(f,t) \approx -\frac{4\pi}{c}fR(t_i;\mathbf{x}) - \frac{4\pi}{c}f_c\dot{R}(t_i;\mathbf{x})(t-t_i), \tag{9}$$

where

$$\dot{R}(t;\mathbf{x}) = x\dot{\theta}\sin\gamma\cos(\dot{\theta}t) - y\dot{\theta}\sin\gamma\sin(\dot{\theta}t) \tag{10}$$

is the range rate of a scatterer at $\mathbf{x}$. Substitution of (5) and (9) into (8) produces

$$g(r,\dot{r}) = BT\int_{\Omega}\sigma(\mathbf{x})\exp\left(-j\frac{4\pi}{c}f_cR(t;\mathbf{x})\right)\operatorname{sinc}\left(\frac{r-R(t_i;\mathbf{x})}{c/2B}\right)\operatorname{sinc}\left(\frac{\dot{r}-\dot{R}(t_i;\mathbf{x})}{\lambda/2T}\right)d\mathbf{x}. \tag{11}$$

As indicated by (11), RDM processing generates an "image" where each scatterer on the object is rendered as 2D sinc function offset by its range and range rate at time t_i. Note that to invoke the approximation in (9), T must be sufficiently small so that higher-order terms in the expansion of t are negligible.

Figure 4a shows a RDM generated from the compact-range dataset using 5° aspect change and 2 GHz bandwidth (scaled to X-band). The RDM reveals the hexagonal bus on the satellite model, with solar panels evident to the left and right of the bus.

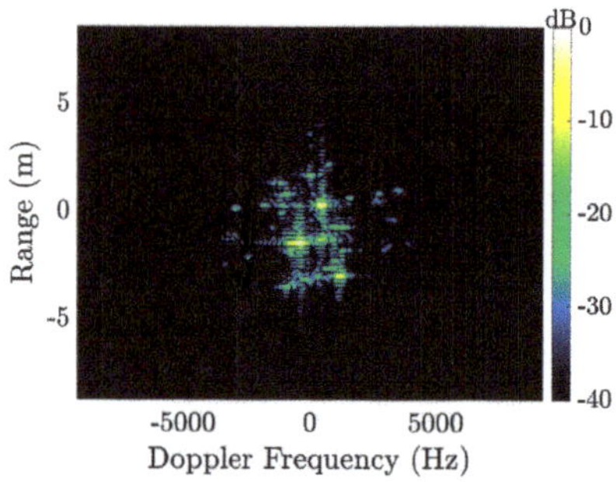

(**a**) Range–Doppler Map

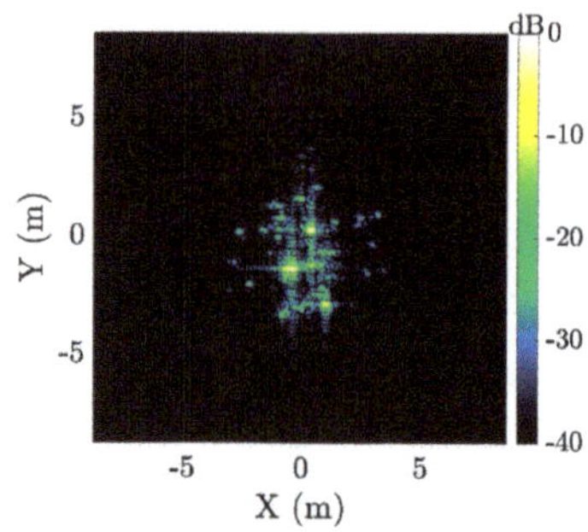

(**b**) ISAR image

Figure 4. (**a**) Range–Doppler Map and (**b**) ISAR image reconstructions of the compact-range satellite model.

2.4.3. Inverse Synthetic Aperture Radar (ISAR) Imaging

The RDM in (11) provides an image of a satellite in the sense that individual scatterers on the object are resolved. For proper image interpretation, we desire a product where scatterers are mapped to their physical positions, and defocusing artifacts due to nonlinearity in the phase are remedied (note that in the approximation in (9), such nonlinearity is assumed to be negligible). Inverse Synthetic Aperture Radar (ISAR) is an approach used to generate a correctly rendered image of the satellite. Like the RDM, an ISAR image is generated by processing a fixed collection window (or integration interval) T centered at a particular time t_i. To derive the ISAR imaging equation, we first express (4) in terms of $\bar{t} = t - t_i$:

$$R(\bar{t};\mathbf{x}) = x\sin\gamma\sin(\dot{\theta}\bar{t} + \theta_i) + y\sin\gamma\cos(\dot{\theta}\bar{t} + \theta_i) + z\cos\gamma,$$

where $\theta_i = \dot{\theta}t_i$ is the angle at the center of the interval. By making use of trigonometric identities and rearranging terms, this expression can be represented as

$$R(\bar{t};\mathbf{x}) = (x\cos\theta_i - y\sin\theta_i)\sin\gamma\sin(\dot{\theta}\bar{t}) + (x\sin\theta_i + y\cos\theta_i)\sin\gamma\cos(\dot{\theta}\bar{t}) + z\cos\gamma. \quad (12)$$

To proceed further, we next consider the special case where $\gamma = \pi/2$, which is the case where the radar LOS is orthogonal to the rotation axis. This removes the dependency upon the height z in (12). Defining the rotated coordinates

$$\begin{aligned} x' &= x\cos\theta_i - y\sin\theta_i, \\ y' &= x\sin\theta_i + y\cos\theta_i, \\ z' &= z, \end{aligned}$$

we have

$$R(\bar{t};\mathbf{x}) = x'\sin(\dot{\theta}\bar{t}) + y'\cos(\dot{\theta}\bar{t}).$$

Insertion into (5) leads to the ISAR imaging equation

$$\begin{aligned} G(f,\bar{t}) &= \int_{\Omega_x}\int_{\Omega_y}\left[\int_{\Omega_z}\sigma(x',y',z')dz'\right]\exp\left(-j2\pi\left[\frac{2f}{c}x'\sin\theta(\bar{t}) + \frac{2f}{c}y'\cos\theta(\bar{t})\right]\right)dx'dy' \\ &= \int_{\Omega_x}\int_{\Omega_y}\bar{\sigma}(x',y')\exp\left(-j2\pi\left[\frac{2f}{c}x'\sin\theta(\bar{t}) + \frac{2f}{c}y'\cos\theta(\bar{t})\right]\right)dx'dy', \end{aligned} \quad (13)$$

where

$$\bar{\sigma}(x,y) = \int_{\Omega_z}\sigma(x,y,z)dz \quad (14)$$

is the projection of the reflectivity function along the z axis.

The phase history data $G(f, \bar{t})$ in (13) has a nice spatial Fourier sampling interpretation. Taking the three-dimensional Fourier transform of the scattering reflectivity function $\sigma(x, y, z)$, we obtain

$$S_\sigma(k_x, k_y, k_z) = \iiint_{\mathbb{R}^3} \sigma(x, y, z) \exp\left(-j2\pi\left[k_x x + k_y y + k_z z\right]\right) dx dy dz. \tag{15}$$

Note that by the Projection Slice Theorem, $S_{\bar{\sigma}}(k_x, k_y) = S_\sigma(k_x, k_y, 0)$, where $S_{\bar{\sigma}}$ is the 2D Fourier transform of $\bar{\sigma}$. Using this fact, and (13), we can relate $(f, \bar{t})$ samples of the phase-history signal to S_σ:

$$G\left(f, \frac{\theta}{\dot{\theta}}\right) = S_\sigma(k_x(f, \theta), k_y(f, \theta), 0), \tag{16}$$

where

$$k_x(f, \theta) = \frac{2f}{c} \sin\theta(\bar{t}),$$
$$k_y(f, \theta) = \frac{2f}{c} \cos\theta(\bar{t}),$$

are spatial frequencies in the x, y, z dimensions, and $\theta(\bar{t}) = \dot{\theta}\bar{t}$. Thus, (16) reveals phase-history samples lie on an annular region on the $k_z = 0$ plane in the Fourier domain. An image can be reconstructed using a polar integral over the following annular region A:

$$\begin{aligned} \mathrm{g}(x, y) &= \iint_A S_\sigma(k_x, k_y, 0) \exp\left(j2\pi\left[k_x x + k_y y\right]\right) dk_x dk_y \\ &= \int_{-\frac{\Delta\theta}{2}}^{\frac{\Delta\theta}{2}} \int_{2(f_c - B/2)/c}^{2(f_c + B/2)/c} S_\sigma(k_r \sin\theta, k_r \cos\theta, 0) \exp\left(j2\pi\left[k_r \sin\theta x + k_r \cos\theta y\right]\right) k_r dk_r d\theta, \end{aligned} \tag{17}$$

where $\Delta\theta = \dot{\theta}T$ is the integration angle. In Appendix A (Appendix A.1), we show that the image has the form

$$\begin{aligned} \mathrm{g}(x, y) \approx {} & \left[\lambda/2\Delta\theta\right]^{-1} \left[c/2B\right]^{-1} \int_{\Omega_x} \int_{\Omega_y} \bar{\sigma}(x', y') \exp\left(\frac{-j4\pi f_c y'}{c}\right) \\ & \times \mathrm{sinc}\left(\frac{x - x'}{\lambda/2\Delta\theta}\right) \mathrm{sinc}\left(\frac{y - y'}{c/2B}\right) dx' dy'. \end{aligned} \tag{18}$$

Note $\mathrm{g}(x, y)$ computed in (18) has the same units as $\bar{\sigma}(x, y)$. Thus, the ISAR image computed in (18) is a two-dimensional projection of the three-dimensional reflectivity function onto the $z' = 0$ plane, where the object is rotated about the rotation axis by θ_i.

For the general case where $\gamma < \pi/2$, the imaging relationship cannot be described exactly by a 2D Fourier transform because of the $z\cos\gamma$ term in (12). When forming an image using (17), the non-zero z term causes a depth-of-focus effect where scatterers with $z \neq 0$ appear defocused. We will analyze this effect in more detail later in the paper when discussing the limitations of the RDT technique. A slant-plane image can be formed when $\gamma < \pi/2$, where the image is a projection of the 3D reflectivity onto the tangent plane of the cone traced out by the radar line of sight. We derive expressions for the slant-plane image in Appendix A.2.

Figure 4 compares and contrasts RDM and ISAR image reconstructions for the same compact-range dataset. Both results shown in Figure 4a,b are generated using 5° worth of aspect change and 2 GHz of bandwidth.

2.4.4. Doppler–Time Intensity (DTI)

Doppler–Time Intensity (DTI) plots provide snapshots of the Doppler spectrum of the object over time. The technique is ideal for cases where the object is unresolved in range but exhibits rotation relative to the radar. In these cases, the trajectories of individual scatterers are traced out over time, in a manner similar to that observed in RTI plots.

Before defining the expression for the DTI data, we first state the condition for an unresolved object in range:

$$|\hat{\mathbf{u}}(t) \cdot \mathbf{x}| < \frac{c}{2B} \text{ for all } \mathbf{x} \in \Omega. \tag{19}$$

This states that all of the scatterers are contained within a single range resolution cell. Such is often the case for narrowband radar (e.g., given 1 MHz bandwidth, the range resolution is 150 m, which is much larger than most satellites). Without loss of generality, we assume that the object return is contained within the zeroth range bin, corresponding to the point tracked by the radar. Given (19), we can express the zeroth bin of the RTI in (7) as a time series:

$$\hat{g}(0,t) \approx B \int_{\Omega} \sigma(\mathbf{x}) \exp\left(-j\frac{4\pi}{c} f_c R(t;\mathbf{x})\right) d\mathbf{x}. \tag{20}$$

Here, we have made use of the approximation

$$\operatorname{sinc}\left(\frac{r - R(t;\mathbf{x})}{c/2B}\right) \approx 1$$

which is valid when $|\hat{\mathbf{u}}(t) \cdot \mathbf{x}| \ll c/2B$.

The DTI data are computed from the time series (20) using

$$\begin{aligned} X_D(\dot{r}, \tau_k) &= \int_{\tau_k - T/2}^{\tau_k + T/2} \hat{g}(0,t) \operatorname{rect}\left(\frac{t - \tau_k}{T}\right) \exp\left(j\frac{4\pi}{c} f_c (t - \tau_k) \dot{r}\right) dt \\ &= \int_{-T/2}^{T/2} \hat{g}(0, \bar{t}_k + \tau_k) \operatorname{rect}\left(\frac{\bar{t}_k}{T}\right) \exp\left(j\frac{4\pi}{c} f_c \bar{t}_k \dot{r}\right) d\bar{t}_k, \end{aligned} \tag{21}$$

where $\bar{t}_k = t - \tau_k$. This expression is equivalent to the Short-Time Fourier Transform (STFT), or spectrogram, of $\hat{g}(0,t)$. Insertion of (20) into (21) produces

$$X_D(\dot{r}, \tau_k) = BT \int_{\Omega} \sigma(\mathbf{x}) \exp\left(-j\frac{4\pi}{c} f_c R(\tau_k;\mathbf{x})\right) \operatorname{sinc}\left(\frac{\dot{r} - \dot{R}(\tau_k;\mathbf{x})}{\lambda/2T}\right) d\mathbf{x}, \tag{22}$$

where from (10)

$$\begin{aligned} \dot{R}(\tau_k;\mathbf{x}) &= x\dot{\theta} \sin\gamma \cos(\dot{\theta}\tau_k) - y\dot{\theta} \sin\gamma \sin(\dot{\theta}\tau_k) \\ &= \rho\dot{\theta} \sin\gamma \cos(\dot{\theta}\tau_k + \varphi), \end{aligned} \tag{23}$$

and $x = \rho\cos\varphi$ and $y = \rho\sin\varphi$. As with the range trajectory (4), the Doppler trajectory takes the form of a sinusoid, but without the height-dependent offset term. Inspection of (22) and the first equality in (23) reveals that the Doppler spectra provide projections of the target reflectivity along the cross-range axis

$$\hat{\mathbf{q}}(t) = \begin{bmatrix} \cos(\dot{\theta}t)\sin\gamma \\ -\sin(\dot{\theta}t)\sin\gamma \\ 0 \end{bmatrix} \tag{24}$$

i.e., $\dot{R}(\tau_k;\mathbf{x}) = \dot{\theta}\hat{\mathbf{q}}(t) \cdot \mathbf{x}$. Using this fact, we will later show that the Doppler spectra can be processed using filtered backprojection techniques to form a resolved tomographic image, an approach termed Doppler tomography.

Figure 5 shows a DTI formed for a narrowband setting ($B \approx 0$ Hz) using data from the compact-range satellite model collect over a single full rotation. The bandwidth is effectively zero such that individual point scatterers are all contained within a single range bin. The DTI shows sinusoidal trajectories of different point scatterers. As observed in

Figure 3, we observe bright specular flashes occur for Doppler profiles computed at aspect angle views $\theta = 90^\circ$ and 270°.

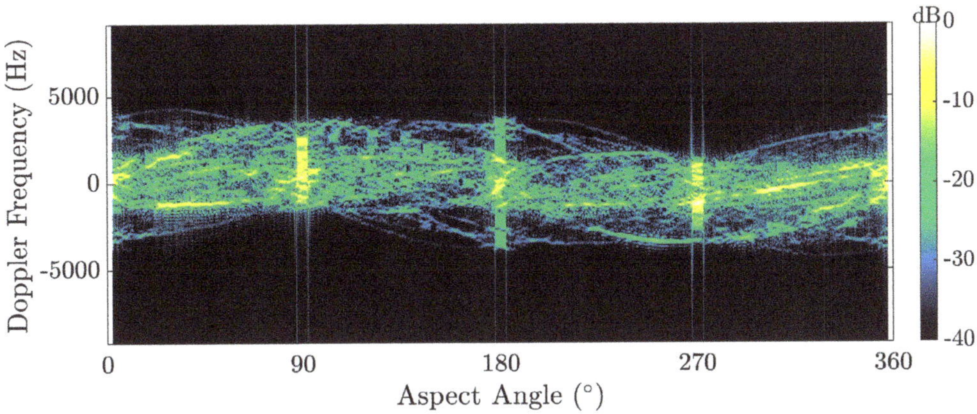

Figure 5. DTI for the compact-range satellite model observed over a full rotation.

3. 3D Range-Doppler-time (RDT) Tensor Technique

In this section, we present the novel 3D RDT tensor processing technique, which provides a generalized approach to Doppler characterization, and detail the enhanced range and Doppler products that can be generated from the RDT tensor.

3.1. Mathematical Description

At the end of Section 2, we explored the case where the radar-signal return from a rotating object was unresolved in range, i.e., the signal was contained within a single range resolution cell. For tracking radars with wider bandwidth (greater than 10 MHz), the target may be partially resolved in range, causing it to occupy multiple range bins. Here, the DTI will not capture the complete trajectory of a scatterer when it migrates out of a range bin. To better characterize the signal in such cases, we can extend the DTI data in (21) to include a third dimension that captures the range evolution of the return:

$$X(r, \dot{r}, \tau_k) = \int_{-T/2}^{T/2} \hat{g}(r, \bar{t}_k + \tau_k)\, \mathrm{rect}\left(\frac{\bar{t}_k}{T}\right) \exp\left(j\frac{4\pi}{c} f_c \bar{t}_k \dot{r}\right) d\bar{t}_k, \tag{25}$$

where $\bar{t}_k = t - \tau_k$. We denote (25) as the range-Doppler-time (RDT) representation of the radar signal. It is convenient to express (25) as a discrete time signal to account for the fact that the radar data are sampled and processed digitally, and also to account for the inherent discrete nature of the data in slow-time arising from using a pulsed radar. This gives rise to the 3D RDT tensor

$$X[n, m, k] = \frac{1}{M} \sum_{l=-M/2}^{M/2-1} \hat{g}[n, l + \bar{l}_k] \exp\left(j\frac{2\pi}{M} lm\right), \tag{26}$$

where $\bar{l}_k = \mathrm{round}(f_p \tau_k)$ and f_p is the pulse-repetition frequency of the radar (alternatively, one could interpolate to a grid centered about τ_k in place of the 'round' operation to achieve a more accurate result). Here, n, m, and k are range, Doppler, and time indices, respectively. Figure 6 illustrates the 3D RDT. In the next section we provide motivation for a number of useful characterization products that can be derived by collapsing (reducing) the dimensionality of the RDT, and by applying scaling/rotation transforms to the RDT. These include Doppler superpulses and range superpulses, which are enhanced versions of the DTI and RTI, respectively, and the power-sum image reconstruction, which is a non-coherent combination of RDMs across one full rotation period.

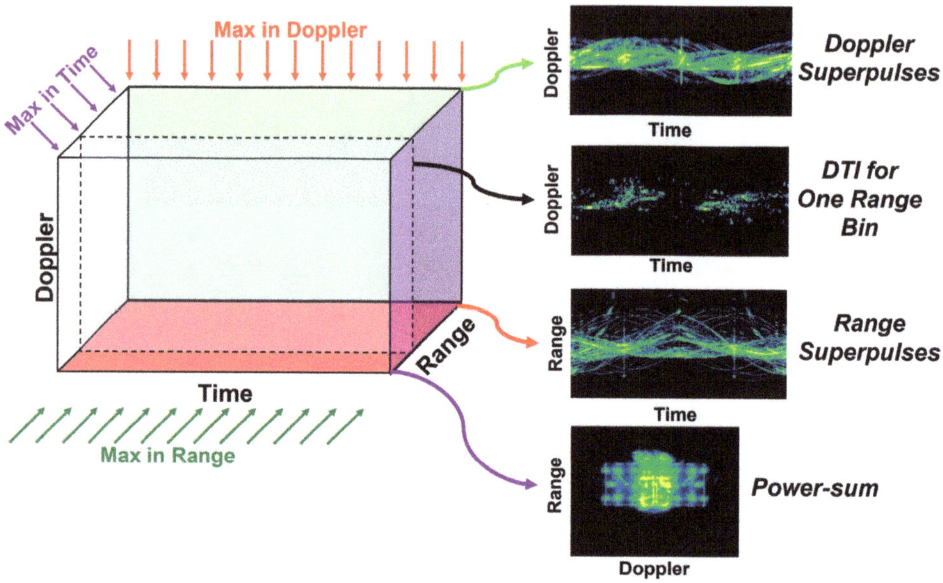

Figure 6. 3D RDT tensor.

Evaluation of (26) for the object model in (5) results in

$$X[n,m,k] = \int_{\Omega} \sigma(\mathbf{x}) \exp\left(-j\frac{4\pi}{c} f_c R(\tau_k;\mathbf{x})\right) \mathrm{sinc}_N\left(\frac{n\delta_R - R(\tau_k;\mathbf{x})}{\delta_R}\right) \mathrm{sinc}_M\left(\frac{m\delta_{\dot{R}} - \dot{R}(\tau_k;\mathbf{x})}{\delta_{\dot{R}}}\right) d\mathbf{x}, \quad (27)$$

where $\delta_R = c/2B$ and $\delta_{\dot{R}} = \lambda/2T$ are the range and Doppler resolution cell sizes, respectively, and sinc_N denotes a periodic sinc function of length N.

We will now examine how each scatterer in (27) evolves as a function of time. We define $\mathbf{h}(\tau_k;\mathbf{x})$ as a 2D vector describing the position of the scatterer in range–Doppler space at each instant of time:

$$\begin{aligned} \mathbf{h}(\tau_k;\mathbf{x}) &= \begin{bmatrix} \dot{R}(\tau_k;\mathbf{x}) \\ R(\tau_k;\mathbf{x}) \end{bmatrix} \\ &= \rho \sin\gamma \begin{bmatrix} \dot{\theta}\cos(\dot{\theta}\tau_k + \varphi) \\ \sin(\dot{\theta}\tau_k + \varphi) \end{bmatrix} + \begin{bmatrix} 0 \\ z\cos\gamma \end{bmatrix} \end{aligned} \quad (28)$$

The expression shows that each scatterer traces out a helix in a 3D range-Doppler-time space. The helix is compressed in Doppler by $\dot{\theta}$ and offset in range by the constant $z\cos\gamma$.

Figure 7 shows an example RTI and RDT generated using simulated radar data for a target scene consisting of two isotropic point scatterers located at $\mathbf{x}_0 = [4.33\text{ m}, 2.5\text{ m}, 3.0\text{ m}]^T$ and $\mathbf{x}_1 = [-2.12\text{ m}, -2.12\text{ m}, -2.0\text{ m}]^T$, where $\gamma = 90^\circ$. Here, the data were simulated at X-band ($f_c = 10$ GHz) with bandwidth $B = 1$ GHz. Both scatterers trace out separate helical trajectories, each described by (28). For this simulation, the rotation period of the target scene is 50 s. In Section 4.6, we highlight an example where $\gamma \neq 90^\circ$, which causes the defocusing of radar-image reconstructions.

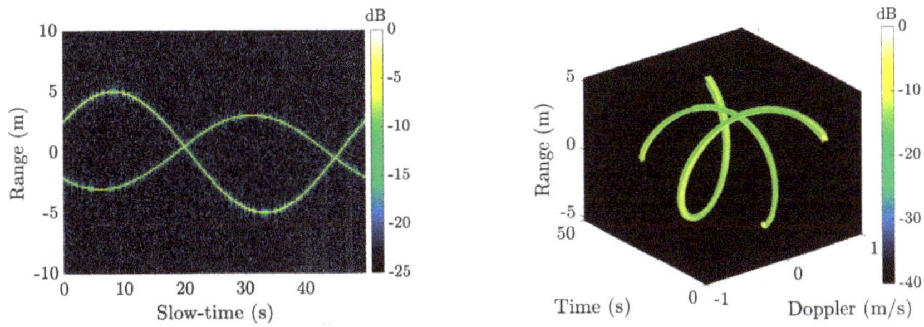

(**a**) RTI (**b**) RDT

Figure 7. Simulated (**a**) RTI and (**b**) RDT for a target scene.

Later in this section, we will define a compressed representation of the 3D scatterer trajectory in (28) based on the pure-tumble motion assumption, and use this representation to derive a signal transform that reduces the rank of the 3D tensor in (27). The reduced-rank tensor provides a path for scatterer estimation, data denoising, and rotation-rate estimation. It also provides a framework describing the relationship between different imaging techniques.

3.2. Range and Doppler Superpulses

In radar-signal analysis, preprocessing is often applied to the RTI data in (6) to accentuate dim scatterers. One such approach is coherent presumming, which applies a boxcar filter in slow-time to coherently sum adjacent pulses. This coherent integration results in a theoretical factor of N improvement to the SNR over a single pulse for an ideal-point scatterer (constant magnitude and phase). Similarly, the generation of the RDM in (8) provides the theoretical scatterer signal-to-noise improvement. This improvement is achieved by use of the (discrete) Fourier transform, which carries out coherent integration under a variety of range-rate hypotheses (i.e., different $\dot{r}$ matched filters in (8)). A limitation of coherent presumming is that it only fully enhances scatterers with zero range rate; scatterers with non-zero range rate experience signal attenuation due to phase variation induced by the scatterer's motion (destructive interference). This can be understood by noting that presumming returns the zero-Doppler bin of the RDM. To maximize SNR, the correct range-rate hypothesis (matched filter) must be applied when combining the pulses.

To provide a more optimal enhanced signal display, we have developed the range-superpulse concept. Here, instead of summing across a sequence of time-shifted boxcar windows, we instead generate a RDM at each window lag and return the Doppler bin with peak intensity for each range bin, thus generating a 1D signal (peak Doppler versus range). This is equivalent to collapsing the 3D RDT in Doppler using the 'max' operation

$$X_{RS}[n,k] = \max_{m} |X[n,m,k]|, \tag{29}$$

where $X_{RS}[n,k]$ denotes the range superpulses.

In a similar fashion, we can apply a max operation to the 3D tensor in range to generate Doppler superpulses:

$$X_{DS}[m,k] = \max_{n} |X[n,m,k]|. \tag{30}$$

Doppler superpulses capture the energy of scatterers that migrate through multiple range resolution cells, allowing DTI techniques such as Doppler tomography to be applied to medium-band and wideband data. In [16], Doppler superpulses are found to provide superior visualizations of the imaged object in cases where the bandwidth is sufficiently large to resolve the object in range.

Figure 8 shows range superpulses generated using (29) for the compact-range dataset. Doppler superpulses for the same dataset are computed using (30) and visualized in

Figure 9. Comparing and contrasting Figures 3 and 8, we immediately see observed improvements in range characterization. Similarly, we observe finer Doppler characterization using superpulse processing after comparing and contrasting Figures 5 and 9. Superpulse processing best serves to enhance the SNR of individual point scatterers that migrate nonlinearly across range and Doppler. These RDT-derived products enhance the sinusoidal trajectories of tumbling targets, which enables higher fidelity rotation-period estimation.

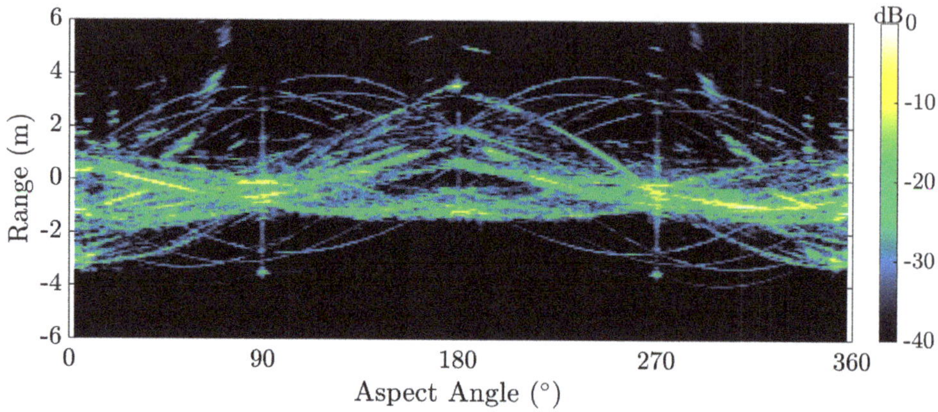

Figure 8. Range superpulses, $X_{RS}[n,k]$, for the compact-range satellite model observed over a full rotation.

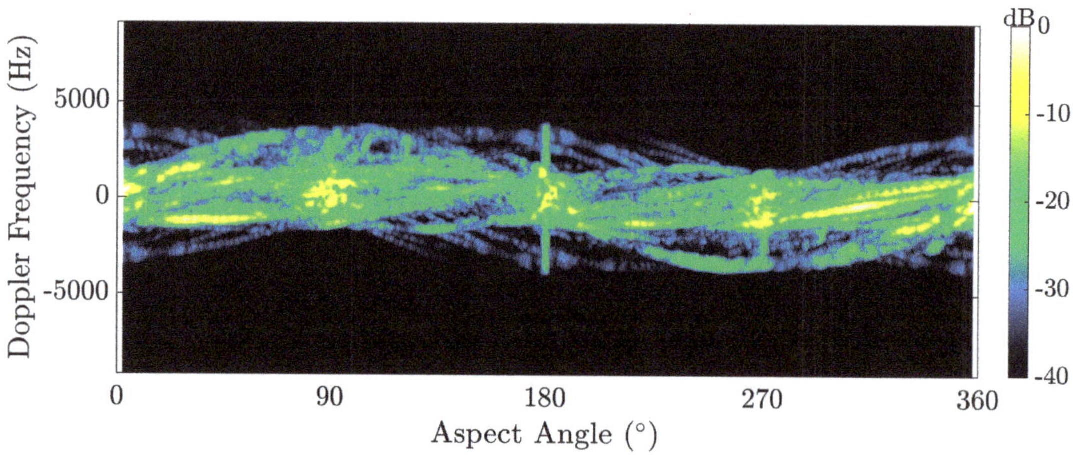

Figure 9. Doppler superpulses, $X_{DS}[m,k]$, for the compact-range satellite model observed over a full rotation.

3.3. Compressed Trajectory Representation and Rank-Reducing Transform

As established by (28), each scatterer traces out a helix in a 3D tensor space. When the rotation rate $\dot{\theta}$ is known, it is possible to "unwind" the helices and map each scatterer to a spatial position. We define the transformation matrix

$$\mathbf{T}_{\dot{\theta},\gamma}(\tau_k) \triangleq \mathbf{R}(\tau_k;\dot{\theta})\mathbf{diag}([(\dot{\theta}\sin\gamma)^{-1},(\sin\gamma)^{-1}]) \tag{31}$$

where $\mathbf{diag}(\mathbf{v})$ is a diagonal matrix with $\mathbf{v}$ along the diagonal, and

$$\mathbf{R}(\tau_k;\dot{\theta}) = \begin{bmatrix} \cos(\dot{\theta}\tau_k) & \sin(\dot{\theta}\tau_k) \\ -\sin(\dot{\theta}\tau_k) & \cos(\dot{\theta}\tau_k) \end{bmatrix}. \tag{32}$$

Starting with (28), we define the transformed position

$$\mathbf{p}(\tau_k;\mathbf{x},\dot{\theta},\gamma) = \mathbf{T}_{\dot{\theta},\gamma}(\tau_k)\mathbf{h}(\tau_k;\mathbf{x}). \tag{33}$$

This results in

$$\mathbf{p}(\tau_k;\mathbf{x},\dot{\theta},\gamma) = \rho\begin{bmatrix} \cos\varphi \\ \sin\varphi \end{bmatrix} + \frac{z}{\tan\gamma}\begin{bmatrix} \sin(\dot{\theta}\tau_k) \\ \cos(\dot{\theta}\tau_k) \end{bmatrix} \tag{34}$$

The first term in (34) gives the x and y positions of the object in the body frame, while the second term is a depth-of-focus artifact that causes the scatterer to trace out a "donut" centered at x, y with radius $z/\tan\gamma$ over the rotation period. Note that when $\gamma = \pi/2$ (rotation axis orthogonal to the radar line of sight), the second term disappears, and the trajectory collapses to a single point.

We now define a signal transform that will realize the scaling and rotation operation producing the positional mapping in (34). This will be achieved by applying a geometric transformation to the continuous-time tensor data in (25). Note from (33) that

$$\mathbf{h}(\tau_k;\mathbf{x}) = \mathbf{T}^{-1}_{\dot{\theta},\gamma}(\tau_k)\mathbf{p}(\tau_k;\mathbf{x},\dot{\theta},\gamma),$$

where, by (31) and (32),

$$\mathbf{T}^{-1}_{\dot{\theta},\gamma}(\tau_k) = \mathbf{diag}([\dot{\theta}\sin\gamma, \sin\gamma])\mathbf{R}^T(\tau_k;\dot{\theta}). \tag{35}$$

The continuous-time representation of the transformed RDT-tensor signal data is

$$\tilde{X}(p_y, p_x, \tau_k) = X\Big(\mathbf{e}_2^T\mathbf{T}^{-1}_{\dot{\theta},\gamma}(\tau_k)\mathbf{p},\ \mathbf{e}_1^T\mathbf{T}^{-1}_{\dot{\theta},\gamma}(\tau_k)\mathbf{p},\ \tau_k\Big). \tag{36}$$

where $\mathbf{p} = [p_x, p_y]^T$ and $\mathbf{e}_1$ and $\mathbf{e}_2$ are the standard basis vectors of length 2. The discrete-time counterpart of (36) is

$$\tilde{X}[n', m', k] = \tilde{X}(n'\delta y, m'\delta_x, \tau_k), \tag{37}$$

where δ_x and δ_y are the desired x and y pixel sizes. We refer to the matrix in (31) as the rank-reducing transform, and to the transformed tensor in (37) as the aligned RDT. In (37), the data are scaled along the range and Doppler axes to yield true position, and then rotated to a common reference frame. As a result, the helical scatterer trajectories are converted to linear ones, thus reducing the rank of the 3D tensor. If $\gamma < \pi/2$, the trajectories will exhibit some helical motion when $z \neq 0$.

Figure 10a shows an example using simulated data where the rank-reducing transform (31) has been applied to the RDT shown in Figure 7b. Note the helical trajectories of both isotropic scatterers have been rotationally aligned along the slow-time axis. Given that the imaging geometry corresponds to $\gamma = 90°$, the aligned trajectories described by (34) are lines. To implement (36) starting with the (discrete-time) RDT in (26), the sampled data can be interpolated to generate the desired values of the continuous-time function (25). Bilinear interpolation was used for the results in this paper.

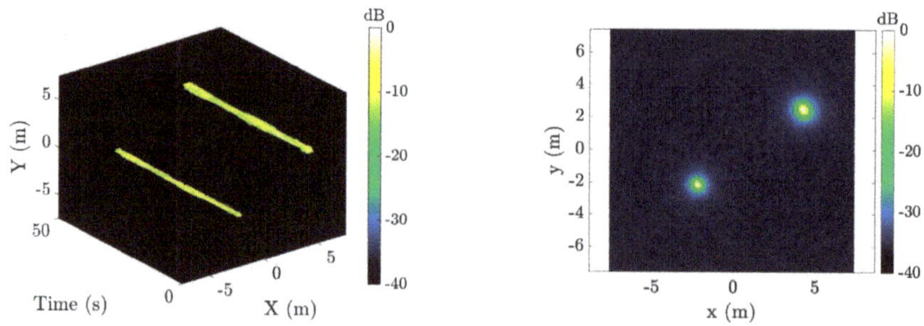

(**a**) Aligned RDT (**b**) Power-sum image

Figure 10. (**a**) Rotationally aligned RDT for target scene in Figure 7, and (**b**) corresponding power-sum image.

The representation in (37) will be useful in establishing the relationship between Doppler tomography and its higher resolution counterpart BENI. As will be discussed later in the paper, the low-rank representation of the signal lends itself well to signal extraction and denoising techniques that make use of the multi-dimensional Singular Value Decomposition (SVD). In addition, (37) provides a parameterized representation of the signal that can be combined with an image-quality metric to yield an optimization-based approach for estimating the rotation rate (in cases where it is not known a priori).

3.4. Power-Sum Image Reconstruction

In addition to superpulse processing, the RDT technique also enables radar-image reconstruction after rotational alignment using the rank-reducing transform. Consider taking the sum of the squared magnitude of the rank-reduced tensor in (37) across the time dimension. This yields

$$\tilde{X}_{PS}[n',m'] = \sum_{k=0}^{K-1} |\tilde{X}[n',m',k]|^2, \tag{38}$$

where $\tilde{X}_{PS}$ denotes the power-sum image. Using (35), (36), and (37), we have

$$\tilde{X}_{PS}[n',m'] = \sum_{k=0}^{K-1} |X(\sin\gamma(x_{m'}\sin(\dot{\theta}\tau_k) + y_{n'}\cos(\dot{\theta}\tau_k)),\ \dot{\theta}\sin\gamma(x_{m'}\cos(\dot{\theta}\tau_k) - y_{n'}\sin(\dot{\theta}\tau_k)),\ \tau_k)|^2, \tag{39}$$

where $x_{m'} = \delta_x m'$ and $y_{n'} = \delta_y n'$.

Figure 10b shows the power-sum image derived from the aligned RDT in Figure 10a. The image accurately reproduces the two scatterers in the x, y plane, which is orthogonal to the rotation axis. An example of a power-sum image generated from the compact-range data is presented in Figure 18a. As we will show in the sections that follow, the power-sum image is important to understanding the relationship between Doppler tomography and BENI, and to interpreting images from each of these techniques.

3.4.1. Relation to Doppler Tomography

Tomographic characterization of rotating objects using Doppler information, also known as Doppler tomography, is built on the principle that point scatterers along a line of constant cross-range u will reflect back radar-return signals that share the same Doppler-shifted frequency f_D in traveling towards the radar [22].

We previously established that each column (i.e., fixed τ_k) of the DTI data in (21), representing a single Doppler spectrum (or Doppler profile), provides a 1D projection of the object reflectivity along the cross-range axis in (24). Thus, the DTI data are analogous to the sinogram-domain projection data used in tomographic imaging. The Doppler tomography image is formed as

$$g_{DT}(x,y) = \int_{-T_p/2}^{T_p/2} \left| X_D^f\left(\dot{\theta}\sin\gamma(x\cos(\dot{\theta}\tau) - y\sin(\dot{\theta}\tau)), \tau\right) \right|^2 d\tau, \quad (40)$$

where

$$X_D^f(\dot{r},\tau) = \int_{-T/2}^{T/2} \hat{X}_D(t,\tau)|t| \exp\left(-j\frac{4\pi}{\lambda}t\dot{r}\right) dt$$

is the ramp-filtered data, as required for filtered backprojection, with $\hat{X}_D(t,\tau)$ denoting the inverse Fourier transform of $X_D(\dot{r},\tau)$ in Doppler [34]. Note that we apply the reconstruction to the power (squared magnitude) of the DTI data. This is due to the phase term in (22), which causes the signal to decohere when combining projections.

Doppler tomography can be viewed within the framework of the RDT technique as a special case of the power-sum reconstruction introduced in Section 3.4. Expression (39) shows how each range-Doppler slice (or RDM) of the RDT contributes to the power-sum image. In the regime where the object is unresolved in range, note that

$$\begin{aligned} &X\left(\sin\gamma(x_{m'}\sin(\dot{\theta}\tau_k) + y_{n'}\cos(\dot{\theta}\tau_k)),\ \dot{\theta}\sin\gamma(x_{m'}\cos(\dot{\theta}\tau_k) - y_{n'}\sin(\dot{\theta}\tau_k)),\ \tau_k\right) \\ &\approx X\left(0,\ \dot{\theta}\sin\gamma(x_{m'}\cos(\dot{\theta}\tau_k) - y_{n'}\sin(\dot{\theta}\tau_k)),\ \tau_k\right) \end{aligned}$$

i.e., the signal does not vary significantly in the range dimension because the object is unresolved for all values of $x_{m'}, y_{n'}$ over the object's support. As a result, each column of the rank-reduced tensor is essentially an unfiltered projection, and the power-sum image is equivalent to the Doppler tomography image in (40) without the ramp-filtering step.

Figure 11 presents an example illustrating Doppler tomography as a limiting case of the power-sum reconstruction when the bandwidth is sufficiently low that the target is unresolved in range. Figure 11a shows the unaligned RDT (i.e., with no rank-reducing transform applied) for the two-scatterer example in Figure 7b, but with the bandwidth reduced from 1 GHz to 10 MHz. Here, each scatterer contributes a 1D projection to the range-Doppler slices of the RDT (note that the discretization in range is due to the range-sample spacing). Figure 11a shows the aligned RDT. The individual RDMs are rotated such that their intersection across time corresponds to the 2D scatterer position. The power-sum reconstruction is displayed in Figure 11c, revealing that both scatterers are correctly rendered.

Note over a full rotation period, the Doppler profiles of a uniformly tumbling object show signs of anisotropic scattering and are not symmetric about a half rotation. Therefore, reconstructions are computed using over $\theta \in [0, 2\pi]$.

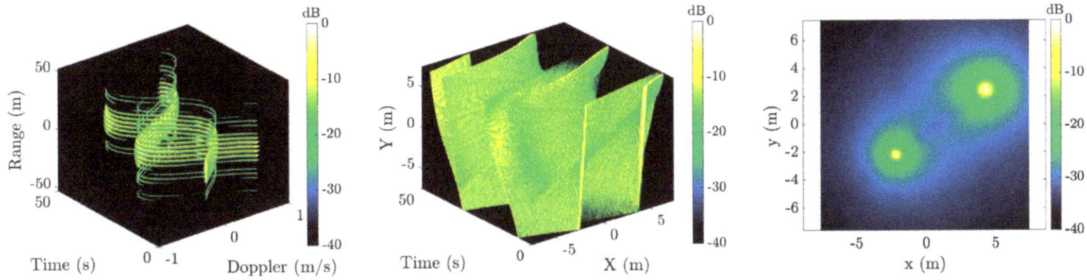

(**a**) Unaligned RDT (**b**) Aligned RDT (**c**) Power-sum reconstruction

Figure 11. Imaging example for the two-scatterer geometry in Figure 7b, but using a narrowband waveform with 10 MHz bandwidth.

3.4.2. Relation to Bandwidth Enhanced Non-Coherent Imaging (BENI)

BENI was designed to improve upon Doppler tomography for cases where some bandwidth is available, but the bandwidth is too small for high-resolution radar imaging. BENI has also been shown to provide exquisite detail with wideband data [28]. The first step in the algorithm is to partition radar data collected over one rotation period into a sequence of time windows of duration T centered at times τ_k, $k = 0, 1, \ldots K-1$, similar to DTI processing. BandWidth Extrapolation (BWE) is applied to each window to improve the range resolution [35]. The data are then resampled in range to generate equal range and cross-range pixel sizes; this step is important because the range resolution is generally much coarser than the cross-range resolution. "Local" Extended Coherent Processing (LECP) is then performed to correct linear range migration and quadratic phase error [36]; these effects arise due to deviations from the first-order approximation assumed in (9). Bandlimited interpolation is applied to the resulting images to generate a finer pixel grid. Lastly, the images are rotated to a common coordinate frame (using knowledge of $\dot{\theta}$), converted to power scale (or log-power scale), and non-coherently combined to form a single output image. The final BENI image can be represented as

$$g_{BENI}(x,y) = \frac{1}{K}\sum_{k=0}^{K-1} \left| g_{LECP,k}\left(x\cos(\dot{\theta}\tau_k) - y\sin(\dot{\theta}\tau_k), x\sin(\dot{\theta}\tau_k) + y\cos(\dot{\theta}\tau_k)\right)\right|^2, \quad (41)$$

where $g_{LECP,k}$ is the k-th LECP image.

Comparison of (39) and (41) reveals that the two expressions have a similar form; note that in (41), $g_{LECP,k}$ is indexed in x, y position rather than in $r, \dot{r}$, and thus, the arguments are switched. Thus, the power sum image formed from $\tilde{X}$ is essentially the BENI image, but without the BWE and motion-compensation steps. Unlike the Doppler tomography case, (41) assumes the object is partially resolved so that the signal varies along the range dimension. In other words, where the projections in Doppler tomography provide no resolving capability in range, the LECP images that BENI combines coarsely resolve the object in range, resulting in a superior reconstruction.

Using the 3D RDT tensor technique, we have established Doppler tomography as a limiting case of BENI when the bandwidth becomes sufficiently low that (19) no longer holds (the object is unresolved in range). Without BWE, BENI in this limiting case generates intermediate LECP images that have rank-1 (prior to rotational alignment). These images are essentially the same as the unfiltered backprojection images that are used for Doppler tomography.

4. Performance Assessment of RDT Tensor Processing

In this section, we analyze the RDT tensor technique in more detail, studying deviations from ideal assumptions, exploring factors that influence image quality, and highlighting challenges and edge cases associated with the imaging process.

4.1. *Influence of Rotation-Rate Estimate*

Successful narrowband radar imaging, where RDT-based power-sum image reconstruction is a particular approach, requires a high-fidelity rotational motion description of the space object. This section explores the influence of rotation-rate estimation errors on relative cross-range position of a single point scatterer on a uniformly rotating object. Using the target's frame of reference, the relative cross-range position of a point scatterer is the projection of the scatterer's position $\mathbf{x_p} = [x_0, y_0, z_0]^T$ onto the unit vector direction:

$$\hat{\mathbf{q}}(t;\dot{\theta}) = \begin{bmatrix} \cos(\dot{\theta}t)\sin\gamma \\ -\sin(\dot{\theta}t)\sin\gamma \\ 0 \end{bmatrix} \tag{42}$$

$$\begin{aligned} u(\mathbf{x_p},t;\dot{\theta}) &= \hat{\mathbf{q}}(\mathbf{t})^T\mathbf{x_p} \\ &= x_0\cos(\dot{\theta}t)\sin\gamma - y_0\sin(\dot{\theta}t)\sin\gamma. \end{aligned} \tag{43}$$

We begin our perturbation analysis by modeling an additive error in our rotation rate estimate, ϵ. Our updated relative cross-range measurement becomes:

$$\begin{aligned} u(\mathbf{x_p},t;\dot{\theta}+\epsilon) &= \hat{\mathbf{q}}^T(t;\dot{\theta}+\epsilon)\mathbf{x_p} \\ &= x_0\cos([\dot{\theta}+\epsilon]t)\sin\gamma - y_0\sin([\dot{\theta}+\epsilon]t)\sin\gamma. \end{aligned} \tag{44}$$

Let Δu denote the change in relative cross-range position resulting from inaccuracies in rotation rate estimation:

$$\Delta u \triangleq u(\mathbf{x_p},t;\dot{\theta}+\epsilon) - u(\mathbf{x_p},t;\dot{\theta}) \tag{45}$$

$$= x_0\sin\gamma\big[\cos([\dot{\theta}+\epsilon]t) - \cos(\dot{\theta}t)\big] - y_0\sin\gamma\big[\sin([\dot{\theta}+\epsilon]t) - \sin(\dot{\theta}t)\big] \tag{46}$$

$$\begin{aligned} &= x_0\sin\gamma\big[\cos(\dot{\theta}t)\cos(\epsilon t) - \sin(\dot{\theta}t)\sin(\epsilon t) - \cos(\dot{\theta}t)\big] \\ &\quad - y_0\sin\gamma\big[\sin(\dot{\theta}t)\cos(\epsilon t) + \cos(\dot{\theta}t)\sin(\epsilon t) - \sin(\dot{\theta}t)\big]. \end{aligned} \tag{47}$$

Using a first-order Taylor expansion around $\epsilon = 0$, we simplify (47) to become the following expression:

$$\begin{aligned} \Delta u &\approx x_0\sin\gamma\big[\cos(\dot{\theta}t)\times 1 - \sin(\dot{\theta}t)\times 0 - \cos(\dot{\theta}t)\big] \\ &\quad - y_0\sin\gamma\big[\sin(\dot{\theta}t)\times 1 + \cos(\dot{\theta}t)\times \epsilon t - \sin(\dot{\theta}t)\big] \\ &= -y_0\sin\gamma\cos(\dot{\theta}t)\epsilon t. \end{aligned} \tag{48}$$

The result shown in (48) highlights cross-range position error is approximately linear both in range position as well as rotation-rate estimation error. Using (33), the 1D error (48) can be mapped to error in the 2D image plane:

$$\Delta\mathbf{p}(t;\mathbf{x},\dot{\theta}) = \begin{bmatrix} \cos(\dot{\theta}t) \\ -\sin(\dot{\theta}t) \end{bmatrix} y_0\cos(\dot{\theta}t)\epsilon t. \tag{49}$$

An upper bound for the norm of the 2D error is

$$\begin{aligned} \|\Delta\mathbf{p}(t;\mathbf{x},\dot{\theta})\|_2 &= |y_0\cos(\dot{\theta}t)\epsilon t| \\ &\leq y_0\epsilon T_p. \end{aligned} \tag{50}$$

If we require the 2D error to be less than one resolution cell in the power-sum image, we can derive the maximum allowable rotation-rate error. Using $d_x = d_y = \lambda/2\Delta\theta$ (i.e., the limiting narrowband case) with the requirement $\|\Delta\mathbf{p}(t;\mathbf{x},\dot{\theta})\|_2 \leq d_x$, we have the result

$$\epsilon \leq \frac{\lambda}{2\Delta\theta T_p y_0}. \tag{51}$$

Thus, the requirement on the max rotation-rate error becomes more stringent with increasing object size (larger y_0), increasing rotation period T_p, and decreasing wavelength λ.

4.2. Influence of Integration Time

Selecting a suitable integration time T_{int} is key to producing well-focused narrowband images. Due to the uniform rotation of the satellite, a range-dependent quadratic phase error is induced, which causes Doppler smearing of a point scatter located at (x_p, y_p) on the radar-image plane. To illustrate this, we can form a second-order Taylor expansion of the phase of an individual point scatterer's return signal around the point $(f,t) = (f_c, 0)$

$$G_p(f,t) \approx \underbrace{\exp\left[\frac{-j4\pi f y_p}{c}\right] \exp\left[-\frac{j4\pi f_c \dot{\theta} x_p t}{c}\right]}_{\text{linear imaging approximation}} \times \underbrace{\exp\left[-\frac{j4\pi (f-f_c)\dot{\theta} x_p t}{c}\right]}_{\text{linear range walk due to rotation}} \times \underbrace{\exp\left[\frac{j2\pi f_c \dot{\theta}^2 y_p t^2}{c}\right]}_{\text{quadratic phase error}} \tag{52}$$

Here, the third term generates range-walk migration. In BENI, this is corrected by applying the Keystone Transform to the data as part of LECP processing. The final term in (52) is a chirp signal. This quadratic phase expression induces smearing along cross-range. This effect worsens with increasing range y_p, and defocuses scatterers far away from the axis of rotation. To characterize the Doppler spread from this quadratic-motion error, we define the spectral width of the chirp term as the extent (in Doppler frequency bins) of the nearly constant magnitude response, where most of the energy of the chirp is contained. The total change in Doppler across the integration time interval can be computed as follows:

$$\ddot{R} = y_p \dot{\theta}^2 \tag{53}$$

$$\Delta \dot{R} = \ddot{R}\left(\frac{T_{int}}{2}\right) - \ddot{R}\left(\frac{-T_{int}}{2}\right) = y_p \dot{\theta}^2 T_{int}. \tag{54}$$

We define spectral width a, as the total change in Doppler divided by Doppler resolution $\delta\dot{R} = \lambda / 2T_{int}$:

$$a \triangleq \frac{\Delta \dot{R}}{\delta \dot{R}} \tag{55}$$

$$= \frac{2 y_p \dot{\theta}^2}{\lambda} T_{int}^2. \tag{56}$$

To ensure that the signal energies of all point scatterers are constrained within one Doppler bin (spectral width, $a \leq 1$), T_{int} should be selected such that:

$$T_{int} \leq \sqrt{\frac{\lambda}{2\dot{\theta}^2 L_{MAX}}}, \tag{57}$$

where L_{MAX} is defined as the object's maximum length extent. When individual point scatterers can be resolved, acceleration error in (52) can be corrected since y_p is known. This is done in BENI as part of LECP. In Doppler tomography, (57) restricts integration time per Doppler profile, which in turn influences the final image resolution.

4.3. Transient Features

Narrowband images degrade with aspect-angle-dependent scattering, as the fidelity of the rotation-rate estimation is impacted and narrowband imaging techniques such as RDT-based power-sum image reconstruction assume scatterers are present for a full rotation. For example, small targets such as cubesats are often only visible when one of their faces is perpendicular to the radar LOS. Another example of anisotropic scattering behavior is bright specular returns. This anisotropic scatterer behavior generates artifacts in the narrowband images and reduces their quality.

To overcome this problem, in Doppler tomography, filtered backprojected images are added using a logarithmic power sum (instead of a linear power sum) due to the presence of bright specular returns. Tomographic reconstructions of the compact range satellite model are shown in Figure 12. Linear-power reconstructions formed using (40) often have degraded image quality due to speculars and large sidelobes as shown in Figure 12a. Notable improvements are achieved when using logarithmic power of Doppler profiles for tomographic reconstruction. In Figure 12b, we observe structural features such as the sets of rivets on the solar panels of the satellite model.

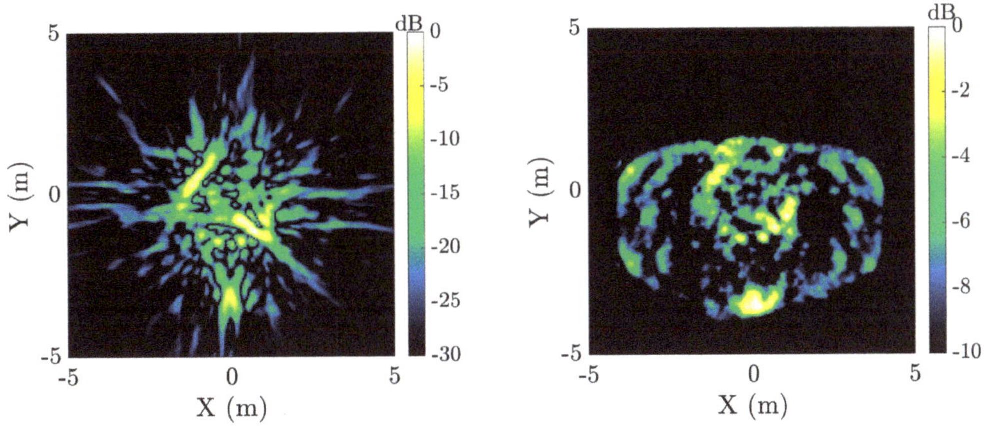

(**a**) Linear-power reconstruction (**b**) Logarithmic-power reconstruction

Figure 12. Doppler tomography reconstructions formed using (**a**) linear power and (**b**) logarithmic power of Doppler profiles of the rotating compact-range satellite model.

Similarly, the final BENI image is calculated using a global logarithmic power sum of the K rotationally aligned ECP images such that

$$g_{BENI}[x_m, y_n] = \frac{20}{K} \sum_{k=1}^{K} \log_{10}[|g_k[x_m, y_n]|]. \tag{58}$$

We can derive a geometric-mean interpretation of the BENI reconstruction where the final image can be interpreted as a geometric average of radar cross section (RCS) over aspect-change intervals:

$$\begin{aligned} g_{BENI}[x_m, y_n] &= 10 \log_{10} \left[\prod_{k=1}^{K} |g_k[x_m, y_n]|^2\right]^{\frac{1}{K}} \\ &= 10 \log_{10} \left[\prod_{k=1}^{K} \sigma_{k,linear}[x_m, y_n]\right]^{\frac{1}{K}}, \end{aligned} \tag{59}$$

where

$$\sigma_{k,linear}[x_m, y_n] = |g_k[x_m, y_n]|^2$$

i.e., the intensity of each pixel corresponds to RCS (for calibrated radar data). From this it can be seen that, for an isotropic point scatterer (where σ_k is constant for all k), BENI image magnitude is equivalent to RCS measured in dBsm.

The expression (59) reveals why the logarithmic sum is effective in suppressing transient features. A transient feature at pixel (x_m, y_m) has a magnitude near zero for at least one value of k. This results in the product of intensities also being near zero, leading to a RCS value approaching $-\infty$ dBsm in the BENI image. In contrast, a persistent scatterer contributes non-zero values for all k, thus generating a detectable feature.

4.4. Resolvability of Features

In standard radar imaging, formal definitions of resolution, which is the minimum separation at which two point scatterers can be distinguished, are well established. Typically, the peak-to-null distance of the point-spread function is used as a measure of resolution.

From (18), the range and cross-range resolution of an ISAR image are $\delta_R = c/2B$ and $\delta_U = \lambda/2\Delta\theta$, respectively. We seek to establish a similar notion of resolution for power-sum imaging. To motivate this, we first examine experimental results applying BENI (a variant of power-sum imaging) to compact-range data with varying aspect change and bandwidth. We then derive analytical expressions that characterize RDT-based power-sum image resolution as a function of bandwidth, center frequency, and aspect change.

In narrowband settings, range resolution is coarse such that the target occupies a small number of range bins. Finer range resolution in BENI image reconstructions can be attributed to two processing steps. The first is Doppler-compressed BWE, which works best when the satellite consists of discrete collections of impulse-shaped features such as edges, rivets, and corner reflectors. In practice, we extrapolate by a factor of 3, $B_{eff} = 3B$. For medium-band settings ($B \propto 100$ MHz), BWE can highlight structural features better. However, when dealing with reduced bandwidth scenarios ($B \propto 10$ MHz) and limited aspect-change diversity, target-scene range characterization using Doppler-compressed BWE may only offer limited improvements. The second step in BENI is globally non-coherent image reconstruction across different rotationally aligned ECP images as shown in (41). Note this image-formation step is implemented using a logarithmic power sum which has a geometric-mean interpretation as described in Section 4.3.

Figure 13 shows a matrix of examples of BENI reconstructions under varying bandwidth and aspect change. It is through exploiting wide aspect diversity that we are able to achieve better effective range resolution for drastically reduced bandwidth scenarios ($B \propto 10$ MHz).

In linear scale, each of the rotationally aligned images are multipled to each other in an element-wise manner prior to evaluation of the logarithm. The remainder of this section will explore the improvement in pixel resolution for a power-sum image formed using two rotationally aligned images corresponding to angle views $\theta_1 = 0°$ and $\theta_2 = 90°$. This allows us to analyze two orthogonal slices through the point-spread frunction to obtain a better understanding of pixel resolution.

Figure 14 illustrates the overlapping rotationally aligned images formed over these two different angle views. Without loss of generality, we assume a point scatterer on the target is centered at the origin such that the continuous, rotationally aligned images corresponding to angle views $\theta_1 = 0°$ and $\theta_2 = 90°$ have the following zero-Doppler cuts:

$$g_1(0, y) = \text{sinc}\left(\frac{y}{\delta R}\right), \tag{60}$$

$$g_2(0, y) = \text{sinc}\left(\frac{y}{\delta \dot{R}/\dot{\theta}}\right), \tag{61}$$

where $\delta R = c/2B$ and $\delta \dot{R} = c/2f_c T_{int}$. We will now take a closer look at the linear scale representation of the zero-Doppler cut of the corresponding power-sum image, which is the element-wise product of 0° and 90° projections as shown in Figure 14. Noting that:

$$\text{sinc}(x) \triangleq \sin(\pi x)/\pi x, \tag{62}$$

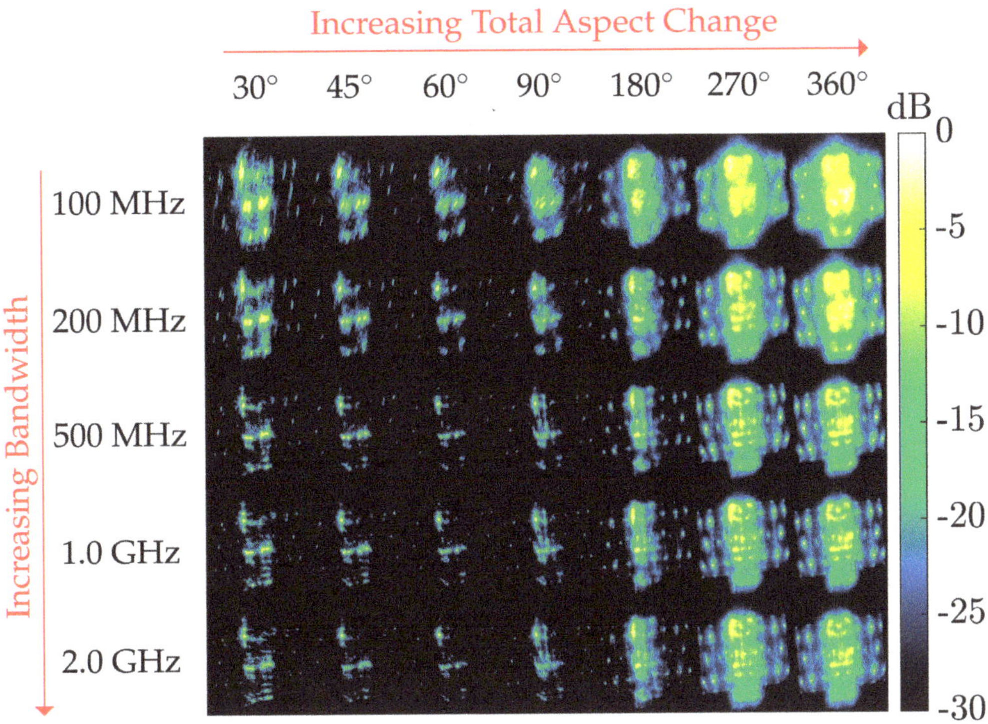

Figure 13. Limited aspect change and reduced bandwidth BENI reconstructions of the compact-range satellite model.

$$g_{PS}(0,y) = \text{sinc}\left(\frac{y}{\delta R}\right)\text{sinc}\left(\frac{y}{\delta \dot{R}/\dot{\theta}}\right) \tag{63}$$

$$= \left(1 - \frac{1}{3!}\left[\frac{\pi y}{\delta R}\right]^2 + \frac{1}{5!}\left[\frac{\pi y}{\delta R}\right]^4 - \cdots\right)\left(1 - \frac{1}{3!}\left[\frac{\pi y}{\delta \dot{R}/\dot{\theta}}\right]^2 + \frac{1}{5!}\left[\frac{\pi y}{\delta \dot{R}/\dot{\theta}}\right]^4 - \cdots\right)$$

$$\approx 1 - \frac{\pi^2 y^2}{6}\left(\left[\frac{1}{\delta R}\right]^2 + \left[\frac{1}{\delta \dot{R}/\dot{\theta}}\right]^2\right), \tag{64}$$

where (64) was obtained using a Taylor expansion of $\sin(\pi x)$. Note the resulting simplified expression is a parabola and the corresponding roots of (64) provide a quantitative measure of range resolution improvement along the y-axis. The following positive root of (64) can be defined as the effective range resolution δy:

$$\begin{aligned}\delta y &= \frac{\sqrt{6}}{\pi}\left(\left[\frac{1}{\delta R}\right]^2 + \left[\frac{1}{\delta \dot{R}/\dot{\theta}}\right]^2\right)^{-1/2} \\ &= \frac{\sqrt{6}}{\pi}\left(\frac{c}{2}\sqrt{\frac{1}{(f_c\Delta\theta)^2 + B^2}}\right).\end{aligned} \tag{65}$$

The expression for δy shows how both aspect change $\Delta\theta$ and bandwidth B contribute to finer characterization along the $y-$axis. Note that (65) can be upper bounded as

$$\delta y = \frac{\sqrt{6}}{\pi}\sqrt{\frac{(\delta R)^2(\delta\dot{R}/\dot{\theta})^2}{(\delta R)^2+(\delta\dot{R}/\dot{\theta})^2}} < \min\{\delta R, \delta\dot{R}/\dot{\theta}\}. \quad (66)$$

This result demonstrates that the power-sum pixel resolution is finer than both δR and $\delta\dot{R}/\dot{\theta}$.

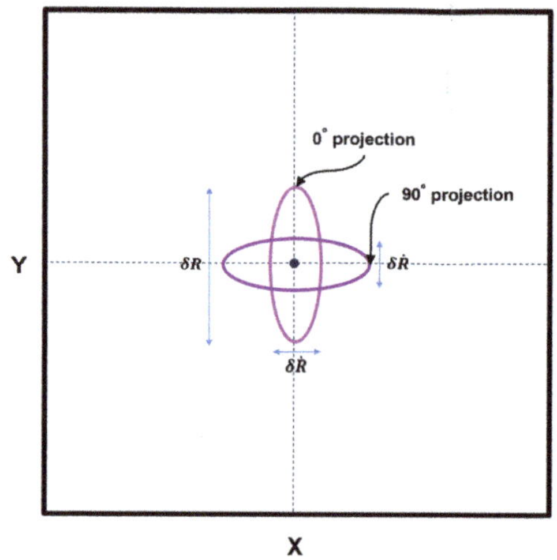

Figure 14. Power-sum reconstruction pixel resolution illustration.

4.5. Dynamic Motion: Precession and Non-Uniform Rotation

Thus far, we have assumed that the imaged object is in a pure tumble motion (uniform rotation along a fixed axis), and that the radar is stationary relative to the center of mass of the object. In practice, these assumptions do not hold perfectly, resulting in deviations from the assumed signal model. When the rotation axis varies over the collection interval, the response of a point scatterer is no longer described by the helical trajectory in (28). One important case of interest is spin-precession motion. Here, the rotation axis precesses about a fixed axis at a rate $\dot{\theta}_P$. The transformation from the object body frame to an inertial frame (such as ECI) can be expressed as

$$\hat{\mathbf{u}}_{ECI}(t) = \mathbf{U}_P\mathbf{R}_z(\dot{\theta}_P t)\mathbf{R}_y(\beta)\mathbf{R}_z(\dot{\theta}_S t)\hat{\mathbf{u}}(t), \quad (67)$$

where $\mathbf{U}_P$ is the precession coordinate frame (where the z-axis is the precession axis), β is the angle between the precession axis and the spin axis, $\dot{\theta}_S$ is the spin rate, $\dot{\theta}_P$ is the precession rate, and $\hat{\mathbf{u}}(t)$ and $\hat{\mathbf{u}}_{ECI}(t)$ denote vectors in the body and inertial frames, respectively. The notation $\mathbf{R}_y(\beta)$ indicates a rotation matrix that applies a rotation of angle β about the y axis. The application of (67) to (3) results in scatterer range and Doppler trajectories that are not purely sinusoidal, but rather, that have a time-varying amplitude modulation, phase angle, and offset. As a result, the techniques in this paper cannot be applied directly.

Another case is where the rotation axis is fixed, but the rotation rate is time varying (non-uniform rotation). If the non-uniformity can be estimated, it is possible to resample the data to equally spaced aspect angles. This enables the generation of Doppler profiles with fixed cross-range sample spacing and extent. It also allows Doppler profiles (or images) to be produced at equal angles, as desired for Doppler tomography and BENI.

Lastly, if there is relative motion between the radar and the target center of mass, as is the case for non-geostationary objects, the radar LOS is variable in the inertial frame. Thus, the model for the body-frame LOS in (1) does not hold.

The problems of estimating rotational motion parameters from the data for the cases above, and of applying these parameters to generate images, are subjects of ongoing research.

4.6. Unfavorable Imaging Geometries

As alluded to in Section 3.3, the tumble angle γ can cause depth-of-focus effects, which degrade the quality of power-sum image reconstructions. To illustrate this, in this section we present simulated radar data modeling two wideband ($B = 1$ GHz, $f_c = 10$ GHz) collects where the same single 0 dBsm peak RCS-isotropic point scatterer undergoing uniform rotation has been observed with two different geometries, where $\gamma = 90^\circ, \gamma = 60^\circ$ for Figure 15 and Figure 16, respectively.

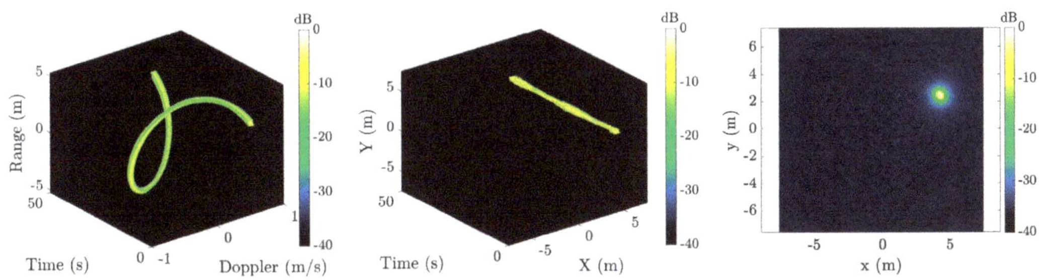

(a) Unaligned RDT (b) Aligned RDT (c) Power-sum reconstruction

Figure 15. Simulated imaging geometry of uniformly rotating point scatterer where $\gamma = 90^\circ$.

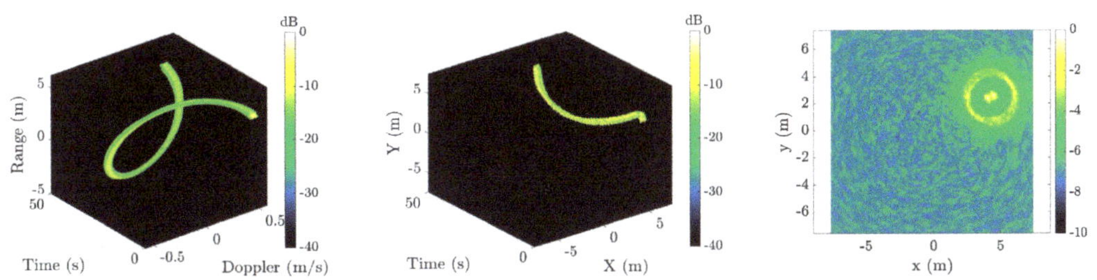

(a) Unaligned RDT (b) Aligned RDT (c) Power-sum reconstruction

Figure 16. Simulated imaging geometry of uniformly rotating point scatterer where $\gamma = 60^\circ$.

The rotationally aligned RDT tensor for the $\gamma = 60^\circ$ scenario is displayed in Figure 16b, revealing that the isotropic point scatterer follows a residual helical trajectory. The radius of this helical path, as described by (34), is equal to $z/tan\gamma \approx 1.73$ m, which closely matches the radius of the circular region in Figure 16c. When $\gamma = 90^\circ$, the aligned trajectory of the isotropic point scatterer is a line as shown in Figure 15b. Knowledge of γ is needed to properly align the RDT in Figure 16a to obtain a high-fidelity power-sum image reconstruction as shown in Figure 15c.

5. Extensions to the RDT Tensor Technique

5.1. Rotational-Rate Estimation

As discussed in Section 4.1, obtaining an accurate estimate of the object's rotation rate is critical for producing well-focused narrowband images. We aim to automate rotation-rate estimation for different target scenarios. Previous work provides a variety of techniques for estimating rotation rate, which can be split into three broad categories: time-based, frequency-based, and image quality analysis. Time-based techniques, such as auto-correlation, have been used for estimating rotational motion parameters in past work [37–39]. Frequency-based techniques, such as cepstrum analysis are also commonly used for identifying periodicity [40–42]. However, both rely on the assumption that at least one full rotation of the target object has been observed. In comparison, image-quality analysis techniques such as SIRTA, which compare image quality across a search space of possible periods, do not assume observation of more than one rotation [43–45].

For automating rotation-rate estimation, we are interested in more than just the ideal scenario where greater than one full rotation has been observed. In future work, we will explore different estimation techniques across a variety of scenarios with a focus on assessing the performance of the methods when only a partial rotation is observed.

5.2. RDT Tensor Denoising Using Higher-Order Singular Value Decomposition (HOSVD)

Noting the rich three-dimensional structure of our RDT data tensor, in this section we investigate the utility of applying Higher-Order Singular Value Decomposition (HOSVD), a generalization of the matrix SVD, for image enhancement via SVD truncation [46]. After performing rotational alignment to our RDT tensor, we recognize isotropic point scatterers on an object that persist over wide-aspect diversity (such as that of a full rotation) are redundant across the slow-time dimension. This redundancy enables us to exploit low-rankness to help denoise our RDT tensor prior to reconstructing power-sum images.

The starting point of HOSVD is forming three separate matrix representations of the RDT tensor. This process is often referred to as matricization or matrix unfolding [46,47]. As visually depicted in Figure 17, we form three matricizations of the rotationally aligned RDT tensor $\mathbf{X} \in \mathbb{R}^{N\times M\times K}$ by rearranging mode-1 (range), mode-2 (Doppler), mode-3 (time) fibers across each dimension to form $\mathbf{X_1} \in \mathbb{R}^{N\times KM}, \mathbf{X_2} \in \mathbb{R}^{M\times KN}, \mathbf{X_3} \in \mathbb{R}^{K\times NM}$, respectively.

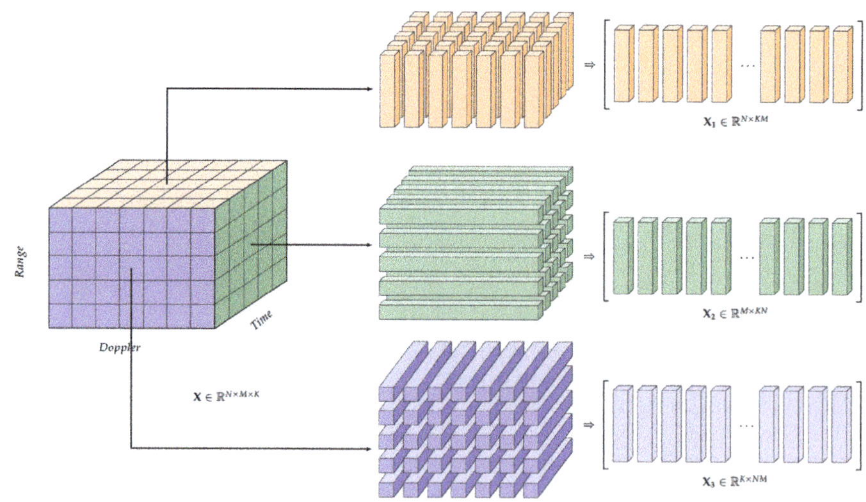

Figure 17. Matricization of the RDT Tensor.

For our application, we use the Tucker decomposition model for the rotationally aligned RDT tensor [47]. That is,

$$X[n,m,k] = \sum_{c}\sum_{b}\sum_{a} G[a,b,c]U_1[n,a]U_2[m,b]U_3[k,c], \tag{68}$$

where $\mathbf{G} \in \mathbb{R}^{N\times M\times K}$ is the three-dimensional core tensor consisting of singular values, and $\mathbf{U_1} \in \mathbb{R}^{N\times N}, \mathbf{U_2} \in \mathbb{R}^{M\times M}, \mathbf{U_3} \in \mathbb{R}^{K\times K}$ are three orthonormal matrices whose columns correspond to left-singular vectors of $\mathbf{X_1}, \mathbf{X_2}, \mathbf{X_3}$, respectively. $\mathbf{U_1}, \mathbf{U_2}, \mathbf{U_3}$ are computed using the standard SVD whereas the core tensor $\mathbf{G}$ is computed using the Tucker's Method 1 or HOSVD algorithm [47].

For our experiments, we computed the core tensor using the HOSVD routine in the MATLAB Tensor Toolbox [48]. We truncated singular values $G[a,b,c]$ such that we kept only nonzero values for $G[a,b,1]$, as shown in (69). This corresponds to a rank-$(N,M,1)$ approximation of $\mathbf{X}$ (rank-1 approximation along the time dimension of our tensor). We formed our denoised rotationally aligned RDT tensor using (70):

$$G_{truncated}[a,b,c] = \begin{cases} G[a,b,c] & \text{if } c = 1 \\ 0 & \text{if } c \neq 1 \end{cases} \tag{69}$$

$$\begin{aligned} X_{denoised}[n,m,k] &= \sum_c \sum_b \sum_a G_{truncated}[a,b,c] U_1[n,a] U_2[m,b] U_3[k,c] \\ &= U_3[k,1] \sum_b \sum_a G[a,b,1] U_1[n,a] U_2[m,b]. \end{aligned} \tag{70}$$

Figure 18 compares and contrasts the full rank-(N,M,K) logarithmic power-sum image reconstructions to that of the reduced rank-$(N,M,1)$ logarithmic power-sum image reconstruction. Note the sharper appearance of rivets and edges on the solar panels of the compact-range satellite. Future work involves a continued investigation into tensor decompositions for radar-image enhancement.

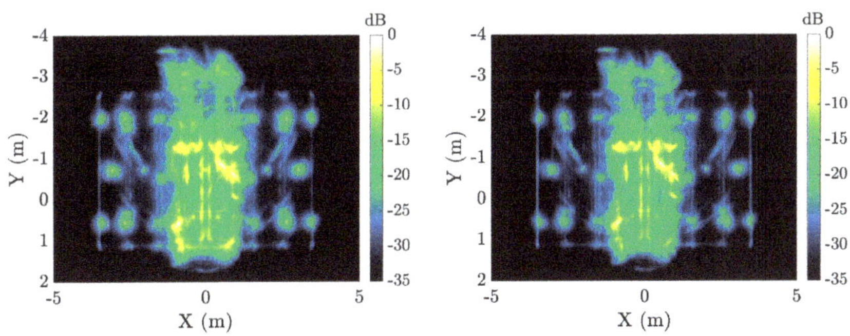

(**a**) Power Sum of $\mathbf{X}$ (**b**) Power Sum of $\mathbf{X_{denoised}}$

Figure 18. Logarithmic power-sum image reconstructions formed using (**a**) full-rank approximation $\mathbf{X}$ and (**b**) HOSVD truncation to form rank-$(N,M,1)$ approximation of $\mathbf{X}$.

6. Conclusions

In this work, we proposed a powerful unified technique for narrowband radar deep-space satellite characterization. Using our three-dimensional RDT tensor representation of radar signals, we generated RDT-derived products such as range superpulses, Doppler superpulses, and power-sum image reconstructions. Utilizing radar data obtained for a compact-range satellite model, we demonstrated that each of these RDT-derived products successfully enhanced range and Doppler characterization. We also established the equivalence of BENI and Doppler tomography narrowband imaging techniques as different versions of power-sum reconstructions using the rank-reduced RDT. In addition, we quantified limitations of these radar imaging techniques as a function of carrier frequency, sensitivity, integration time, and rotation-rate estimate uncertainty. Another accomplished objective of this work included providing interpretability of RDT-based power-sum images with respect to pixel RCS measurements over varying imaging geometries (orientations of the object). For future work, we will continue to address challenges encountered when deviating from ideal radar-signal models, and explore other ways to exploit the rich RDT data structure for radar-image enhancement.

Author Contributions: Conceptualization, A.S., J.C., R.L.M.J. and M.D.A.; formal analysis, A.S. and R.L.M.J.; data curation, A.S. and J.C.; writing—original draft preparation, A.S., J.C., R.L.M.J. and M.D.A.; writing—review and editing, A.S., J.C., R.L.M.J. and M.D.A.; visualization, A.S. and J.C.; supervision, R.L.M.J. and M.D.A. All authors have read and agreed to the published version of the manuscript.

Funding: This research was funded under Air Force Contract No. FA8702-15-D-0001.

Data Availability Statement: Data are contained within the article.

Acknowledgments: The authors acknowledge Randall Seed of MIT Lincoln Laboratory for initial explorations in narrowband-radar Doppler tomography for space-object characterization. In addition, the authors thank Kenneth Cole, also of MIT Lincoln Laboratory, for his feedback in preparing this manuscript.

Conflicts of Interest: The authors declare no conflicts of interest.

Abbreviations

The following abbreviations are used in this manuscript:

BENI	Bandwidth Enhanced Non-Coherent Imaging
BWE	BandWidth Extrapolation
DTI	Doppler-Time Intensity
GEO	Geosynchronous Earth Orbit
HOSVD	Higher-Order Singular Value Decomposition
ISAR	Inverse Synthetic Aperture Radar
LECP	Local Extended Coherent Processing
LEO	Low Earth Orbit
LOS	Line-Of-Sight
MEO	Medium Earth Orbit
RDM	Range–Doppler Map
RDT	range-Doppler-time Tensor
RTI	Range–Time Intensity
SDA	Space Domain Awareness
SVD	Singular Value Decomposition

Appendix A. Notation and Terminology

Appendix A.1. Derivation of ISAR Imaging Expression

An image can be reconstructed using a polar integral over the following annular region A:

$$\begin{aligned} g(x,y) &= \iint_A S_\sigma(k_x,k_y,0)\exp\left(j2\pi\left[k_x x + k_y y\right]\right)dk_x dk_y \\ &= \int_{\frac{-\Delta\theta}{2}}^{\frac{\Delta\theta}{2}} \int_{2(f_c-B/2)/c}^{2(f_c+B/2)/c} S_\sigma(k_r\sin\theta, k_r\cos\theta, 0) \\ &\qquad \times \exp\left(j2\pi\left[k_r\sin\theta x + k_r\cos\theta y\right]\right)k_r dk_r d\theta, \end{aligned} \tag{A1}$$

where $\Delta\theta = \dot{\theta}T$ is the integration angle. In practice, ISAR images are formed using a linear-imaging approximation over $(f,\bar{t}) = (f_c,0)$,

$$
\begin{aligned}
G\left(f, \frac{\theta}{\dot{\theta}}\right) &\approx \int_{\Omega_x} \int_{\Omega_y} \bar{\sigma}(x', y') \exp\left(\frac{-j4\pi f_c x'\theta}{c}\right) \exp\left(\frac{-j4\pi f y'}{c}\right) dx' dy' \\
k_x(f, \theta) &\approx \frac{2f_c}{c}\theta \\
k_y(f, \theta) &\approx \frac{2f}{c}.
\end{aligned}
$$

After performing a change of variables $(dk_x = 2f_c d\theta/c,\ dk_y = 2df/c)$, we can approximate (17) as an integration over a rectangular region parametrized by $f \in [f_c - \frac{B}{2}, f_c + \frac{B}{2}]$, $\theta \in [-\frac{\Delta\theta}{2}, \frac{\Delta\theta}{2}]$

$$
\begin{aligned}
g(x, y) &\approx \frac{4f_c}{c^2} \int_{\frac{-\Delta\theta}{2}}^{\frac{\Delta\theta}{2}} \int_{f_c - B/2}^{f_c + B/2} G\left(f, \frac{\theta}{\dot{\theta}}\right) \exp\left(\frac{j4\pi f_c x\theta}{c}\right) \exp\left(\frac{j4\pi f y}{c}\right) df d\theta \qquad \text{(A2)} \\
&= \frac{4f_c}{c^2} \int_{\Omega_x} \int_{\Omega_y} \bar{\sigma}(x', y') \left[\int_{\frac{-\Delta\theta}{2}}^{\frac{\Delta\theta}{2}} \exp\left(\frac{-j4\pi f_c (x - x')\theta}{c}\right) d\theta \right] \\
&\qquad \times \left[\int_{f_c - B/2}^{f_c + B/2} \exp\left(\frac{-j4\pi f (y - y')}{c}\right) df \right] dx' dy' \\
&= \frac{4f_c}{c^2} B\Delta\theta \int_{\Omega_x} \int_{\Omega_y} \bar{\sigma}(x', y') \exp\left(\frac{-j4\pi f_c y'}{c}\right) \operatorname{sinc}\left(\frac{x - x'}{\lambda/2\Delta\theta}\right) \operatorname{sinc}\left(\frac{y - y'}{c/2B}\right) dx' dy' \\
&= [\lambda/2\Delta\theta]^{-1} [c/2B]^{-1} \int_{\Omega_x} \int_{\Omega_y} \bar{\sigma}(x', y') \exp\left(\frac{-j4\pi f_c y'}{c}\right) \operatorname{sinc}\left(\frac{x - x'}{\lambda/2\Delta\theta}\right) \qquad \text{(A3)} \\
&\qquad \times \operatorname{sinc}\left(\frac{y - y'}{c/2B}\right) dx' dy'.
\end{aligned}
$$

Table A1. Key notation in paper.

$\vec{\omega}$	Angular velocity vector (rad/s)
$\mathbf{r}_{TX}(t)$	Radar position relative to target center of mass (m)
$\mathbf{x}$	3D scatterer (scattering element) position (m)
$\sigma(\mathbf{x})$	Complex-valued radar scattering reflectivity density
$\Omega \subset \mathbb{R}^3$	3D volumetric support of target
$\hat{\mathbf{u}}(t)$	Radar line-of-sight unit vector
$\hat{\mathbf{q}}(t)$	Cross-range axis unit vector
T_p	Tumble period (s)
$\dot{\theta} = 2\pi/T_p$	Rotation rate (rad/s)
γ	Tumble angle between line of sight and angular velocity $\vec{\omega}$ (rad)
$R(t;\mathbf{x})$	Relative range of scatterer at $\mathbf{x}$ (m)
$\dot{R}(t;\mathbf{x})$	Range rate of scatterer at $\mathbf{x}$ (m/s)
$G(f,t)$	Phase history data of the received signal
$f = f_c + f_B$	Transmission frequency (Hz)
f_B	Baseband frequency (Hz)
f_c	Carrier frequency (Hz)
$\lambda = c/f_c$	Wavelength (m)
B	Bandwidth (Hz)
T	Integration interval (s)
$\Delta\theta = \dot{\theta}T$	Integration angle (rad)
τ_k	Center of k-th Doppler-processing interval (s)
$\delta_R = c/2B$	Range resolution (m)
$\delta_{\dot{R}} = \lambda/2T$	Doppler resolution (m/s)
$\hat{g}(r,t)$	Range–Time Intensity (RTI)
$g(r,\dot{r})$	Range–Doppler Map (RDM)

Table A1. *Cont.*

$g(x,y)$	2D ISAR image
$X_D(\dot{r},\tau_k)$	Doppler–Time Intensity (DTI)
$X(r,\dot{r},\tau_k)$	3D range-Doppler-time (RDT) tensor (continuous time)
$X[n,m,k]$	3D range-Doppler-time (RDT) tensor (discrete time)
$X_{RS}[n,k]$	Range superpulses
$X_{DS}[m,k]$	Doppler superpulses
$\tilde{X}[n',m',k]$	Rank-reduced RDT tensor
$\tilde{X}_{PS}[n',m']$	Power-sum image
$g_{DT}(x,y)$	Doppler tomography image
$g_{BENI}(x,y)$	BENI image
$\mathbf{h}(\tau_k;\mathbf{x})$	Position of scatterer $\mathbf{x}$ in RDT space (m, m/s)
$\mathbf{p}(\tau_k;\mathbf{x},\dot{\theta},\gamma)$	Transformed position after rank-reducing transform (m, m)
$\mathbf{T}_{\dot{\theta},\gamma}(\tau_k)$	Rank-reducing transform matrix

Appendix A.2. Derivation of Slant-Plane Image

A slant-plane image can be formed when $\gamma < \pi/2$ by invoking the first-order expansion

$$\sin(\dot{\theta}\bar{t}) \approx \dot{\theta}\bar{t}$$
$$\cos(\dot{\theta}\bar{t}) \approx 1.$$

Substitution into (12) yields

$$R(\bar{t};\mathbf{x}) \approx (x\cos\theta_i - y\sin\theta_i)\dot{\theta}_{\text{eff}}\bar{t} + (x\sin\theta_i + y\cos\theta_i)\sin\gamma + z\cos\gamma, \tag{A4}$$

where

$$\dot{\theta}_{\text{eff}} = \|\vec{\omega} \times \hat{\mathbf{u}}(t)\| = (\sin\gamma)\dot{\theta}$$

is the effective rotation rate observed by the radar, i.e., the component of rotation orthogonal to the radar line of sight. The conversion between body-frame coordinates $\mathbf{x}$ and slant-plane coordinates $\mathbf{x}''$ at angle θ_i is

$$\begin{bmatrix} x'' \\ y'' \\ z'' \end{bmatrix} = \begin{bmatrix} 1 & 0 & 0 \\ 0 & \sin\gamma & \cos\gamma \\ 0 & -\cos\gamma & \sin\gamma \end{bmatrix} \begin{bmatrix} \cos\theta_i & -\sin\theta_i & 0 \\ \sin\theta_i & \cos\theta_i & 0 \\ 0 & 0 & 1 \end{bmatrix} \begin{bmatrix} x \\ y \\ z \end{bmatrix} \tag{A5}$$

Applying (A5) to (A4) produces

$$R(\bar{t};\mathbf{x}'') \approx x''\dot{\theta}_{\text{eff}}\bar{t} + y''.$$

Applying the linear-imaging approximation, which includes the approximations used when producing the RDM in (8), we generate the slant-plane image as

$$g_S(x,y) = \int_{-T/2}^{T/2}\int_{-B/2}^{B/2} G(f,\bar{t}) \exp\left(j\frac{4\pi}{c}f_B y\right) \exp\left(j\frac{4\pi}{c}f_c\dot{\theta}_{\text{eff}}\bar{t}x\right) df_B d\bar{t}. \tag{A6}$$

The resulting image is

$$g_S(x,y) \approx BT \int_{\Omega_y}\int_{\Omega_x} \bar{\sigma}_S(x'',y'') \exp\left(-j\frac{4\pi}{c}f_c y''\right) \operatorname{sinc}\left(\frac{x - x''}{\lambda/(2\sin\gamma\Delta\theta)}\right) \times \operatorname{sinc}\left(\frac{y-y''}{c/2B}\right) dx''dy'', \tag{A7}$$

where $\bar{\sigma}_S(x'',y'')$ is the 2D projection of the 3D reflectivity onto the slant plane.

References

1. Sridharan, R.; Pensa, A. *Perspectives in Space Surveillance*; MIT Press: Cambridge, MA, USA, 2017.
2. Walker, J. Range-Doppler imaging of rotating objects. *IEEE Trans. Aerosp. Electron. Syst.* **1980**, *AES-16*, 23–52. [CrossRef]
3. Munson, D.; O'Brien, D.; Jenkins, W. A tomographic formulation of spotlight-mode synthetic aperture radar. *Proc. IEEE* **1983**, *71*, 917–925. [CrossRef]
4. Swenson, A.T.; Nebelecky, C.K.; Wilkinson, D.; Crassidis, J.L. Resident Space Object Shape and Material Estimation using Polarimetric Data. In Proceedings of the AAS Guidance, Navigation and Control (GN&C Conference), Sopot, Poland, 12–16 June 2023; pp. 23–140.
5. Dianetti, A.; Crassidis, J. Resident Space Object Characterization Using Polarized Light Curves. *J. Guid. Control Dyn.* **2022**, *46*, 1–18. [CrossRef]
6. Anderson, J.D.; Anderson, A.J.; Zuehlke, D.A.; Canales, D.; Lovell, T.A. Resident Space Object Identification in Arbitrary Unresolved Space Images. In Proceedings of the Proceedings of the 33rd AAS/AIAA Space Flight Mechanics Meeting], AAS/AIAA, Austin, TX, USA, 15–19 January 2023.
7. Suthakar, V.; Sanvido, A.A.; Qashoa, R.; Lee, R.S.K. Comparative Analysis of Resident Space Object (RSO) Detection Methods. *Sensors* **2023**, *23*, 9668. [CrossRef] [PubMed]
8. Wu, X.; Sahoo, D.; Hoi, S.C. Recent advances in deep learning for object detection. *Neurocomputing* **2020**, *396*, 39–64. [CrossRef]
9. Massimi, F.; Ferrara, P.; Benedetto, F. Deep Learning Methods for Space Situational Awareness in Mega-Constellations Satellite-Based Internet of Things Networks. *Sensors* **2023**, *23*, 124. [CrossRef]
10. Linares, R.; Furfaro, R. Space Object classification using deep Convolutional Neural Networks. In Proceedings of the 2016 19th International Conference on Information Fusion (FUSION), Heidelberg, Germany, 5–8 July 2016; pp. 1140–1146.
11. Jia, B.; Pham, K.; Blasch, E.; Wang, Z.; Shen, D.; Chen, G. Space object classification using deep neural networks. In Proceedings of the 2018 IEEE Aerospace Conference, Big Sky, MT, USA, 3–10 March 2018; pp. 1–8. [CrossRef]
12. Linares, R.; Furfaro, R.; Reddy, V. Space Objects Classification via Light-Curve Measurements Using Deep Convolutional Neural Networks. *J. Astronaut. Sci.* **2020**, *67*, 1063–1091. [CrossRef]
13. Qashoa, R.; Lee, R. Classification of Low Earth Orbit (LEO) Resident Space Objects' (RSO) Light Curves Using a Support Vector Machine (SVM) and Long Short-Term Memory (LSTM). *Sensors* **2023**, *23*, 6539. [CrossRef] [PubMed]
14. Furfaro, R.; Linares, R.; Gaylor, D.; Jah, M.; Walls, R. Resident space object characterization and behavior understanding via machine learning and ontology-based bayesian networks. In Proceedings of the Advanced Maui Optical and Space Surveillance Technologies Conference, Maui, HI, USA, 20–23 September 2016; p. 35.
15. AlDahoul, N.; Karim, H.A.; De Castro, A.; Tan, M.J.T. Localization and classification of space objects using EfficientDet detector for space situational awareness. *Sci. Rep.* **2022**, *12*, 21896. [CrossRef]
16. Serrano, A.; Morrison, R.L. Doppler Superpulse Processing for Improved Tomographic Characterization of Space Objects. In Proceedings of the 2023 IEEE International Applied Computational Electromagnetics Society Symposium (ACES), Monterey, CA, USA, 26–30 March 2023; pp. 1–2.
17. Chua, C. Doppler-Only Synthetic Aperture Radar. Master's Thesis, Naval Postgraduate School, Monterey, CA, USA, 2006.
18. Androsov, A.; Vygon, V.; Ustinov, N. Reconstruction of images of rotating bodies of arbitrary angular dimensions. I. Structure of Doppler spectra and reconstruction of images from projections. *Sov. J. Quantum Electron.* **1985**, *15*, 168–172. [CrossRef]
19. Marino, R.; Capes, R.; Keicher, W.; Kulkarni, S.; Parker, J.; Swezey, L. Tomographic image reconstruction from laser radar reflective projections. In Proceedings of the Proc. SPIE 0999; Laser Radar III; SPIE: Bellingham, WA, USA, 1989; pp. 248–268.
20. Chen, V. Analysis of radar micro-Doppler with time-frequency transform. In Proceedings of the Tenth IEEE Workshop on Statistical Signal and Array Processing, Pocono Manor, PA, USA, 16 August 2000; pp. 463–466.
21. Coetzee, S.; Baker, C.; Griffiths, H. Narrow band high resolution radar imaging. In Proceedings of the 2006 IEEE Conference on Radar, Verona, NY, USA, 24–27 April 2006; pp. 622–625.
22. Mensa, D.L.; Halevy, S.; Wade, G. Coherent Doppler tomography for microwave imaging. *Proc. IEEE* **1983**, *71*, 254. [CrossRef]
23. Das, Y.; Boerner, W. On radar target shape estimation using algorithms for reconstruction from projections. *IEEE Trans. Antennas Propag.* **1978**, *AP-26*, 274–279. [CrossRef]
24. Bai, X.; Sun, G.; Wu, Q.; Xing, M.; Bao, Z. Narrow-band radar imaging of spinning targets. *Sci. China Inf. Sci.* **2011**, *54*, 873–883. [CrossRef]
25. Lanterman, A.; Munson, D.; Wu, Y. Wide-angle radar imaging using time–frequency distributions. *IEE Proc.-Radar Sonar Navig.* **2003**, *150*, 203–211. [CrossRef]
26. Sun, H.; Feng, H.; Lu, Y. High resolution radar tomographic imaging using single-tone CW signals. In Proceedings of the 2010 IEEE Conference on Radar, Lecce, Italy, 21–25 June 2010; pp. 975–980.
27. McCoy, J.; Magotra, N.; Chang, B. Coherent Doppler tomography—A technique for narrow band SAR. *IEEE Aerosp. Electron. Syst. Mag.* **1991**, *6*, 19–22. [CrossRef] [PubMed]
28. Serrano, A.; Morrison, R.L. BENI: Bandwidth Enhanced Noncoherent Imaging of Rotating Objects. In Proceedings of the 2022 IEEE Radar Conference (RadarConf22), New York, NY, USA, 21–25 March 2022; pp. 1–6. [CrossRef]
29. Serrano, A.; Kobsa, A.; Uysal, F.; Cerutti-Maori, D.; Ghio, S.; Kintz, A.; Morrison, R.L., Jr.; Welch, S.; van Dorp, P.; Hogan, G.; et al. Long baseline bistatic radar imaging of tumbling space objects for enhancing space domain awareness. *IET Radar Sonar Navig.* **2023**. [CrossRef]

30. Uysal, F.; Dorp, P.v.; Serrano, A.; Kobsa, A.; Ghio, S.; Kintz, A.; Bassa, C.; Garrington, S.; Cuenca, M.C.; Otten, M.; et al. Large Baseline Bistatic Radar Imaging for Space Domain Awareness. In Proceedings of the 2023 IEEE International Radar Conference (RADAR), Sydney, Australia, 6–10 November 2023; pp. 1–6. [CrossRef]
31. Benson, C.J.; Naudet, C.J.; Lowe, S.T. Radar Study of Inactive Geosynchronous Earth Orbit Satellite Spin States. *Interplanet. Prog. Rep.* **2020**, *42*, 1–14.
32. Coulombe, M.J.; Horgan, T.; Waldman, J.; Neilson, J.; Carter, S.; Nixon, W. A 160 GHz Polarimetric Compact Range for Scale Model RCS Measurements. In Proceedings of the Antenna Measurements and Techniques Association (AMTA), Seattle, WA, USA, 30 September–3 October 1996; p. 239.
33. High Fidelity Satellite Model in STL Radar Range. Available online: https://www.uml.edu/research/stl/ (accessed on 30 September 2023).
34. Kak, A.; Slaney, M. *Principles of Computerized Tomographic Imaging*; IEEE Press: New York, NY, USA, 1988.
35. Borison, S.; Bowling, S.; Cuomo, K. Super-resolution methods for wideband radar. *Linc. Lab. J.* **1992**, *5*, 441–461.
36. Ausherman, D.; Kozma, A.; Walker, J.; Jones, H.; Poggio, E. Developments in radar imaging. *IEEE Trans. Aerosp. Electron. Syst.* **1984**, *AES-16*, 363–398. [CrossRef]
37. Xing, Y.; You, P.; Yong, S. Parameter Estimation of Micro-Motion Targets for High-Resolution-Range Radar Using Online Measured Reference. *Sensors* **2018**, *18*, 2773. [CrossRef]
38. Yang, Q.; Deng, B.; Wang, H.; Qin, Y. Parameter Estimation and Image Reconstruction of Rotating Targets with Vibrating Interference in the Terahertz Band. *J. Infrared Millim. Terahertz Waves* **2017**, *38*, 909–928. [CrossRef]
39. Zhang, W.; Li, K.; Jiang, W. Parameter Estimation of Radar Targets with Macro-Motion and Micro-Motion Based on Circular Correlation Coefficients. *IEEE Signal Process. Lett.* **2015**, *22*, 633–637. [CrossRef]
40. Bogert, B.P.; Healy, M.J.; Tukey, J.W. The quefrency analysis of time series for echoes: Cepstrum, pseudo-autocovariance, cross-cepstrum and saphe cracking. In Proceedings of the Symposium on Time Series Analysis, Providence, RI, USA, 11–14 June 1963.
41. Oppenheim, A.; Schafer, R. From frequency to quefrency: A history of the cepstrum. *IEEE Signal Process. Mag.* **2004**, *21*, 95–106. [CrossRef]
42. Lee, J.K.; Kabrisky, M.; Oxley, M.E.; Rogers, S.K.; Ruck, D.W. The complex cepstrum applied to two-dimensional images. *Pattern Recognit.* **1993**, *26*, 1579–1592. [CrossRef]
43. Ghio, S.; Martorella, M. Estimation of Rotating RSO Parameters using Radar Data and Joint Time-Frequency Transforms. In Proceedings of the 7th Eur Conf. Space Debris, Darmstadt, Germany, 18–21 April 2017.
44. Ghio, S.; Martorella, M.; Staglianò, D.; Petri, D.; Lischi, S.; Massini, R. Practical implementation of the spectrogram-inverse Radon transform based algorithm for resident space objects parameter estimation. *IET Sci. Meas. Technol.* **2019**, *13*, 1254–1259. [CrossRef]
45. Ghio, S.; Martorella, M.; Staglianò, D.; Petri, D.; Lischi, S.; Massini, R. Experimental Comparison of Radon Domain Approaches for Resident Space Object's Parameter Estimation. *Sensors* **2021**, *21*, 1298. [CrossRef] [PubMed]
46. De Lathauwer, L.; De Moor, B.; Vandewalle, J. A Multilinear Singular Value Decomposition. *SIAM J. Matrix Anal. Appl.* **2000**, *21*, 1253–1278. [CrossRef]
47. Kolda, T.G.; Bader, B.W. Tensor Decompositions and Applications. *SIAM Rev.* **2009**, *51*, 455–500. [CrossRef]
48. Kolda, T.G.; Bader, B.W.; Acar Ataman, E.N.; Dunlavy, D.; Bassett, R.; Battaglino, C.J.; Plantenga, T.; Chi, E.; Hansen, S. Tensor Toolbox for MATLAB v. 3.5; MathWorks Company: Portola Valley, CA, USA, 2023.

remote sensing

MDPI

Article

Space Target Tracking with the HRRP Characteristic-Aided Filter via Space-Based Radar

Shuyu Zheng, Libing Jiang, Qingwei Yang, Yingjian Zhao and Zhuang Wang *

National Key Laboratory of Science and Technology on Automatic Target Recognition, College of Electronic Science and Technology, National University of Defense Technology (NUDT), Changsha 410073, China; zhengshuyu21@nudt.edu.cn (S.Z.); jianglibing@nudt.edu.cn (L.J.); yangqingwei18@nudt.edu.cn (Q.Y.); drztsp@163.com (Y.Z.)

* Correspondence: wangzhuang@nudt.edu.cn; Tel.: +86-139-7516-3062

Abstract: Approaching space target tracking is a typical and challenging mission in the space situational awareness (SSA) field. As the space-based radar is able to monitor the space targets of interest full-weather all-time, the space-based radar system is utilized in this paper. However, most multi-target tracking (MTT) filters in target tracking studies merely utilize the location or narrow measurements, and many potentially valuable electromagnetic scattering characteristics are missed, which leads to space target false tracking problems. The space-based radar transmits a wide-band signal, and the measured high-resolution range profile (HRRP) information is an effective characteristic for different target discrimination. Therefore, the HRRP characteristics of space targets are implemented into the update recursion of the MTT filter, which can be utilized to improve the tracking performance. Then, to predict the target HRRP sequence, the geometrical theory of diffraction (GTD) model is utilized. Additionally, a modified spatial spectrum method with a novel covariance matrix is designed to improve the scattering parameter estimation accuracy. Finally, an adapting threshold is devised for merging the Gaussian mixture (GM) components weights. The proposed threshold is on the basis of the proposed HRRP characteristic-aided probability hypothesis density (PHD) filter, and it can tackle the problem of space target discrimination. Simulation results validate the effectiveness and robustness of the proposed probability hypothesis density (HGI-PHD) filter aided by HRRP information and improved with GM weights.

Keywords: space-based radar; space target tracking; high-resolution range profile (HRRP); the geometrical theory of diffraction (GTD) model; adapting Gaussian mixture (GM) weights; merging threshold

1. Introduction

1.1. Background and Problem Statement

The radar system in this paper is the space-based radar, and the reasons for utilizing a space-based radar to detect or track space targets are analyzed as follows. Note that with the rapid growth of stealth technology and space activity, the existing ground-based equipment fails to monitor our space targets of interest all-weather full-time. Compared with ground-based equipment and space-based optical equipment, a space-based radar can provide better coverage, and it is not constrained by the weather or the curvature of the Earth. Moreover, it possesses a stronger early warning ability for high-speed space targets. Therefore, the space-based radar is utilized to track space targets in this paper, and some valuable problems are solved under space-based observing scenarios.

The problems that exist in the above-mentioned space-based observing scenarios are essential for perceiving the orbital state of the space targets of concern in real-time accurately, which is a challenging mission for space situational awareness (SSA) [1,2]. In general, the SSA surveillance tasks conclude the effective observation of our space targets of interest in the surveillance region, i.e., how the space target orbits change with varying time in order to estimate the individual characteristics or states (such as their positions and velocities in the reference coordinates). It should be mentioned that the number of space targets will change as time varies, and it is not known as a priori knowledge; therefore, the SSA tasks can be regarded as classical multiple target tracking (MTT) problems.

However, MTT will encounter some specific and typical problems when faced with the SSA surveillance tasks. For example, the space-based radar can only cover a limited tracking range due to its relatively smaller radar transmitting power compared to the ground-based radar system. Furthermore, it is challenging to track space targets accurately when they approach each other or the distance between them is close. What is more, all target state components sometimes are not provided thoroughly in a space-based radar system, and these characteristics make it more difficult and complicated for MTT problems.

An appealing and effective way to solve the aforementioned problems is using the random finite set (RFS) theory and finite set statistics (FISST). It is known that FISST is an elegant Bayesian formulation for the description of MTT based on the RFS theory [3,4]. Efficient solutions such as the probability hypothesis density (PHD) [5,6] and cardinalized PHD (CPHD) [7–9] have emerged in the MTT region in recent years. The core goal for the aforementioned filters is to update the first-order multi-target moments of the posterior probability density and update the global posterior probability density rather than recurrence, which is able to reduce the computational complexity to a large degree. Moreover, with the aim of alleviating poor cardinality estimation performance accuracy of the standard PHD filter, the CPHD filter is proposed to ease the above tension. It should be noted that the sequential Monte Carlo (SMC) model [10] and the Gaussian mixtures (GM) model [11] are the two most typical implementations of the above-mentioned Bayesian-based filters.

Note that among the various MTT filters, the PHD filter and the CPHD filter have concise formulation expressions and a slightly small computational burden, which is feasible for engineering applications. However, due to the "spooky action" in the process of PHD filter recursion, some limits exist as we propagate these two filters' legacy PHD. We would like to mention that it is difficult to derive a theoretical mathematical expression for the posterior probability density. Therefore, to tackle this problem, the multi-target cardinality-balanced multi-Bernoulli (CBMeMer) filter [12] is developed by propagating the posterior PDF. But under high-clutter density scenarios, the tracking performance of the CBMeMer filter will deteriorate sharply. In recent years, the generalized labeled multi-Bernoulli (GLMB) filter [13] and its computational simplified form, the labeled multi-Bernoulli (LMB) filter [14], have been proposed sequentially. These two filters are able to distinguish various target trajectories more accurately. However, considering the moment approximate operation, it is hard for the LMB filter to effectively avoid missed detection.

The space-based radar system is able to transmit a wide-band signal, and, thus, it can provide targets additional electromagnetic information like the high-resolution range profile (HRRP). Motivated by the above, we designed a filter that can update the target HRRP pseudo-likelihood. In order to predict the HRRP sequence and employ it in update recursion, the geometrical theory of diffraction (GTD) model is utilized in this paper. Being a classical scattering center model, the GTD model is effective in describing the radar targets' electromagnetic characteristics at high frequencies. In recent years, the GTD scattering center model has been widely applied in lots of radar domains, such as automatic target recognition (ATR), radar cross-section (RCS) extrapolation and interpolation, radar target three-dimensional (3D) reconstruction, and so on. Once the GTD scattering center is reconstructed, we can obtain the targets' electromagnetic scattering data, and the HRRP sequence can be predicted as well. Therefore, it is vitally important to construct a precise GTD scattering model and estimate the scattering parameters accurately. Thus, many effective methods, such as estimating signal parameters via the rational invariant technique (ESPRIT) algorithm [15–18], the multiple signal classification (MUSIC) algorithm [19–22], and the matrix enhancement and matrix pencil (MEMP) algorithm [23–25], have been proposed to estimate the scattering parameters from the back-scattered electromagnetic data.

Based on the aforementioned background as well as to deeply analyze the scientific problems in the process of tracking space targets based on a space-based radar, we separately analyzed the specific problems in the observation scene of a space-based radar and provided specific solutions so as to effectively improve the tracking accuracy of space targets.

1.2. Problem Analysis and Contributions

It is worth stressing that some electromagnetic scattering characteristics remain in radar signals, while most MTT filters fail to utilize them properly. Therefore, some drawbacks existed during the tracking recursion as follows:

1. Generally, most MTT filters merely utilize target information such as distance and orientation but fail to use some targets' additional characteristics that a wide-band radar can provide, for example, the HRRP information.
2. It is difficult to track space targets accurately when they are closely spaced with each other. So how can we improve the tracking performance when space targets are in dense clutter environments (due to thermal noise in space-based platforms, space debris, or other space targets)?
3. If the HRRP characteristics are utilized in the MTT tracking recursion process, then how can we predict the HRRP sequence, and how can we compute the HRRP pseudo-likelihood for different targets?
4. A more accurate GTD model is important for providing precise electromagnetic information for tracking update procedures. However, due to the thermal noise in a space-based platform, the signal-to-noise ratio (SNR) will be lower than the ground-based radar, which will lead to inaccuracy in scattering center parameter estimation. Therefore, it is vitally important to design a novel scattering parameter estimation method that can perform well in low-SNR scenarios.

The standard PHD filters do not utilize the targets' potential electromagnetic information abundantly, and they will encounter the problem of poor tracking performance under low-SNR scenarios or dense clutter environments. To improve the aforementioned problems, a modified PHD filter with HRRP characteristics is proposed in this paper, and the main contributions are shown as follows.

1. To fully utilize the electromagnetic scattering characteristics of space targets, HRRP information is utilized in the update recursion of our proposed MTT filter. The HRRP for various targets and clutter differs significantly from each other as tracking time varies; thus, if the HRRP information is fully utilized, the difference between different targets or clutter will be enlarged, which is beneficial for improving the tracking accuracy.
2. To solve the aforementioned second problem, an adaptive threshold for merging the Gaussian components is designed in this paper. The designed adaptive threshold is relevant to the noise covariance, and it is able to improve the tracking performance of the MTT filter. Furthermore, the discrimination ability between closely spaced targets can be enhanced at the same time.
3. In order to predict the HRRP sequence, the GTD scattering center model is constructed in this paper. Then, the similarity rate between different HRRP sequences is denoted to describe the HRRP pseudo-likelihood.
4. To solve the fourth problem mentioned above with the aim of improving the GTD scattering parameter estimation accuracy in low-SNR scenarios, a modified 3D-ESPRIT algorithm with a novel correlation matrix is proposed in this paper. Furthermore, by squaring the total covariance matrix, the parameter estimation accuracy can be further improved.

The rest of this paper is organized as follows. Section 2 investigates a review of the PHD filter and the GTD scattering center model. Furthermore, to obtain the time-varying observing angle, the observation geometry transformation for a space-based radar is also given in Section 2. Section 3 presents our novel scattering parameter estimation method and the analytic implementation of our proposed PHD (HGI-PHD) filter aided by HRRP characteristics and improved with Gaussian mixture components. The simulation and performance evaluation are illustrated in Section 4. Finally, conclusions are drawn in Section 5.

2. Technical Foundation

In this section, we first introduce the basic theory of PHD filtering, which provides a foundation for space target tracking. Then, to predict the HRRP sequence of space targets, the GTD scattering center model is demonstrated in this section. Finally, the observation geometry transformation for the space-based radar is derived to compute the time-varying observing angle.

2.1. PHD Filtering

In order to ease the computational intractability of the Bayesian multiple-target filter, the PHD filter is proposed by using the first-order moment of an RFS, which can be called the intensity function, and it is used to approximate the posterior multiple-target state density and propagate posterior intensity. The intensity of an RFS is defined as follows:

$$v_k(x|Z_{1:k}) = \int \delta_X(x) f(X|Z_{1:k}) \delta X \tag{1}$$

where $\delta_X(x) = \sum_{w \in X} \delta_w(x)$ and $\delta_w(x)$ represent the Dirac delta function. X and Z represent the states and measurements, respectively. Here, $Z_{1:k}$ is omitted from the intensity to notate it more simply. The integral of the PHD is not a constant, and the expected number of targets is expressed as

$$\int v_k(x) \mathrm{d}x = \lambda_k \tag{2}$$

The PHD filter approximates the density of multiple targets by using a Poisson process as follows:

$$f_{k|k}(X) = e^{-\lambda_k} \prod_{x \in X} \lambda_k v_k(x) \tag{3}$$

Then, the PHD prediction recursion is computed as

$$v_{k|k-1}(x) = \int p_{S,k}(\zeta) f(x|\zeta) v_{k-1}(\zeta) \mathrm{d}\zeta + \gamma_k(x) \tag{4}$$

where $v_{k|k-1}(x)$ represents the predicted PHD density of time k, $p_{S,k}(\zeta)$ is the target survival probability, $\gamma_k(x)$ is the target birth intensity function, and $f(x|\zeta)$ denotes the Markov transition density with state x.

Finally, the PHD update step can be summarized as

$$v_k(x) = [1 - p_{D,k}(x)] v_{k|k-1}(x) + \sum_{z \in Z} \frac{p_{D,k}(x) g_z(z|x) v_{k|k-1}(x)}{\kappa(z) + \int p_{D,k}(\xi) g_z(z|\xi) v_{k|k-1}(\xi) \mathrm{d}\xi} \tag{5}$$

where at time k, $p_{D,k}(x)$ is the detection probability of the target, $g_z(z|x)$ is the measurement likelihood of the target, and $\kappa(z)$ is the clutter RFS intensity function.

2.2. GTD Scattering Center Model

As one of the typical scattering center models, the geometric theory of diffraction (GTD) model enables us to describe the electromagnetic characteristics of radar targets effectively. Therefore, in a high-frequency region, the GTD scattering center model can be expressed as [26]

$$\begin{aligned} E(f_m, \theta_n, \varphi_k) &= \sum_{i=1}^{I} A_i \left(\mathrm{j}\frac{f_m}{f}\right)^{\alpha_i} \exp[-4\pi \mathrm{j} f_m (x_i \cos\theta \cos\varphi + y_i \sin\theta \cos\varphi + z_i \sin\varphi)/\mathrm{c}] + \omega(f_m, \theta_n, \varphi_k) \\ &= \sum_{i=1}^{I} A_i \left(\mathrm{j}\frac{f_0 + m\Delta f}{f}\right)^{\alpha_i} \exp[-4\pi \mathrm{j} f_m (x_i \cos\theta \cos\varphi + y_i \sin\theta \cos\varphi + z_i \sin\varphi)/\mathrm{c}] + \omega(f_m, \theta_n, \varphi_k) \end{aligned} \tag{6}$$

where $E(f_m, \theta_n, \varphi_k)$ represents the back-scattering data of radar targets, I denotes the total scattering centers, and $\{A_i, \alpha_i, x_i, y_i, z_i\}$ denote the scattering intensity, scattering type, transversal position parameter, and longitudinal position parameter and vertical position parameter of the i-th scattering center, respectively. $f_m = f_0 + m\Delta f$ represents the operating frequency, and $\{f_0, \Delta f, m\}$ represent the initial frequency, the frequency step, and the frequency index, respectively. In similarity, θ_n is equal to $\theta_0 + n\Delta\theta$; φ_k is equal to $\varphi_0 + k\Delta\varphi$; θ_0 and φ_0 are the initial azimuth angle and the initial elevation angle, respectively; and $n\Delta\theta$ and $k\Delta\varphi$ are the radar rotation angles, which are relatively small. $\mathrm{c} = 3 \times 10^8$ m/s represents the electromagnetic wave propagation speed, and $\omega(f_m, \theta_n, \varphi_k)$ denotes the Gaussian white noise. The type scattering parameter α_i of typical scattering structures is given in [26].

Then, according to Ref. [27], the scattering center parameters α_i, x_i, y_i, and z_i can be estimated as

$$\alpha_i = \frac{(|P_{xi}| - 1) f_0}{\Delta f} \tag{7}$$

$$x_i = \frac{-\mathrm{angle}(P_{xi}) \times \mathrm{c}}{4\pi\Delta f_x} \tag{8}$$

$$y_i = \frac{-\mathrm{angle}(P_{yi}) \times \mathrm{c}}{4\pi\Delta f_y} \tag{9}$$

$$z_i = \frac{-\mathrm{angle}(P_{zi}) \times \mathrm{c}}{4\pi\Delta f_z} \tag{10}$$

where $P_{xi} = (1 + \alpha_i \frac{\Delta f_x}{f_{x0}}) \exp(-4\pi \mathrm{j} \Delta f_x x_i / c)$, $P_{yi} = \exp(-4\pi \mathrm{j} \Delta f_y y_i / c)$, and $P_{zi} = \exp(-4\pi \mathrm{j} \Delta f_z z_i / c)$.

Finally, the intensity parameters $\tilde{A}$ can be estimated by using the least square method as follows

$$\tilde{A} = (\mathbf{G}^{\mathrm{H}} \mathbf{G})^{-1} \mathbf{G}^{\mathrm{H}} \mathbf{E}_{\mathrm{k}} \tag{11}$$

where

$$\mathbf{G} = [\mathbf{a}_1, \ldots, \mathbf{a}_I] \tag{12}$$

$$\mathbf{a}_i = \begin{bmatrix} a_i(0,0,0), \ldots, a_i(M-1,0,0), a_i(0,1,0), \ldots, \\ a_i(M-1,1,0), \ldots, a_i(M-1,N-1,0), \\ a_i(0,0,1), \ldots, a_i(M-1,N-1,K-1) \end{bmatrix}^{\mathrm{T}} \tag{13}$$

$$a_i(m,n,k) = \mathrm{j}\left(\frac{f_m}{f_0}\right)^{\alpha_i} \exp\left[\frac{-4\pi \mathrm{j} f_m}{\mathrm{c}} (x_i \cos\theta_n \cos\varphi_k + y_i \sin\theta_n \cos\varphi_k + z_i \sin\varphi_k)\right] \tag{14}$$

$$\mathbf{E}_k = \begin{bmatrix} E(f_0, \theta_0, \varphi_0), \ldots, E(f_{M-1}, \theta_0, \varphi_0), a_i(f_0, \theta_1, \varphi_0), \ldots, \\ a_i(f_{M-1}, \theta_1, \varphi_0), \ldots, a_i(f_{M-1}, \theta_{N-1}, \varphi_0), \\ a_i(f_0, \theta_0, \varphi_1), \ldots, a_i(f_{M-1}, \theta_{N-1}, \varphi_{K-1}) \end{bmatrix} \tag{15}$$

$[]^{\mathrm{T}}$ represents the transpose operation.

2.3. Observation Geometry Transformation for Space-Based Radar

In this section, the observation geometry transformation for a space-based radar is derived to obtain accurate observing angles, which are significant for GTD scattering center estimation and the HRRP prediction process.

As depicted in Figure 1, the Earth-centered, Earth-fixed (ECEF) coordinate system $O_e - X_e Y_e Z_e$ and the radial tangential normal (RTN) coordinate system $O_t - X_t Y_t Z_t$ are marked in black lines and in red lines, respectively. The origin of the ECEF coordinate system is at the center of the earth, and it is a non-rotating right-handed Cartesian coordinate system. The direction of its X_e axis points to the equator, and the prime meridian intersection and the Z_e axis pass through the Earth's north pole. As for the RTN coordinate system, the X_t axis is parallel to the direction of the velocity, the Z_t axis is along the main body of the space target, and the Y_t axis is determined according to the right-hand theorem.

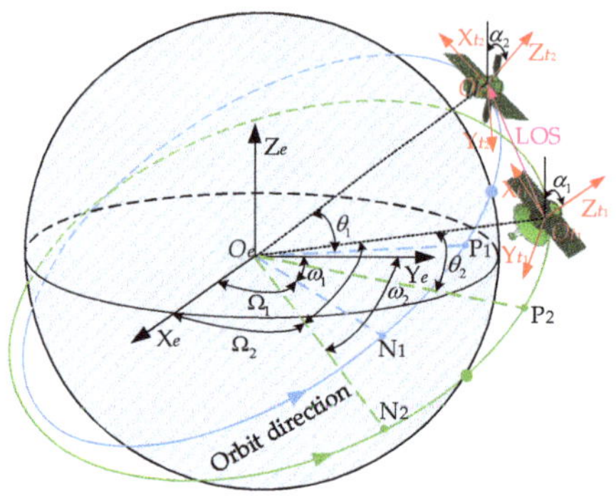

Figure 1. Schematic of the coordination system transformation.

The orbit of the space target is determined by six orbital elements, which are derived from a priori two-line elements (TLEs). To be specific, these elements are expressed as h (semi-major axis), α (inclination), Ω (right ascension of the ascending node), e (eccentricity), ω (argument of perigee), and θ (true anomaly). Thereafter, the observation geometry transformation between the ECEF coordinate system and the RTN coordinate system can be derived according to the given six orbital elements.

Then, the positions of the space target and the space-based radar in the ECEF coordinate system are given as

$$\begin{aligned} \boldsymbol{O}_{t_1 ECEF} &= \boldsymbol{R}_z(\Omega_1)\boldsymbol{R}_x(\alpha_1)\boldsymbol{R}_z(\omega_1) \\ &\cdot \begin{bmatrix} h_1 \frac{e_1+\cos\theta_1}{1+e_1\cos\theta_1} & \left(h_1 \frac{1-e_1{}^2}{1+e_1\cos\theta_1}\right)\sin\theta_1 & 0 \end{bmatrix}^{\mathrm{T}} \end{aligned} \tag{16}$$

$$\begin{aligned} \boldsymbol{O}_{t_2 ECEF} &= \boldsymbol{R}_z(\Omega_2)\boldsymbol{R}_x(\alpha_2)\boldsymbol{R}_z(\omega_2) \\ &\cdot \begin{bmatrix} h_2 \frac{e_2+\cos\theta_2}{1+e_2\cos\theta_2} & \left(h_2 \frac{1-e_2{}^2}{1+e_2\cos\theta_2}\right)\sin\theta_2 & 0 \end{bmatrix}^{\mathrm{T}} \end{aligned} \tag{17}$$

where $\boldsymbol{R}_x$, $\boldsymbol{R}_y$, and $\boldsymbol{R}_z$ represent the rotation matrix around the x axis, the y axis, and the z axis, respectively.

Also, the rotation matrices can be expressed as follows:

$$\boldsymbol{R}_z(\Omega) = \begin{bmatrix} \cos\Omega & -\sin\Omega & 0 \\ \sin\Omega & \cos\Omega & 0 \\ 0 & 0 & 1 \end{bmatrix} \tag{18}$$

$$\boldsymbol{R}_x(\alpha) = \begin{bmatrix} 1 & 0 & 0 \\ 0 & \cos\alpha & -\sin\alpha \\ 0 & \sin\alpha & \cos\alpha \end{bmatrix} \tag{19}$$

$$\boldsymbol{R}_z(\omega) = \begin{bmatrix} \cos\omega & -\sin\omega & 0 \\ \sin\omega & \cos\omega & 0 \\ 0 & 0 & 1 \end{bmatrix} \tag{20}$$

Thus, the coordinate transformation matrix between the ECEF coordinate system and the RTN coordinate system can be expressed as

$$\boldsymbol{M}_{ECEF-OF} = [\boldsymbol{R}_z(\Omega)\boldsymbol{R}_x(\alpha)\boldsymbol{R}_z(\omega+\theta)]^{-1} \tag{21}$$

So, the positions of the space target and space-based radar in the RTN coordinate system are given as

$$\boldsymbol{O}_{t_1OF} = \boldsymbol{M}_{ECEF-OF}(\Omega_1, \alpha_1, \omega_1, \theta_1)\boldsymbol{O}_{t_1ECEF} \tag{22}$$

$$\boldsymbol{O}_{t_2OF} = \boldsymbol{M}_{ECEF-OF}(\Omega_1, \alpha_1, \omega_1, \theta_1)\boldsymbol{O}_{t_2ECEF} \tag{23}$$

Also, the line of sight (LOS) of the radar in the RTN reference coordinate system is computed as

$$\boldsymbol{LOS} = \boldsymbol{O}_{t_1OF} - \boldsymbol{O}_{t_2OF} = (x_\Delta, y_\Delta, z_\Delta) \tag{24}$$

$$\theta_\Delta = \arctan\left(\frac{x_\Delta}{y_\Delta}\right) \tag{25}$$

$$\varphi_\Delta = \arctan\left(\frac{z_\Delta}{\sqrt{x_\Delta^2 + y_\Delta^2 + z_\Delta^2}}\right) \tag{26}$$

where $x_\Delta, y_\Delta, z_\Delta$ represent the location vector of the LOS in the RTN coordinate system and θ_Δ and φ_Δ are the observing azimuth angle and elevation angle in the RTN coordinate system, respectively.

By deriving the observation geometry transformation for the space-based radar, the inherent motion of space targets can be eliminated from consideration in this paper.

3. Proposed Method

In Section 3.1, to predict the HRRP sequence more accurately, we first introduce a novel scattering parameter estimation method based on an improved 3D-ESPRIT algorithm. Then, in Section 3.2, HRRP characteristics obtained via the wide-band radar and HRRP pseudo-likelihood predicted by using the GTD scattering center model are employed in the PHD filter to fully utilize space targets' electromagnetic characteristics. Finally, the main steps of our proposed HGI-PHD filter are given in Section 3.3. To be specific, the flowchart of our proposed HGI-PHD filter is shown in Figure 2.

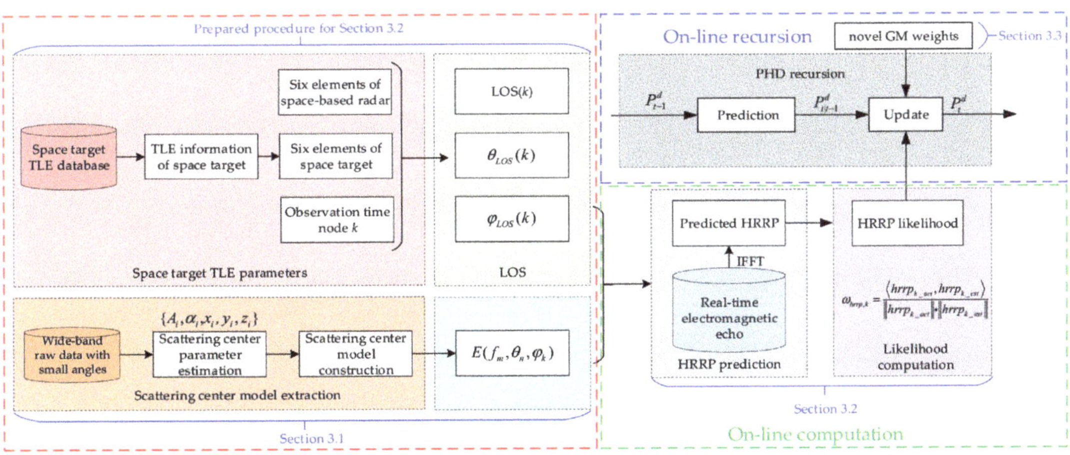

Figure 2. Flowchart of our proposed HGI-PHD filter.

3.1. Parameter Estimation of Scattering Centers

In this section, in order to predict the HRRP sequence more accurately, an improved 3D-ESPRIT algorithm with reconstructed covariance matrices is proposed to estimate the 3D-GTD model parameters, and the main contribution of our proposed 3D-ESPRIT algorithm is shown as follows.

Firstly, by performing the spatial smoothing operation in the x direction, a Hankel matrix pencil $\boldsymbol{X}^x$ can be given by reference [27]

$$\boldsymbol{X}^x = \begin{bmatrix} \boldsymbol{X}_0^x & \boldsymbol{X}_1^x & \cdots & \boldsymbol{X}_{M-P}^x \\ \boldsymbol{X}_1^x & \boldsymbol{X}_2^x & \cdots & \boldsymbol{X}_{M-P+1}^x \\ \vdots & \vdots & \cdots & \vdots \\ \boldsymbol{X}_{P-1}^x & \boldsymbol{X}_P^x & \cdots & \boldsymbol{X}_{M-1}^x \end{bmatrix} \tag{27}$$

where

$$\boldsymbol{X}_m^x = \begin{bmatrix} \boldsymbol{x}(m,0) & \boldsymbol{x}(m,1) & \cdots & \boldsymbol{x}(m,K-L) \\ \boldsymbol{x}(m,1) & \boldsymbol{x}(m,2) & \cdots & \boldsymbol{x}(m,K-L+1) \\ \vdots & \vdots & \cdots & \vdots \\ \boldsymbol{x}(m,L-1) & \boldsymbol{x}(m,L) & \cdots & \boldsymbol{x}(m,K-1) \end{bmatrix} \tag{28}$$

$$\boldsymbol{x}(m,k) = \begin{bmatrix} E(m,0,k) & E(m,1,k) & \cdots & E(m,N-Q,k) \\ E(m,1,k) & E(m,2,k) & \cdots & E(m,N-Q+1,k) \\ \vdots & \vdots & \cdots & \vdots \\ E(m,Q-1,k) & E(m,Q,k) & \cdots & E(m,N-1,k) \end{bmatrix} \tag{29}$$

where $P \in [I+1, M-I+1]$, $Q \in [I+1, N-I+1]$, $L \in [I+1, K-I+1]$, and M, N, K, and I represent the frequency steps, azimuth angle steps, elevation angle steps, and the scattering center numbers, respectively.

Then, a permutation matrix $\boldsymbol{J}$ is defined as

$$\boldsymbol{J} = \begin{bmatrix} 0 & \cdots & 0 & 1 \\ \vdots & 0 & 1 & 0 \\ 0 & \cdot^{\cdot^{\cdot}} & \ddots & \vdots \\ 1 & 0 & \cdots & 0 \end{bmatrix}_{PQL \times PQL} \tag{30}$$

By combining the permutation matrix $\boldsymbol{J}$ and the original back-scattering electromagnetic data $\boldsymbol{X}^x$, a new matrix $\boldsymbol{E}_{conj}$ containing the covariance information of the original scattered data $\boldsymbol{X}^x$ is obtained by

$$\boldsymbol{E}_{conj} = \boldsymbol{J} \cdot \boldsymbol{X}^x \tag{31}$$

Then, we construct the following three covariance matrices to fully utilize the electromagnetic scattering characteristic of radar targets:

$$\begin{cases} \boldsymbol{R}_{\boldsymbol{X}_x\boldsymbol{X}_x} = \boldsymbol{X}_x{\boldsymbol{X}_x}^{\mathrm{H}} \\ \boldsymbol{R}_{\boldsymbol{E}_{conj}\boldsymbol{E}_{conj}} = \boldsymbol{E}_{conj}{\boldsymbol{E}_{conj}}^{\mathrm{H}} \\ \boldsymbol{R}_{\boldsymbol{X}_x\boldsymbol{E}_{conj}} = \boldsymbol{X}_x{\boldsymbol{E}_{conj}}^{\mathrm{H}} \end{cases} \tag{32}$$

Next, by averaging the above three covariance matrices, a novel covariance matrix $\boldsymbol{R}$ is proposed in this paper, which is shown as follows:

$$\boldsymbol{R} = \frac{\boldsymbol{R}_{\boldsymbol{X}_x\boldsymbol{X}_x} + \boldsymbol{R}_{\boldsymbol{E}_{conj}\boldsymbol{E}_{conj}} + \boldsymbol{R}_{\boldsymbol{X}_x\boldsymbol{E}_{conj}}}{3} \tag{33}$$

Note that $\boldsymbol{R}$ is a Hermitten matrix, and it satisfies $\boldsymbol{R} = \boldsymbol{R}^{\mathrm{H}}$. Thus, by squaring matrix $\boldsymbol{R}$, we have the final covariance matrix $\boldsymbol{R}_1$ as follows:

$$\boldsymbol{R}_1 = \boldsymbol{R}\boldsymbol{R}^{\mathrm{H}} = \boldsymbol{R}^2 \tag{34}$$

Also, the following relation satisfies

$$\begin{cases} \lambda_{1i} = \lambda_i^2 \\ \mathbf{\Lambda}_{1i} = \mathbf{\Lambda}_i \end{cases} \quad i = 1, \ldots, I \tag{35}$$

where λ_{1i} and $\mathbf{\Lambda}_{1i}$ are the eigenvalue and the eigenvector of $\boldsymbol{R}_1$, respectively, and λ_i and $\mathbf{\Lambda}_i$ are the eigenvalue and the eigenvector of $\boldsymbol{R}$, respectively.

Note that from (34), the eigenvalues of $\boldsymbol{R}_1$ are twice those of $\boldsymbol{R}$. Therefore, the differences between the noise eigenvalues and the signal eigenvalues can be broadened by constructing $\boldsymbol{R}_1$.

To analyze our motivation for constructing the aforementioned covariance matrix mathematically, the estimated parameters variance is derived as follows:

$$E\left\{(\hat{\xi} - \xi)^2\right\} = \frac{\frac{\sigma^2}{2MNK} \sum\limits_{i=1}^{I} -\frac{\gamma_i}{(\sigma^2 - \gamma_i)^2} \left| \boldsymbol{G}^{\mathrm{H}} \cdot (v_i)^{\mathrm{H}} \right|^2}{\sum\limits_{i=I+1}^{MNK} -\left| \left[d\boldsymbol{G}^{\mathrm{H}} / d(z) \right] \cdot (v_i)^{\mathrm{H}} \right|^2} \tag{36}$$

where $E\{(\hat{\xi} - \xi)^2\}$ represents variance of the estimated parameters; $\hat{\xi}$ and ξ represent the estimated parameter and the original parameter, respectively; σ^2 and γ_i denote eigenvalues of noises and eigenvalues of signals, respectively; $\boldsymbol{v}_i = \gamma_i \boldsymbol{I} - \boldsymbol{X}^x$ represents the eigenmatrix of γ_i; and $\boldsymbol{I}$ denotes a identify matrix.

It can be seen from (36) that $E\{(\hat{\xi} - \xi)^2\}$ decreases when σ^2 and γ_m differ greatly from each other, which will bring out more accurate scattering center parameters. So, constructing $\boldsymbol{R}$ can broaden the differences between σ^2 and γ_m, which can estimate the GTD model parameters more accurately.

Then, the scattering parameters $\{\alpha_i, x_i, y_i, z_i\}$ are estimated according to reference [27], and the intensity parameters $\tilde{\boldsymbol{A}}$ can be obtained by using the least square method in (27). It should be noticed that for the sake of readability, the detailed CRB derivation of the GTD scattering center parameter is given in Appendix A.

3.2. HRRP Characteristic-Aided Filter

Once we have estimated the scattering center model parameters, the HRRP sequence can be predicted at a relatively small observing angle. Thus, the predicted HRRP information can be utilized in the update recursion of the PHD filter. In this section, we will introduce the specific implementation of our proposed HRRP characteristic-aided filter.

Let us denote the surveillance region as three-dimensional; then, the target state can be expressed as

$$\boldsymbol{x}_k = \left[l_{x,k}, v_{x,k}, l_{y,k}, v_{y,k}, l_{z,k}, v_{z,k} \right]^{\mathrm{T}} \tag{37}$$

where at time k, $l_{x,k}$, $l_{y,k}$, and $l_{z,k}$ are the three-dimensional locations of targets. $v_{x,k}$, $v_{y,k}$, and $v_{z,k}$ are the three-dimensional velocities of targets.

Thereafter, as the HRRP characteristics are employed, the measurement state of targets at time k yields

$$z_k = \left[z_{l,k}, z_{hrrp,k} \right] \tag{38}$$

where at time k, $z_{l,k}$ and $z_{hrrp,k}$ denote the location measurements and the HRRP measurements, respectively.

The specific prediction and update recursion of the proposed HGI-PHD filter are discussed as follows.

***Prediction*:**

The prediction recursion of the proposed filter is written as

$$D_{k|k-1}(\boldsymbol{x}_k) = \gamma_k(\boldsymbol{x}_k) + \int \left(p_{s,k}(\boldsymbol{x}_{k-1}) f_{k|k-1}(\boldsymbol{x}_k \middle| \boldsymbol{x}_{k-1}) + b_{k|k-1}(\boldsymbol{x}_k \middle| \boldsymbol{x}_{k-1}) D_{k-1|k-1}(\boldsymbol{x}_{k-1}) \right) d\boldsymbol{x}_{k-1} \tag{39}$$

where at time k, $\gamma_k(\boldsymbol{x}_k)$ denotes the intensity of the birth RFS, $p_{s,k}(\boldsymbol{x}_{k-1})$ is the survival probability, $f_{k|k-1}(\boldsymbol{x}_k | \boldsymbol{x}_{k-1})$ is the Markov transmission density, $b_{k|k-1}(\boldsymbol{x}_k | \boldsymbol{x}_{k-1})$ represents the density of the spawned at time k under the condition of the previous state at time $k-1$, and $D_{k-1|k-1}(\boldsymbol{x}_{k-1})$ is the density of the PHD filter at time $k-1$.

***Update*:**

After employing the HRRP characteristics in the standard PHD filter, the update recursion of the proposed HGI-PHD filter can be given by

$$D_{k|k}(\boldsymbol{x}_k) = L_{\boldsymbol{Z}}(\boldsymbol{x}_k) \cdot D_{k|k-1}(\boldsymbol{x}_k) \tag{40}$$

Assuming the HRRP characteristic of targets is independent of location states, the likelihood for targets and for clutter can be denoted as

$$g(\boldsymbol{z} | \boldsymbol{x}) = g((z_{l,k}, z_{hrrp,k}) | \boldsymbol{x}) = g_l(z_l | \boldsymbol{x}) g_{hrrp}(hrrp) \tag{41}$$

$$c(\boldsymbol{z} | \boldsymbol{x})) = c_z(z | \boldsymbol{x}) c_{hrrp}(hrrp) \tag{42}$$

where $g_l(z_l|x)$ and $g_{hrrp}(hrrp)$ are the state location likelihood and HRRP pseudo-likelihood functions for targets, respectively. $c_z(z|\boldsymbol{x})$ and $c_{hrrp}(hrrp)$ are the state location likelihood and HRRP pseudo-likelihood functions for clutter, respectively.

However, how do we predict the HRRP sequence as time varies, and how do we compute the HRRP pseudo-likelihood for various targets and clutter? These two problems are core points to be solved in this paper, and we give the detailed process as follows.

(1) ***HRRP sequence prediction***

Note that the radar is able to provide range and bearing measurements of the target's center, and the observation model of kinematic measurement is denoted as follows:

$$\boldsymbol{M}_k = [R_k, \beta_k] = h(\boldsymbol{l}_k) + \boldsymbol{w}_k \tag{43}$$

where at time k, $h(\cdot)$ denotes the kinematic observation function; R_k and β_k represent the target range measurements and bearing measurements, respectively; and $\boldsymbol{w}_k$ represents the zero-mean observation noise matrix.

For the space target, due to its previously known orbital information and low maneuverability, the velocity direction is almost aligned with the axial direction of the target's main body, which is shown in Figure 3.

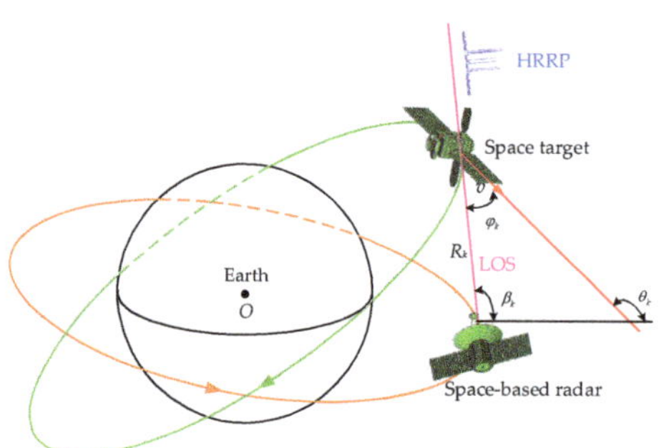

Figure 3. Schematic of target observing angle.

Therefore, the observing angle can be computed as

$$\varphi_k = \theta_k - \beta_k = \cos^{-1}\left(\frac{x_k v_{x,k} + y_k v_{y,k}}{\sqrt{x_k^2 + y_k^2}\sqrt{v_{x,k}^2 + v_{y,k}^2}}\right) \tag{44}$$

where $\theta_k = \tan^{-1}\left(\frac{v_{y,k}}{v_{x,k}}\right)$ denotes the heading angle for space targets.

Once the specific observing angles are given as time varies, different back-scattering echoes can be obtained via the 3D scattering center model in (6). Then, the frequency response of the m-th frequency point can be written as $E_m = E(f_m, \theta_n, \varphi_k)$, and by taking an inverse discrete Fourier transform (IDFT) on frequency response sequence $\boldsymbol{E} = E_0, E_1, \ldots E_M$, the HRRP sequence of space targets can be predicted eventually, which is denoted as $hrrp_{k\_est}$ in this paper.

(2) ***HRRP pseudo-likelihood evaluation***

At time k, we can acquire the actual HRRP sequence of different space targets or clutter, which is denoted as $hrrp_{k\_act}$. Then, the HRRP pseudo-likelihood can be computed as

$$g_{hrrp,k} = \frac{\langle hrrp_{k\_act}, hrrp_{k\_est}\rangle}{\|hrrp_{k\_act}\|\cdot\|hrrp_{k\_est}\|} \tag{45}$$

where $\langle a, b\rangle = \int a(x)b(x)dx$ represents the inner product operation and $\|h\|$ is the l_2 norm function.

Note that the posterior intensity at time $k-1$ is in the form of a Gaussian mixture; then, we have

$$v_{k|k-1}(x) = \sum_{i=1}^{J_{k|k-1}} w_{k|k-1}^i \mathcal{N}\left(x; \boldsymbol{m}_{k-1}^i, \boldsymbol{P}_{k-1}^i\right) \tag{46}$$

where $w_{k|k-1}^i$ is the i-th Gaussian mixture weight at time $k-1$; $\boldsymbol{m}_{k-1}^i$ and $\boldsymbol{P}_{k-1}^i$ denote the mean vector and covariance vector of the i-th Gaussian mixture weight at time $k-1$, respectively; and $J_{k|k-1}$ represents the number of Gaussian mixture weights at time $k-1$.

Substituting (46) into (40), we have the posterior intensity at time k as

$$D_{k|k}(\boldsymbol{x}_k) = [1 - p_d] D_{k|k-1}(\boldsymbol{x}_k) + \sum_{i=1}^{J_{k|k-1}} \sum_{z_k \in \boldsymbol{Z}} \widetilde{w}_{k|k-1}^i \mathcal{N}\left(\boldsymbol{x}_k; \boldsymbol{m}_{k|k-1}^i, \boldsymbol{P}_{k|k-1}^i\right) \tag{47}$$

where the normalized GM weight is denoted as

$$\widetilde{w}_{k|k-1}^j(z_k) = \frac{p_{D,k}(x) g_z(z|x) g_{hrrp}(hrrp) w_{k|k-1}^j \boldsymbol{q}_k^j(z_k)}{\kappa(z) c_{hrrp}(hrrp) + p_{D,k}(x) g_z(z|x) g_{hrrp}(hrrp) \sum_{i=1}^{J_{k|k-1}} w_{k|k-1}^i \boldsymbol{q}_k^i(z_k)} \tag{48}$$

where $\begin{cases} \boldsymbol{q}_k^i(z_k) = N\left(z_k, \boldsymbol{\eta}_{k|k-1}^i, \boldsymbol{S}_{k|k-1}^i\right) \\ \boldsymbol{\eta}_{k|k-1}^i = \boldsymbol{H}_k \boldsymbol{m}_{k|k-1}^i \\ \boldsymbol{S}_{k|k-1}^i = \boldsymbol{H}_k \boldsymbol{P}_{k|k-1}^i \boldsymbol{H}_k^{\mathrm{T}} + \boldsymbol{R}_k \end{cases}$.

3.3. Improved Gaussian Component Weight Merging Principal

By utilizing (43), the posterior intensity of targets can be obtained, and different intensities are represented by different constituent Gaussian component weights. With the aim of tracking space targets with high clutters effectively, label κ is denoted in this paper. Then, we have the set $\boldsymbol{v}_k = \left\{w_k, \boldsymbol{x}_k^i, \boldsymbol{P}_k^i, \kappa_k^i\right\}$, and κ_k^i denotes the label of the kth target at time i.

Following the RFS scheme, the constituent Gaussian mixture component weights are modified as follows.

(1) First, the index of the highest weight component in the target posterior density is selected as follows:

$$i_{\max} = \underset{i \in I}{\operatorname{argmax}}(w_k^i) \tag{49}$$

where I is the index set.

(2) Then, the set of component orders that are most similar to the largest weighted component in the target posterior intensity at time k is represented as

$$\Phi_k = \left\{ i \mid \left(\left(x_k^i - x_k^{i_{\max}} \right)^{\mathrm{T}} \left(\boldsymbol{P}_k^i \right)^{-1} \left(x_k^i - x_k^{i_{\max}} \right) \leq U_{mm} \right) \right\} \tag{50}$$

$$U_{mm} = \left(1 + w_k^i \right) \sigma_s \tag{51}$$

where σ_s is the variance of the measured Gaussian white noise.

(3) If the Gaussian component satisfies Equation (50) and the sum of the component weights fails to exceed twice the state extraction threshold, then a novel set v_k^τ, which contains components, can be obtained as follows:

$$\widetilde{l}_k^\tau = i_k^{i_{\max}}, \widetilde{w}_k^\tau = \sum_{i \in \Phi_k} w_k^i, \boldsymbol{x}_k^\tau = \frac{1}{\widetilde{w}_k^\tau} \sum_{i \in \Phi_k} w_k^i \boldsymbol{x}_k^i \tag{52}$$

While $\sum_{k=1}^{M_k} \widetilde{w}_k^\tau > 1$, the sub-weights of the target are multiplied by a penalized factor except the highest weight $i_{\max}$. Therefore, the novel weights yield the following:

$$w_{novel}^k = \begin{cases} \widetilde{w}_k^\tau, if\ k = i_{\max} \\ \beta \widetilde{w}_k^\tau, if\ k \neq i_{\max}, k \in [1, \ldots, M_k] \end{cases} \tag{53}$$

where $\beta = \alpha(1 - w_k^{i_{\max}})$ denotes the penalized factor and $0 \leq \alpha \leq 1$ is a constant that can determine the penalized degree.

(4) Substituting (53) into (48) yields

$$\overline{w}_{novel}^{i,n} = \begin{cases} \dfrac{w_k^{i,n}}{\kappa_k(\lambda_c) + w_k^{i_{\max}} \beta + \sum\limits_{i=1}^{N_{k|k-1}} w_k^{i,n}}, i \neq i_{\max}, i \in [1, M_k], n \in [1, N_{k|k-1}] \\ \dfrac{w_k^{i_{\max},n} \beta}{\kappa_k(\lambda_c) + w_k^{i_{\max}} \beta + \sum\limits_{i=1}^{N_{k|k-1}} w_k^{i,n}}, i = i_{\max}, n \in [1, N_{k|k-1}] \end{cases} \tag{54}$$

where $w_k^{n,m} = p_{D,k}(x) g_{rcs}(rcs) g_d(d) w_{k|k-1}^j q_k^j(z_k)$ and $\kappa_k(\lambda_c) = \lambda_c c_{rcs}(rcs) c_d(d)$.

(5) Finally, the novel posterior density is computed as follows:

$$v(x) = \sum_{J_k}^{\tau=1} w_k^\tau N(x; x_k^\tau, p_k^\tau) \tag{55}$$

$$\widetilde{\boldsymbol{P}}_k^\tau = \frac{1}{x_k^\tau} \sum_{i \in \Phi_k} w_k^i \left(\boldsymbol{P}_k^i + \left(x_k^\tau - x_k^i \right) \left(x_k^\tau - x_k^i \right)^{\mathrm{T}} \right) \tag{56}$$

We can observe from (51) that U_{mm} is an adaptive threshold varying with the measured noise in tracking scenarios. Therefore, the threshold can be dynamically adjusted during the whole tracking process and has a better tracking performance for the closely spaced targets.

3.4. Key Steps of the Proposed HGI-PHD Filter

The main steps of our proposed methods are given as follows. For the sake of readability, the corresponding pseudo-code of our proposed HGI-PHD filter is provided in Appendix B.

Step 1 Calculating the time-varying line-of-sight (LOS) and observing angles.

Step 2 Estimating the GTD scattering parameter via our proposed improved 3D-ESPRIT algorithm and reconstructing the scattering center model.

Step 3 Combining the time-varying observing angles with the scattering center model to predict the real-time scattering echo of space targets and applying an IFFT transform to the real-time electromagnetic echo to obtain the predicted HRRP sequence.

Step 4 Computing the HRRP likelihood for different targets and clutter.

Step 5 Improving the constituent Gaussian component weights by using the method introduced in Section 3.3.

Step 6 Employing the HRRP likelihood and the improved Gaussian component weight merging principal (e.g., $\overline{w}^{i,n}_{novel}$) in the update recursion of the PHD filter.

4. Simulation Results

In this paper, a merging PHD (HGI-PHD) filter aided by HRRP characteristics and improved with Gaussian mixture components is proposed, which can be applied to track space targets accurately. The time-varying HRRP pseudo-likelihood is utilized in the GM process of the standard PHD filter. The performances of our proposed HGI-PHD filter are validated by the following simulation experiments. Furthermore, the capability of the proposed HGI-PHD filter to discriminate approaching space targets is also demonstrated by utilizing the star-link satellites' real raw data.

4.1. Performance Evaluation of Scattering Center Estimation

In this section, the performance of the modified algorithm with respect to the SNR is compared with the other two methods. We set the scattering center parameters as shown in Table 1; the matrix beam parameters P, Q, and L are all set as 6; the SNR varies from 0 dB to 30 dB with an interval of 1 dB; and 200 Monte Carlo trials are performed for each fixed SNR. To simplify, here, we just provide the mean RMSE of the scattering centers obtained by using different methods. Simulation results are depicted in Figure 4. It can be observed from Figure 4a–e that all of the scattering parameter estimation accuracies obtained via different methods increase as SNR increases, and our proposed improved 3D-ESPRIT method has a better parameter estimation performance than the traditional 3D-ESPRIT algorithm, the method in reference [15], and the method in reference [27]. Furthermore, note that the location parameter estimation accuracy is better than that of the type parameter and intensity parameter due to the latter two parameters being dependent on the aforementioned location parameters.

Table 1. Scattering parameters.

Scattering Centers	$x_i(m)$	$y_i(m)$	$z_i(m)$	α_i	A_i
Scattering center 1	1.202	1.080	1.320	1.0	6.2
Scattering center 2	1.353	1.263	1.541	0.5	5.6
Scattering center 3	1.534	1.702	2.520	0	4.7
Scattering center 4	1.892	2.322	3.210	1.0	3.4

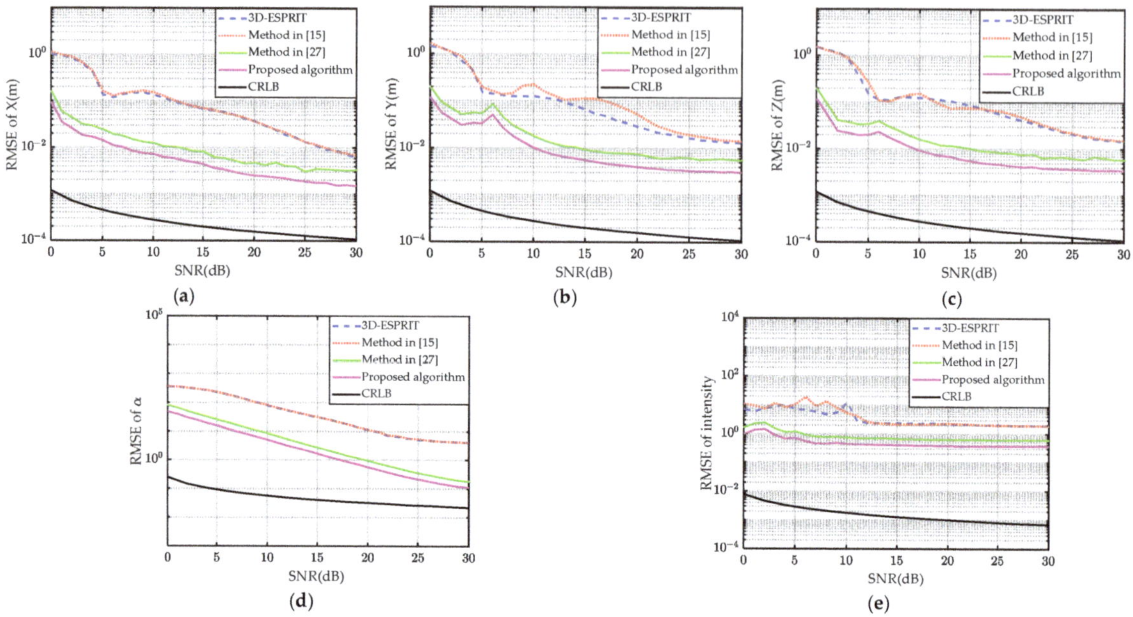

Figure 4. Comparisons of scattering parameter estimation accuracy between different methods. (**a**) X, (**b**) Y, (**c**) Z, (**d**) Type, and (**e**) Intensity.

4.2. HRRP Sequence Reconstructed by the GTD Model

To evaluate the reconstruction and parameter estimation performance of the proposed improved 3D-ESPRIT algorithm with simulated data, the back-scattering electromagnetic echo of three typical targets is calculated using electromagnetic computing software. To be specific, the shape and geometrical size of these three typical targets are depicted in Figure 5. In this simulation, we set the frequency number as 101 points with a range of 15.7~17.7 GHz and a frequency step of 20 MHz. Here, the incident azimuth angle and the elevation angle are set as 0° and 90°, respectively.

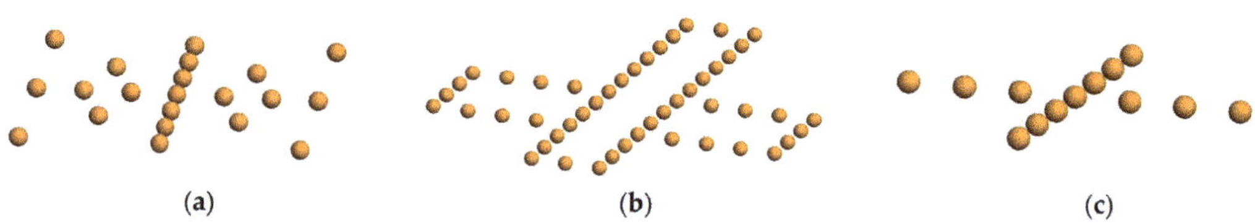

Figure 5. Scattering center model. (**a**) Target 1, (**b**) Target 2, and (**c**) Target 3.

It can be observed from Figure 6 that for these three typical targets, there are 3, 3, and 2 sharp peaks in the HRRP curve of the original data. Note that the reconstructed HRRP curve of these three targets obtained by using our improved 3D-ESPRIT algorithm matches the whole trend well, and it can also describe the fluctuation characteristics accurately. The above simulation results validate the effectiveness of our improved 3D-ESPRIT method and the feasibility of HRRP reconstruction via the GTD scattering center model.

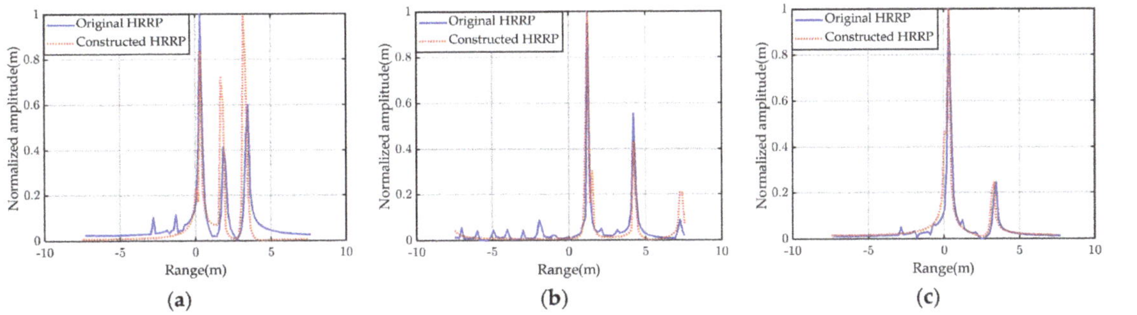

Figure 6. Reconstructed HRRP of three typical space targets with the GTD scattering center model when the azimuth angle is 0 and the elevation angle is 90. (**a**) Target 1, (**b**) Target 2, and (**c**) Target 3.

4.3. HRRP Similarity Rate Comparison

Firstly, the following three-dimensional (3D) simulation scenario is set to obtain the dynamic HRRP of three typical space targets, as depicted in Figure 7. From Figure 7, it can be seen that there are three typical space targets in the surveillance tracking region of the space-based radar.

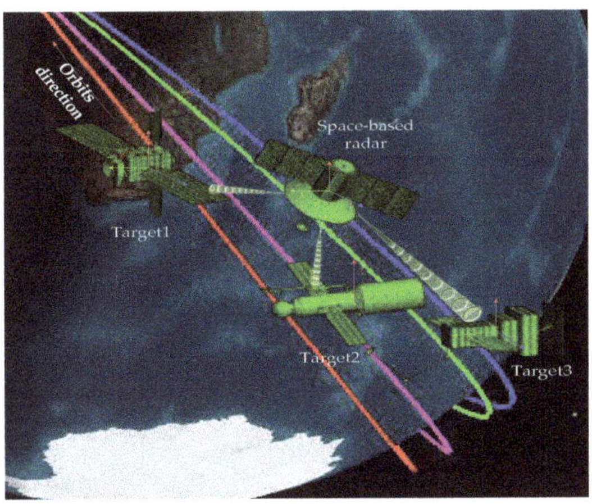

Figure 7. Three-dimensional simulation scenario.

In most applications, LOS parameter sequences are acquired via different methods. In this paper, we solve satellite targets with TLE ephemeris to compute their positions in the SGP4 model, and then their positions relative to the space-based radar can be calculated eventually. Hence, the LOS parameters can be calculated by using the spacecraft orbit computing software, which is shown in Figure 8. Table 2 gives the specific simulation parameters of the space radar and the three typical satellite targets in the spacecraft orbit computing software. Finally, combining the predicted HRRP sequence and the theoretical HRRP sequence of different targets and clutter, we obtain the dynamic coefficients among different targets and clutter as shown in Figure 9. It can be obviously seen that the predicted HRRP and the theoretical HRRP of the same target have the highest similarity rate than those of different targets or clutter. That means the same target has the largest HRRP pseudo-likelihood with its theoretical HRRP sequence, which can be utilized to improve the update accuracy and space target tracking performance.

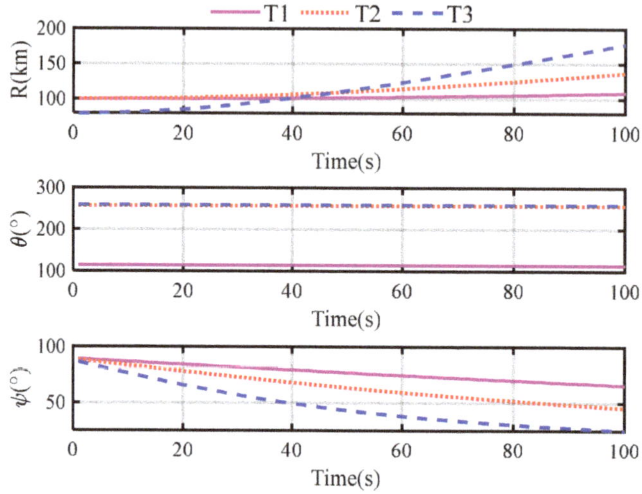

Figure 8. LOS parameters of three typical space targets. (T1: Target 1, T2: Target 2, T3: Target3).

Table 2. Simulation parameters in the spacecraft orbit computing software.

Parameters	Value
Range of detection R	200 km
Orbit altitude of space-based radar H_0	500 km
Orbit altitude of target 1 H_1	400 km
Orbit altitude of target 2 H_2	500 km
Orbit altitude of target 3 H_3	480 km
Orbit inclination of space-based radar ϑ_0	45°
Orbit inclination of target 1 ϑ_1	48°
Orbit inclination of target 2 ϑ_2	55°
Orbit inclination of target 2 ϑ_3	60°

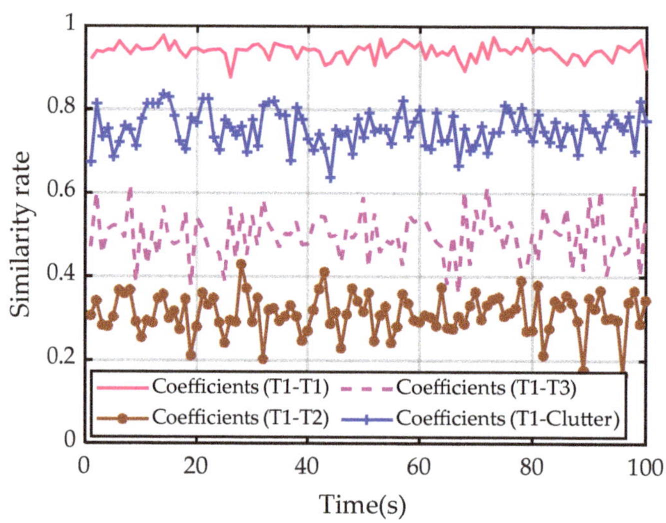

Figure 9. Dynamic coefficients among different targets and clutter.

4.4. Comparisons of OSPA Errors

Suppose there are three space targets existing in the surveillance region with a detection range of [0, 100]km × [0, 100]km × [0, 100]km. As for these three space targets, all of them obey the variable acceleration motion model, and the kinematic state vector includes the positions, velocities, and accelerations as $z_k = \left[x, y, z, v_x, v_y, v_z, a_x, a_y, a_z\right]^{\mathrm{T}}$. Moreover, it should be noticed that the measurement vector $\tilde{x}_k = [x_k, HRRP_k]^{\mathrm{T}}$ includes the position measurements and HRRP characteristic measurements.

Let us set the duration for the whole surveillance simulation as 100 s and the survival probability for independent space targets as 0.98. As for the three space targets, they are born at 1s, and the second target vanishes at 70s, while the third one is present throughout the whole tracking process. Then, the pruning procedure is employed with varying time by utilizing the Gaussian weight threshold as $T_{th} = 10^{-5}$ and the maximum number of the mixture components as 100, and we fix the cardinality distribution over 100 terms. Furthermore, the detection probability is set as $P_d = 0.98$ for all filters, $V = 8 \times 10^6$ m^3 represents the whole space surveillance region volume, and measurements for detected targets are immersed with clutter, which obeys a Poisson RFS distribution and has an intensity of 1.4×10^{-5} m^{-3} (i.e., 112 false alarms over space surveillance region per frame). Notice that the HRRP probability density functions of clutter are obtained by adding white Gaussian noise to the HRRP probability density functions of targets.

Here, the optimal sub-pattern assignment (OSPA) metric is employed for the detection performance evaluation metric, and its specific form is given as follows [28]:

$$D_c^p(X, Y) \triangleq \left[\frac{1}{N}\left(\min_{\ell \in \Pi_N^M} \sum_{i=1}^{M} d_c(x_i, y_{\ell(i)})^p + c^p(N - M)\right)\right]^{\frac{1}{p}} \tag{57}$$

where $d_c(x_i, y_{\ell(i)}) \triangleq \min(c, \|x_i - y_{\ell(i)}\|)$ is the cut-off distance with parameter $c > 0$, $\|\cdot\|$ denotes the Euclidean distance, and Π_N^M represents the set of permutations of cardinality M on $\{1, 2, \ldots, N\}$. p is the order parameter of the OSPA metric with a range of $1 \leq p \leq \infty$, and the OSPA distance is given by $d_c^p(X, Y) = d_c^p(Y, X)$, which corresponds to the case of $M > N$. Then, the average OSPA results after 100 Monte Carlo simulations for detection probabilities with parameters set as $p = 1$ and $c = 200$ m are shown in Figure 10, which is able to demonstrate the performance metrics for the accuracy of target location estimation as well as target number estimation.

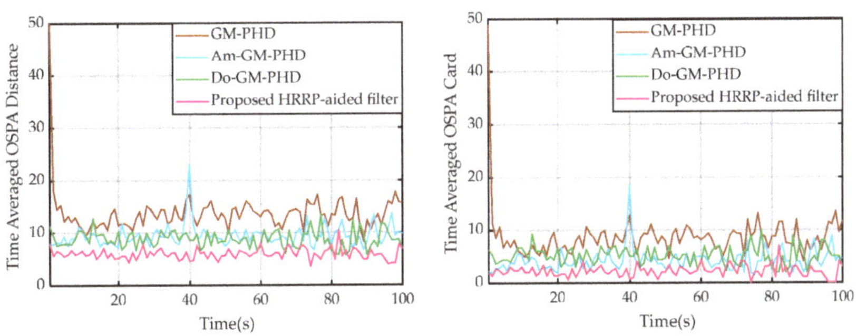

(a) (b)

Figure 10. Average of OSPA errors over 100 Monte Carlo trials. (**a**) Mean OSPA distance. (**b**) Mean OSPA cardinality.

It can be seen from Figure 10a,b that the GM-PHD filter has the highest location error and average OSPA cardinality among the four methods in general. Note that for most of the observing time, the amplitude-aided GM-PHD (Am-GM-PHD) filter and the Doppler-aided

GM-PHD (Do-GM-PHD) filter have better location estimation accuracy and cardinality estimation accuracy than the conventional GM-PHD filter. We would like to mention that the estimated OSPA location obtained by the Am-GM-PHD filter fluctuates sharply as the time sequence varies. Furthermore, it can be observed that our proposed HRRP-aided filter has lower estimation OSPA results than the other three methods as depicted in Figure 11. We can also notice that our proposed filter fits better with the ground truth data than the other three methods.

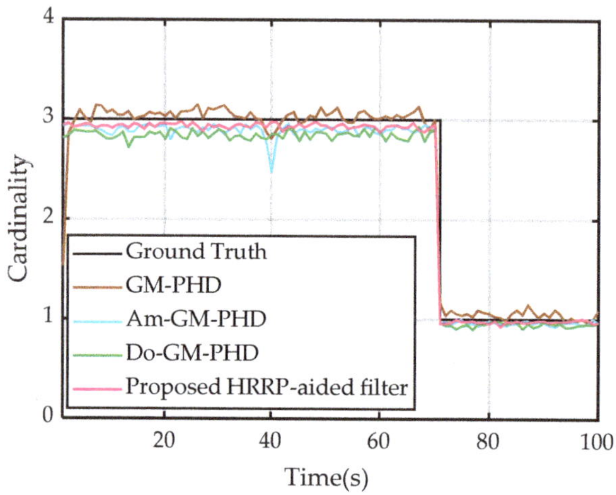

Figure 11. Cardinality estimation results.

The explanation is that, as expected, the proposed filter utilizes the HRRP characteristics of space targets obtained via the wide-band space-based radar system, which is more delicate in describing the target scattering features than the aforementioned ones (e.g., amplitude, etc.). Simulation results show that the proposed HRRP-aided filter outperforms the GM-PHD filter, the Am-GM-PHD filter, and the Do-GM-PHD filter, especially in dense clutter detection scenarios.

4.5. Effective Tracking of Two Closely Spaced Targets

To further exploit the performance of our proposed HGI-PHD filter to track closely spaced multiple targets, two closely spaced targets are chosen for tracking in this subsection. Note that due to Space-X satellites having similar orbit characteristics, two star-link satellites with serial numbers 55713U and 55715U are selected as the closely spaced targets to validate the discrimination functionality of the proposed filter.

The TLE information and observing duration for the two selected star-link satellites are depicted in Table 3, and the observing duration is from 3 Mar 2023 01:55:39.000 to 3 Mar 2023 01:57:19.000. Moreover, the 3D observation diagram in the spacecraft orbit computing software is shown in Figure 12.

Table 3. TLE information of the two selected star-link satellites.

Targets	TLE Information
Satellite 1	1 55713U 23026U 23066.91667824 .00087040 00000-0 95406-3 0 9990 2 55713 43.0022 235.7401 0001445 276.7526 56.7822 15.62262486 1301
Satellite 2	1 55715U 23026W 23066.91667824 .00053204 00000-0 58720-3 0 9995 2 55715 41.0058 235.6894 0002065 272.0442 84.7138 15.62266923 1303

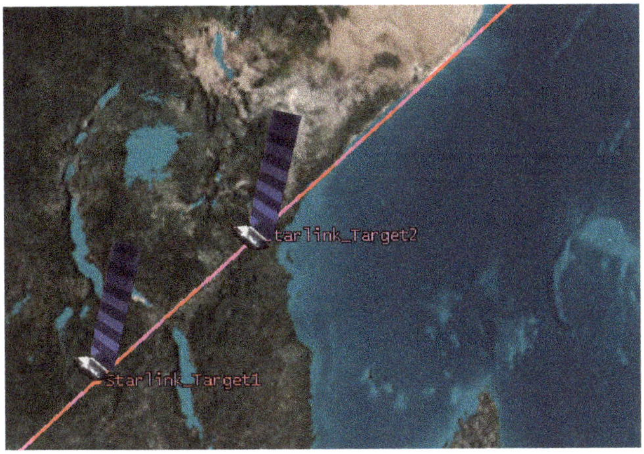

Figure 12. Three-dimensional observation diagram of two closely spaced targets.

After obtaining the trajectories of the two space targets, the tracking performance of various MTT filters can be compared. Figure 13a,b show the time-averaged OSPA distance and the time-averaged OSPA cardinality over 200 Monte Carlo trials, respectively. Figure 14 shows the mean deviations of the cardinality distributions. As shown in Figures 13 and 14, the proposed HGI-PHD filter produces the best results in the simulation not only in OSPA distance but in cardinality estimation as well. In contrast, the worst performance is from the standard GM-PHD filter, i.e., both its OSPA distance and OSPA cardinality are high in the initial stage of our simulation. Furthermore, it can be also seen that the results obtained from the Am-GM-PHD filter and the Do-GM-PHD filter fluctuate more violently than those obtained from the HGI-PHD filter. Hence, we are convinced by comparing Figures 13 and 14 that the proposed HGI-PHD filter outperforms the other three filters for resolving closely spaced targets.

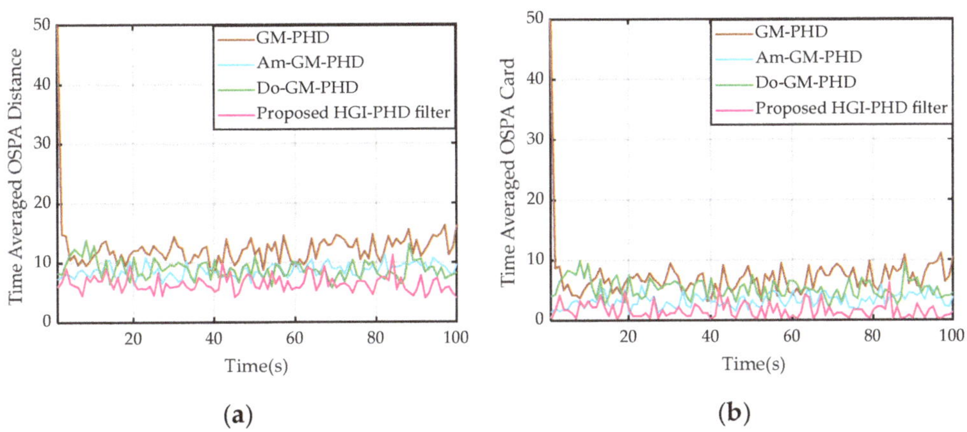

Figure 13. Tracking performance for two closely spaced targets. (**a**) Mean OSPA distance. (**b**) Mean OSPA cardinality.

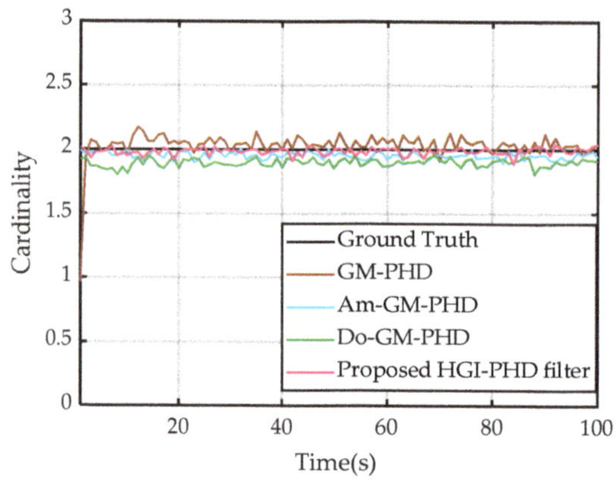

Figure 14. Cardinality estimation results of two closely spaced targets.

4.6. Computational Time Analysis

To analyze the computational complexity of different filters, the computational time is compared with an Intel(R) Core(TM) i7-8700 CPU @ 3.20GHz PC with 16 GB RAM and MATLAB R2021b. The core number we have utilized is 6, and the operating system is Windows 10. And the average computational time of the four filters is given in Table 4 after 100 Monte Carlo trials. It can be observed the proposed filter has a little heavier computational burden than the Am-GM-PHD filter under two clutter scenarios, which are 3.005 s and 3.086 s corresponding to two different clutter rates. The Do-GM-PHD filter and the standard GM-PHD filter have similar computational costs. In fact, due to computing the prior knowledge of the time-varying HRRP pseudo-likelihood, the proposed filter has a similar computational cost to the Am-GM-PHD filter theoretically, and the results in Table 4 validate it quantitatively.

Table 4. Comparisons of Average Computational Time.

Clutter λ_c	GM-PHD	Am-GM-PHD	Do-GM-PHD	The Proposed Filter
50	0.725 s	3.249 s	0.780 s	3.005 s
100	0.815 s	3.425 s	1.075 s	3.086 s

5. Conclusions

In the standard PHD filter, the tracking performance deteriorates sharply when clutter rates are high, detection probabilities are low, or targets approach each other. To deal with the aforementioned problems, an adapted PHD-based filter aided by HRRP characteristics and with an adapting threshold for the GM component merging procedure was proposed in this paper. By fully using the HRRP information, the electromagnetic characteristics of space targets were utilized in the update recursion of the standard PHD filter. We have also derived the mathematical proof for the GM component merging procedure.

Our proposed HGI-PHD method can be applied to the space situational awareness region. It was shown that our proposed HGI-PHD filter outperformed the GM-PHD filter, the Am-GM-PHD filter, and the Do-GM-PHD filter, especially in scenarios where the detection probabilities are low, clutter rates are high, or there are closely spaced targets. the robustness and effectiveness of the proposed HGI-PHD filter were validated through simulation experiments and using real trajectories. Future work includes the development of efficient real-time algorithms for SAA tasks and the improvement of tracking accuracy for space targets.

Author Contributions: S.Z. proposed the conceptualization and methodology; Q.Y. wrote the draft manuscript; Z.W. and L.J. supervised the experimental analysis and revised the manuscript; Y.Z. proof-read and revised the first draft. All authors have read and agreed to the published version of the manuscript.

Funding: This research received no external funding.

Data Availability Statement: Not applicable.

Acknowledgments: The authors would like to thank the editors and all the reviewers for their very valuable and insightful comments during the revision of this work.

Conflicts of Interest: The authors declare no conflict of interest.

Appendix A

Derivation of CRLB of the GTD Model

Here, we derive the deterministic Cramer–Rao lower bound (CRLB) of the GTD scattering center model parameters. For simplicity, we give the CRLB derivation of the one-dimensional GTD model, and the CRB for the 3D-GTD model can be obtained by expansion. The one-dimensional GTD model can be written as

$$\begin{aligned} E(f_m) &= \sum_{i=1}^{I} A_i \exp(-\mathrm{j}4\pi f_0 r_i/\mathrm{c})(1+\tfrac{m\Delta f}{f_0})^{\alpha_i} \exp(-\mathrm{j}4\pi m\Delta f r_i/\mathrm{c}) + \omega(f_m) \\ &= \sum_{i=1}^{I} a_i(1+\tfrac{m\Delta f}{f_0})^{\alpha_i} \exp(-\mathrm{j}m\varpi_i) + \omega(f_m) \end{aligned} \tag{A1}$$

where $\{r_i, a_i, A_i\}$ represent the position parameter, scattering intensity, and scattering type of the i-th scattering center, respectively. And other parameters have been defined in (6). Note that a_i is equal to $A_i \cdot \exp(-\mathrm{j}4\pi f_0 r_i/\mathrm{c}) = a_{R_i} + \mathrm{j}a_{I_i} = |a_i|e^{\mathrm{j}\varphi_i}$, and ϖ_i is equal to $-4\pi\Delta f r_i/\mathrm{c}$.

Thereafter, the estimated scattering parameters in (A1) can be expressed as follows:

$$\zeta = \begin{bmatrix} \sigma^2 & \zeta_{a\mathrm{Re}}^{\mathrm{T}} & \zeta_{a\mathrm{Im}}^{\mathrm{T}} & \zeta_{P}^{\mathrm{T}} & \zeta_{\varpi}^{\mathrm{T}} \end{bmatrix}^{\mathrm{T}} \tag{A2}$$

where σ^2 denotes variance of Gaussian white noise, $\zeta_{a\mathrm{Re}}^{\mathrm{T}} = [a_{\mathrm{Re}1}, \ldots, a_{\mathrm{Re}I}]$, $\zeta_{a\mathrm{Im}}^{\mathrm{T}} = [a_{\mathrm{Im}1}, \ldots, a_{\mathrm{Im}I}]$, $\zeta_{P}^{\mathrm{T}} = [p_1, \ldots, p_I]$, and $\zeta_{\varpi}^{\mathrm{T}} = [\varpi_1, \ldots, \varpi_I]$.

When the relative operating bandwidth of radar satisfies $\gamma = \frac{N\Delta f}{f_0} = \frac{B}{f_0} \ll 1$, then we can obtain the following approximation:

$$\left(1+m\frac{\Delta f}{f_0}\right)^{\alpha_i} = \exp\left(\alpha_i \cdot \ln(m\frac{\Delta f}{f_0})\right) \approx \exp\left(\alpha_i \cdot m\frac{\Delta f}{f_0}\right) \tag{A3}$$

So, the 1D-GTD model is transformed as the damped exponential (DE) model, which is shown as

$$E_{DE}(f_m) = \sum_{i=1}^{I} A_i \exp(m\alpha_i\Delta f/f_0)\exp(-4\pi\mathrm{j}f_m r_i/\mathrm{c}) + \omega(f_m) = \sum_{i=1}^{I} A_i v_i^m + \omega(f_m) \tag{A4}$$

where $v_i^m = p_i \cdot \exp(-4\pi\mathrm{j}f_m r_i/\mathrm{c})$ and $p_i = \exp(m\alpha_i\Delta f/f_0)$.

In [29], it has been verified that the CRLB of the DE model and the GTD scattering center model are substitutable by both theoretical derivations and simulation experiments. Therefore, the CRLB of the DE model is derived here to substitute that of the GTD model.

Due to being the Gaussian white noise, the CRLB matrix of the DE model can be computed as

$$\boldsymbol{CRLB}_{DE} = \begin{bmatrix} \frac{\sigma^4}{N} & \boldsymbol{0}_{1\times 4\mathrm{I}} \\ \boldsymbol{0}_{4\mathrm{I}\times 1} & \frac{\sigma^2}{2}\boldsymbol{F} \end{bmatrix} \tag{A5}$$

$$F = \begin{bmatrix} \mathrm{Re}\{E + EB\Delta_1^{-1}BE\} & -\mathrm{Im}\{E + EB\Delta_1^{-1}BE\} & -\mathrm{Re}\{EBA\Delta^{-1}P\} & \mathrm{Im}\{EBA\Delta^{-1}\} \\ \mathrm{Im}\{E + EB\Delta_1^{-1}BE\} & \mathrm{Re}\{E + EB\Delta_1^{-1}BE\} & -\mathrm{Im}\{EBA\Delta^{-1}P\} & -\mathrm{Re}\{EBA\Delta^{-1}\} \\ -\mathrm{Re}\{P\Delta^{-1}A^{H}BE\} & \mathrm{Im}\{P\Delta^{-1}A^{H}BE\} & \mathrm{Re}\{P\Delta^{-1}P\} & -\mathrm{Im}\{P\Delta^{-1}\} \\ -\mathrm{Im}\{\Delta^{-1}A^{H}BE\} & -\mathrm{Re}\{\Delta^{-1}A^{H}BE\} & \mathrm{Im}\{\Delta^{-1}P\} & \mathrm{Re}\{\Delta^{-1}\} \end{bmatrix} \tag{A6}$$

where $\boldsymbol{0}$ denotes the zero matrix, $\boldsymbol{\Delta} = \boldsymbol{A}^{H}\boldsymbol{\Delta}_1\boldsymbol{A}$, $\boldsymbol{A} = \mathrm{diag}(a_1, a_2, \ldots, a_I)$, $\boldsymbol{\Delta}_1 = \boldsymbol{B}_2 - \boldsymbol{BEB}$, $\boldsymbol{B}_2 = \boldsymbol{V}^{H}\boldsymbol{N}^2\boldsymbol{V}$, $\boldsymbol{B} = \boldsymbol{V}^{H}\boldsymbol{NV}$, $\boldsymbol{V} = [\boldsymbol{v}_1, \boldsymbol{v}_2, \ldots, \boldsymbol{v}_I]$, $\boldsymbol{v}_i = v_i^l \cdot [1, v_i, \ldots, v_i^{N-1}]^{T}$, $l = \begin{cases} -(N-1)/2, & N \text{ is odd} \\ -N/2+1, & N \text{ is even} \end{cases}$, $\boldsymbol{N} = \mathrm{diag}(-(N-1)/2, \ldots, (N-1)/2)$, and $\boldsymbol{P} = \mathrm{diag}(p_1, p_2, \ldots, p_I)$.

According to (A5) and (A6), it can be observed that it is complicated to obtain the Cramer–Rao lower bound of the DE model. However, for most wide-band radars, we have $\Delta f / f_0 \ll 1$, which is equivalent to $p_i \approx 1$. Therefore, by simplifying the CRLB matrix in (A5), the CRLB of the DE model can be obtained as

$$CRLB_{a_i} \approx \frac{\sigma^2}{2N} \tag{A7}$$

$$CRLB_{v_i} \approx \frac{6\sigma^2}{|a_i|^2 N^3} \tag{A8}$$

$$CRLB_{p_i} \approx \frac{6\sigma^2 p_i^2}{|a_i|^2 N^3} \tag{A9}$$

Based on (A4), the exact relations between the DE model parameters and the GTD model parameters are given by

$$r_i = -\frac{c}{4\pi\Delta f} \cdot \omega_i \tag{A10}$$

$$\alpha_i = \frac{f_0}{\Delta f} \cdot \ln p_i \tag{A11}$$

$$A_i = a_i \cdot \exp(\mathrm{j}4\pi f_0 r_i / c) \tag{A12}$$

According to Equations (A7)–(A12), we obtain

$$\mathrm{var}\{r_i\} \geq \left(\frac{c}{4\pi\Delta f}\right)^2 \cdot \frac{6\sigma^2}{|a_i|^2 N^3} = \frac{3}{2\pi^2} \cdot \frac{1}{SNR_i} \tag{A13}$$

$$\mathrm{var}\{\alpha_i\} \geq \left(\frac{f_0}{\Delta f}\right)^2 \cdot \frac{6\sigma^2 p_i^2}{|a_i|^2 N^3} = \frac{6}{\gamma^2 \cdot SNR_i} \tag{A14}$$

$$\mathrm{var}\{|A_i|\} \geq \frac{\sigma^2}{2N} \tag{A15}$$

where SNR_i denotes the peak signal-to-noise ratio of the i-th scattering center and $SNR_i \approx \frac{N|a_i|^2}{\sigma^2}$.

Finally, by extending the 1D-GTD model to the 3D-GTD model, we have

$$\mathrm{var}\{\hat{x}_i\} = var\{\hat{y}_i\} = var\{\hat{z}_i\} \geq \frac{3}{2\pi^2 SNR_{ii}} \tag{A16}$$

$$\mathrm{var}\{\hat{\alpha}_i\} \geq \frac{6}{\gamma^2 SNR_{ii}} \tag{A17}$$

$$\mathrm{var}\{\hat{A}_i\} \geq \frac{1}{2SNR_{ii}} \tag{A18}$$

$$SNR_{ii} \approx \frac{MNK|A_i|^2}{\sigma^2} \tag{A19}$$

where σ^2 denotes the white Gaussian noise variance.

Appendix B

Pseudo-code for the HGI-PHD filter

given $\left\{\overline{w}^i_{novel,k-1}, \boldsymbol{m}^i_{k-1}, \boldsymbol{P}^i_{k-1}\right\}_{i=1}^{M_{k|k-1}}$, the measurement set $\boldsymbol{Z}_k$
step 1 Prediction of the HGI-PHD filter
$i = 0.$
for $m = 1, \ldots, M_{k-1}$
$i = i + 1.$
$\overline{w}^i_{novel,k|k-1} = p_{s,k}\overline{w}^m_{novel,k-1},$
$\boldsymbol{m}^m_{k-1} = \boldsymbol{F}_{k-1}\boldsymbol{m}^m_{k|k-1}, \boldsymbol{S}^m_k = \boldsymbol{R}_k + \boldsymbol{H}_{k|k-1}\boldsymbol{P}^m_{k|k-1}\boldsymbol{H}^{\mathrm{T}}_k.$
$\boldsymbol{K}^m_k = \boldsymbol{P}^m_{k|k-1}\boldsymbol{H}^{\mathrm{T}}_k\left[\boldsymbol{S}^m_k\right]^{-1}, \boldsymbol{P}^m_{k|k} = \left[\boldsymbol{I} - \boldsymbol{K}^m_k\boldsymbol{H}_k\right]\boldsymbol{P}^m_{k|k-1}.$
end
$M_{k|k-1} = i.$
step 2 Construction of the HGI-PHD update components
for $m = 1, \ldots, M_{k|k-1}$
$\boldsymbol{\eta}^m_{k|k-1} = \boldsymbol{H}_k\boldsymbol{m}^m_{k|k-1}, \boldsymbol{S}^m_k = \boldsymbol{R}_k + \boldsymbol{H}_k\boldsymbol{P}^m_{k|k-1}\boldsymbol{H}^{\mathrm{T}}_k,$
$\boldsymbol{K}^m_k = \boldsymbol{P}^m_{k|k-1}\boldsymbol{H}^{\mathrm{T}}_k\left[\boldsymbol{S}^m_k\right]^{-1}, \boldsymbol{P}^m_{k|k} = \left[\boldsymbol{I} - \boldsymbol{K}^m_k\boldsymbol{H}_k\right]\boldsymbol{P}^m_{k|k-1}.$
end
step 3 Update of the HGI-PHD filter
for $m = 1, \ldots, M_{k|k-1}$
$\overline{w}^m_{novel,k} = (1 - p_{D,k})\overline{w}^m_{novel,k|k-1},$
$\boldsymbol{m}^m_k = \boldsymbol{m}^m_{k|k-1}, \boldsymbol{P}^m_k = \boldsymbol{P}^m_{k|k-1}.$
end
$n = 0.$
for each measurement $z \in \boldsymbol{Z}_k$
$n = n + 1.$
For $m = 1, \ldots, M_{k|k-1}$
$w^{nM_{k|k-1}+m}_{novel,k} = \dfrac{p_{D,k}(x_k)g_{hrrp}(hrrp_k)w^n_{k|k-1}\mathbf{q}^n_k(\mathbf{Z}_k)}{\lambda_c c_{hrrp}(hrrp_k)c_d(d_k) + p_{D,k}(x_k)g_{hrrp}(hrrp_k)\sum\limits_{i=1}^{N_{k|k-1}} w^n_{k|k-1}\mathbf{q}^n_k(\mathbf{Z}_k)}$
$\boldsymbol{m}^{nM_{k|k-1}+m}_k = \boldsymbol{m}^m_{k|k-1} + \boldsymbol{K}^m_k(\boldsymbol{z} - \boldsymbol{\eta}^m_{k|k-1}),$
$\boldsymbol{P}^{jM_{k|k-1}+m}_k = \boldsymbol{P}^m_{k|k}.$
end
$m_{\max} = \underset{m \in I}{\operatorname{argmax}}(w^i_k), m \in \left[1, nM_{k|k-1} + M_{k|k-1}\right], \Re = \kappa_k(\lambda_c) + w^{m_{\max}}_{novel,k}\beta + \sum\limits_{i=1}^{N_{k|k-1}} w^{jM_{k|k-1}+m}_{novel,k},$
$\overline{w}^{nM_{k|k-1}+m}_{novel,k} = \begin{cases} \dfrac{w^{nM_{k|k-1}+m}_{novel,k}}{\Re}, m \neq m_{\max}, m \in [1, M_k], n \in [1, N_{k|k-1}] \\ \dfrac{w^{nM_{k|k-1}+m}_{novel,k}\beta}{\Re}, m = m_{\max}, n \in [1, N_{k|k-1}] \end{cases}.$
end
$nM_{k|k-1} + M_{k|k-1}.$
output $\left\{\overline{w}^i_{novel,k}, \boldsymbol{m}^i_k, \boldsymbol{P}^i_k\right\}_{i=1}^{M_{k|k}}.$

References

1. Emmanuel, D.; Carolin, F.; Jose, F.; Jérémie, H.; Daniel, C. Novel Multi-Object Filtering Approach for Space Situational Awareness. *J. Guid. Control Dyn.* **2017**, *41*, 59–73.
2. Du, Y.; Jiang, Y.; Wang, Y.; Zhou, W.; Liu, Z. ISAR imaging for low Earth-orbit target based on coherent integrated smoothed generalized cubic phase function. *IEEE Trans. Geosci. Remote Sens.* **2020**, *58*, 1205–1220. [CrossRef]
3. Faber, W.; Mishra, U.; Chakravorty, S.; Hussein, I. Application of a randomized-finite set statistics technique (R-FISST) to space situational awareness. *J. Astronaut. Sci.* **2022**, *69*, 1149–1178. [CrossRef]
4. Li, Y.X.; Xiao, H.T.; Wu, H.; Hu, R.; Fu, Q. Posterior Cramer-Rao lower bounds for complicated multi-target tracking with labeled FISST based filters. *Signal Process.* **2016**, *127*, 156–167. [CrossRef]

5. Garc´ıa-Fernandez, A.; Svensson, L.; Morelande, M. Multiple target tracking based on sets of trajectories. *IEEE Trans. Aerosp. Electron. Syst.* **2020**, *56*, 1685–1707. [CrossRef]
6. Mahler, R. Multi-target Bayes filter via first-order multi-target moments. *IEEE Trans. Aerosp. Electron. Syst.* **2003**, *39*, 1152–1178. [CrossRef]
7. Vo, B.T.; Vo, B.N.; Cantoni, A. Analytic implementations of the cardinalized probability hypothesis density filter. *IEEE Trans. Signal Process.* **2007**, *55*, 3553–3567. [CrossRef]
8. Li, C.; Wang, W.; Kirubarajan, T.; Sun, J.; Lei, P. PHD and CPHD Filtering with Unknown Detection Probability. *IEEE Trans. Aerosp. Electron. Syst.* **2018**, *66*, 3784–3798. [CrossRef]
9. Ángel, F.; Fernánde, G.; Svensson, L. Trajectory PHD and CPHD Filters. *IEEE Trans. Signal Process.* **2017**, *67*, 5702–5714.
10. Vo, B.N.; Singh, S.; Doucet, A. Sequential Monte Carlo methods for multi-target filtering with random finite sets. *IEEE Trans. Aerosp. Electron. Syst.* **2005**, *41*, 1224–1245.
11. Vo, B.N.; Ma, W.K. The Gaussian mixture probability hypothesis density filter. *IEEE Trans. Signal Process.* **2006**, *54*, 4091–4104. [CrossRef]
12. Vo, B.T.; Vo, B.N.; Cantoni, A. The cardinality balanced multi-target multi-Bernoulli filter and its implementations. *IEEE Trans. Signal Process.* **2009**, *57*, 409–423.
13. Vo, B.T.; Vo, B.N. Labeled random finite sets and multi-object conjugate priors. *IEEE Trans. Signal Process.* **2013**, *61*, 3460–3475. [CrossRef]
14. Reuter, S.; Vo, B.T.; Vo, B.N.; Dietmayer, K. The labeled multiBernoulli filter. *IEEE Trans. Signal Process.* **2014**, *62*, 3246–3260.
15. Stephanie, R.; Mohamed, N. Estimation of frequencies and damping factors by two-dimensional ESPRIT type methods. *IEEE Trans. Signal Process.* **2001**, *49*, 237–245.
16. Souleymen, S.; Konstantin, U.; Pierre, C. Multidimensional ESPRIT for damped and undamped signals: Algorithm, computations, and perturbation analysis. *IEEE Trans. Signal Process.* **2017**, *65*, 5897–5910.
17. Zhang, W.; Zhang, X.F.; Sun, H.P. Non-circular generalised-Esprit algorithm for direction of arrival estimation. *IET Radar Sonar Navig.* **2017**, *11*, 736–744.
18. Cheng, Q. A simple modification of Esprit. *IEEE Signal Process. Lett.* **2018**, *25*, 1256–1260.
19. Zhang, H.M.; Zhang, H.Y. Research on DOA estimation method of sonar radar target based on Music algorithm. *J. Phys. Conf. Ser.* **2019**, *1176*, 032001. [CrossRef]
20. Chen, J.; Guan, D.; Tong, Y. Two-dimensional direction of arrival estimation for improved archimedean spiral array with MUSIC algorithm. *IEEE Access* **2018**, *6*, 49740–49745. [CrossRef]
21. Anderew, L.K.; Inder, J.G. A modified MUSIC algorithm for direction of arrival estimation in the presence of antenna array manifold mismatch. *IEEE Trans. Antennas Propag.* **2016**, *64*, 4836–4847.
22. Zhang, X.F.; Xu, L.Y.; Xu, L. Direction of departure (DOD) and direction of arrival(DOA) estimation in MIMO radar with reduced dimension MUSIC. *IEEE Commun. Lett.* **2010**, *14*, 1161–1163. [CrossRef]
23. Li, H.W.; Zhang, L.; Jiang, C.Q. Joint TOA and DOA estimation based on improved matrix pencil method. In Proceedings of the 2018 IEEE 4th International Conference on Computer and Communication, Chengdu, China, 7–10 December 2018; pp. 763–768.
24. Huang, J.Q.; Gumpper, K.; Chi, Y.J. Fast two-dimensional super-high resolution image reconstruction algorithm for ultra-high emitter density. *Optic-Lett.* **2015**, *40*, 2989–2992. [CrossRef] [PubMed]
25. Tang, W.G.; Jiang, H.; Pang, S.X. Grid-free DOD and DOA estimation for MIMO radar via duality-based 2D atomic norm minimization. *IEEE Access* **2019**, *7*, 60827–60836. [CrossRef]
26. Xu, X.J. *New Techniques for Radar Target Scattering Signature Measurement and Processing*; National Defence Industry Press: Beijing, China, 2018; pp. 258–264.
27. Zheng, S.Y.; Zhang, X.K.; Zhao, W.C.; Zhou, J.X.; Zong, B.F.; Xu, J.H. Parameter estimation of GTD model and RCS extrapolation based on a modified 3D-ESPRIT algorithm. *J. Syst. Eng. Electron.* **2020**, *31*, 1206–1215.
28. Vo, B.N.; Vo, B.T.; Hoang, H.G. An efficient implementation of the generalized labeled multi-Bernoulli filter. *IEEE Trans. Signal Process.* **2017**, *65*, 1975–1987. [CrossRef]
29. Zhou, J.X. Theory and Technology on Reconstructing 3D Scattering Centers of Radar Targets in Optical Region. Ph.D. Thesis, National University of Defense Technology, Changsha, China, 2006; pp. 21–32.

remote sensing

MDPI

Article

Numerical Evaluation of Planetary Radar Backscatter Models for Self-Affine Fractal Surfaces

Anne Virkki [1,2]

[1] Department of Physics, University of Helsinki, Gustaf Hällströminkatu 2, 00560 Helsinki, Finland; anne.virkki@helsinki.fi

[2] Finnish Geospatial Research Institute, National Land Survey, Vuorimiehentie 5, 02150 Espoo, Finland

Abstract: Numerous analytical radar-scattering laws have been published through the past decades to interpret planetary radar observations, such as Hagfors' law, which has been commonly used for the Moon, and the cosine law, which is commonly used in the shape modeling of asteroids. Many of the laws have not been numerically validated in terms of their interpretation and limitations. This paper evaluates radar-scattering laws for self-affine fractal surfaces using a numerical approach. Traditionally, the autocorrelation function and, more recently, the Hurst exponent, which describes the self-affinity, have been used to quantify the height correlation. Here, hundreds of three-dimensional synthetic surfaces parameterized using a root-mean-square (rms) height and a Hurst exponent were generated, and their backscattering coefficient functions were computed to evaluate their consistency with selected analytical models. The numerical results were also compared to empirical models for roughness and radar-scattering measurements of Hawaii lava flows and found consistent. The Gaussian law performed best at predicting the rms slope regardless of the Hurst exponent. Consistent with the literature, it was found to be the most reliable radar-scattering law for the inverse modeling of the rms slopes and the Fresnel reflection coefficient from the quasi-specular backscattering peak, when homogeneous statistical properties and a ray-optics approach can be assumed. The contribution of multiple scattering in the backscattered power increases as a function of rms slope up to about 20% of the backscattered power at normal incidence when the rms slope angle is 46°.

Keywords: radar; microwave scattering; fractal surfaces; planetary science

1. Introduction

The characterization of planetary surfaces relies dominantly on remote-sensing techniques due to the high cost of in situ measurement opportunities provided by spacecraft missions. Visual and near-infrared (VNIR) observations can be utilized for photometric, polarimetric, and spectroscopic characterization of the physical properties from the mineralogy of regolith to convex-hull shape estimates of SSOs. The grain sizes can be estimated as well, but the estimates are constrained by the size scale of the utilized wavelength. In order to characterize the planetary surfaces on a larger scale, such as the topography and regolith particles on a centimeter-to-meter scale, radar observations are arguably the best remote-sensing tool for the purpose. In typical radar observations, a high-power microwave signal is transmitted, and the scattered echo is received using the same (monostatic) or a different antenna (bistatic). The comparison of the received signal properties to those of the transmitted signal in terms of the expected echo power, wavelength, and polarization provides various ways for the characterization of the regolith as well as the target as a whole. In fact, the radar delay-Doppler images provide the finest mapping resolution for Earth-based observations of SSOs: the echo power mapped as a function of the round-trip time (delay) measured in fractions of microseconds, and thus providing a range resolution as fine as a few meters, and the Doppler frequency, which arises from the rotation of the object. For the Moon, fully spatial synthetic aperture radar (SAR) images are an option; for example, the Lunar Reconnaissance Orbiter's Mini-RF instrument has observed numerous regions of the Moon using the SAR technique, e.g., [1]), and more recently, also the dual-frequency L-band and S-band synthetic aperture radar (DFSAR) instrument on Chandrayaan-2 has contributed a wealth of data [2].

This article is focused on the characterization methods of the topographic properties of planetary surfaces dominated by fine-grained regolith (grain sizes much smaller than the wavelength). The first analytical radar backscattering laws for this purpose were derived in the 1950s and 1960s and inspired by the first radar observations of the Moon and Venus (e.g., [3,4]). Radar scattering laws refer to the backscattering coefficient (BSC) functions as a function of the incidence angle, i.e., the angular distribution of the area-normalized backscattering reflection coefficient as a function of the angle between the direction of the incident illumination and the normal of a surface element. The planetary surfaces have been traditionally parameterized using the root-mean-square (rms) height and the autocorrelation length, which is the average vertical offset between any two arbitrary points on the surface separated by a horizontal distance (lag) at which the height correlation has decreased to e^{-1} (about 37%).

The first analytical laws assumed the Gaussian height probability density function and height autocorrelation function because of mathematical convenience [5]. Hagfors [3] showed that a Gaussian height autocorrelation function did not provide a good fit to the Lunar radar data but that an exponential function provided a better fit. Thus, Hagfors' law (see Section 2) became the most commonly used radar scattering law since 1964, as it provided the best fit to the data; however, not without controversy, because the exponential height autocorrelation function is non-physical, as the law does not conserve energy. Also, Hagfors' law was not a good fit for many asteroids' radar-scattering profiles. For example, Mitchell et al. [6] used only the Gaussian law, cosine law [7,8], and a "flat law" [9], for attempts to fit radar scattering by (1) Ceres, (2) Pallas, and (4) Vesta. The cosine law remains the most used scattering law option in the shape modeling of asteroids based on radar data (e.g., [10–13]); however, the physical interpretation of the fit parameters of the cosine law is not well constrained as will be discussed further below.

Close to the year 2000, several radar scattering laws based on self-affine fractal surface properties emerged [14–16]. Self-affine fractal and Brownian surfaces are physically more realistic than surfaces described by a single, specific height autocorrelation function. Instead of the autocorrelation length, the height elements' correlation in self-affine fractal surfaces is better described using the Hurst exponent, H, which is related to the fractal dimension $D = 3 - H$, so that $H = 1$ represents an ideally fractal topography, $z(x, y)$ [17]. For a two-dimensional surface profile, $z(x)$, the Hurst exponent is $H = 2 - D$. Turcotte [17] defined the Hurst exponent as the power-law index for the increase of the height standard deviation (traditionally called the rms height) as a function of the sample length, L:

$$h(L) = h_0 \left(\frac{L}{L_0} \right)^H, \tag{1}$$

where h_0 is measured for a sample length L_0. As the equation makes clear, $h(L)$ does not remain stationary through the sample but depends on the selected length; therefore, no single value of h can uniquely describe the surface roughness. A similar equivalence can be found for the rms (Allan) deviation as a function of the sampling interval, Δr, (hereafter, lag):

$$\nu(\Delta r) = \nu_0 \left(\frac{\Delta r}{\Delta r_0} \right)^H. \tag{2}$$

The rms slope $s(\Delta r) = \nu(\Delta r)(\Delta r)^{-1}$ and is thus proportional to $(\Delta r)^{H-1}$. It is related to the slope angle so that $\theta_s = \arctan(s)$.

Optionally, the topography can be characterized using the power spectrum, which is related to the Fourier transform of the topography profile so that the power spectrum defines the amplitude of a given spatial frequency [17,18]. The two methods of characterization do not, however, always provide consistent results. Austin et al. [19] reported that the fractal dimensions derived using the power-spectrum method are biased toward greater Hurst exponents (lower fractal dimension), attributing it to spectral leakage toward higher frequencies, and further suggesting that filtering out the highest frequencies (Hanning filter) may improve the results. According to Shepard et al. [18], the deviogram (a logarithmic graph of the rms deviation as a function of lag) does not require filtering and is less noisy than the power spectrum, so characterizing the topography using the deviogram is often preferred. However, generating synthetic surfaces using the power spectrum can be faster and more consistent (free from artifacts) than spatial methods ([18] and references therein).

Furthermore, it is important to remember that no physically realistic surface is infinitely fractal due to structural limitations, but the length-scale limits of the self-affinity depend on, for example, the mineralogical composition or the formation mechanisms. A length-scale limit of self-affinity in the power spectrum is typically called a roll-off frequency or wavevector; it appears as a change of gradient in the power spectrum, or even a plateau. Similarly, it changes the gradient of a deviogram. The fractality has a lower limit determined by the grain size for particular surfaces or the mineralogical structure for solid materials, and also the roughness below the Rayleigh limit is dominated by the wavelength scale and larger height undulations in terms of contribution to the echo power, so the roughness larger than the Rayleigh limit is more relevant to the scattering profile. For polarization studies, the wavelength-scale and sub-wavelength-scale roughness play a greater role than they do for the intensity of the echo. In the larger size scales, there is periodicity beyond the upper size scale; however, a finite measurement length can also have a similar effect, as will be discussed in detail in this paper. Similarly, in numerical models, the discretized surfaces are always limited by a discretization parameter (i.e., the size of the element) and a finite size of the topography model. The mathematical consequence of this fact is that although integration of the power spectra or the autocorrelation functions through the range of scales may converge from zero to infinity, and thus be analytically convenient, it is not physically correct and thus risks the realism of the models.

In order to evaluate the scattering laws, I conducted numerical simulations of the BSC functions of rough surfaces using different levels of self-affinity and evaluated to what extent the analytically derived scattering laws are able to predict the rms slopes. The benefit of the numerical approach is the ability to compare the consistency between the analytical and numerical models, to investigate possible constraints in each approach, and to quantify the effects of shadowing and multiple scattering, which are challenging analytical problems, especially for a three-dimensional surface. Shepard and Campbell [20] have investigated comprehensively shadowing models for self-affine fractal surfaces, which is why shadowing effects are addressed here only briefly. This work takes a step forward from the existing numerical models by, for example, Ulaby et al. [21], who have provided a variety of numerical methods to simulate microwave scattering; however, their methods are limited to the autocorrelation function approach and a length-independent rms height. Although the numerical models have shown to provide good fits to measured data, they cannot provide a straightforward model of the BSC functions when the surface statistics are measured and described using the Hurst exponent instead of the autocorrelation function. For example, Campbell [22] measured the topographic properties and the microwave backscattering coefficients of Hawaii lava flows, which can be considered analogous to all basaltic planetary surfaces. Only empirical modeling was presented in [22] with a call for numerical computations as future work.

2. Methods

2.1. The Generation of the Synthetic Surfaces

I generated synthetic rough surfaces using a Python translation of the Matlab code "Surface generator: artificial randomly rough surfaces", which is based on theory presented by Persson [23], and freely available through MathWorks (https://www.mathworks.com/

matlabcentral/fileexchange/60817-surface-generator-artificial-randomly-rough-surfaces; accessed on 16 June 2023). The algorithm is described in detail in Appendix A. The surface model is defined using a Hurst exponent, an edge length (L_e), an rms height (h_L), a number of elements per edge (m), and a roll-off wavevector index (q_r), which defines the roll-off wavevector (to be explained in further detail in Section 2.3). The physical interpretation of the roll-off length is to quantify the spatial wavelength limit where the self-affinity properties of the surface change. The rms height is equal to the height standard deviation of the whole topography model. The number of elements could be different for each side but here I use only square areas discretized to m elements per side. The element size is then defined as $d_e = L_e m^{-1}$, which can be set to be, for example, the wavelength of the measurements. Also, an electric permittivity, ϵ, is required for computing the scattering properties.

After, the vertex coordinates were generated so that their height correlation with respect to the horizontal scale was consistent with the given Hurst exponent, the vertices were connected into triangular facets (as common in 3D models), and could be saved in a wavefront file format when needed. The facet size has to be at least one wavelength wide for the ray-optics scattering regime, and thus, for the Fresnel scattering to remain physically valid. A second requirement for physical validity—especially relevant to self-affine fractal surfaces with low Hurst exponents—is the rms height within each element: if the element-scale rms height is below the Rayleigh limit, i.e., $h < \lambda(8\cos\theta_i)^{-1}$, the element can be considered electromagnetically "smooth", whereas if the rms height is much greater than the Rayleigh limit, the validity of Fresnel scattering becomes questionable. This fact will be considered in the analysis where relevant in the evaluation of the scattering laws.

It is also important to note that because the generation of the sample surfaces has stochastic elements, individual topography models are not necessarily fully isotropic, but an ensemble average over several sample surfaces is required. Figure 1 visualizes example surfaces using different surface-roughness parameters. Once generated, the topography statistics (total and projected area and the statistical roughness properties) of the surface models were recorded and analyzed.

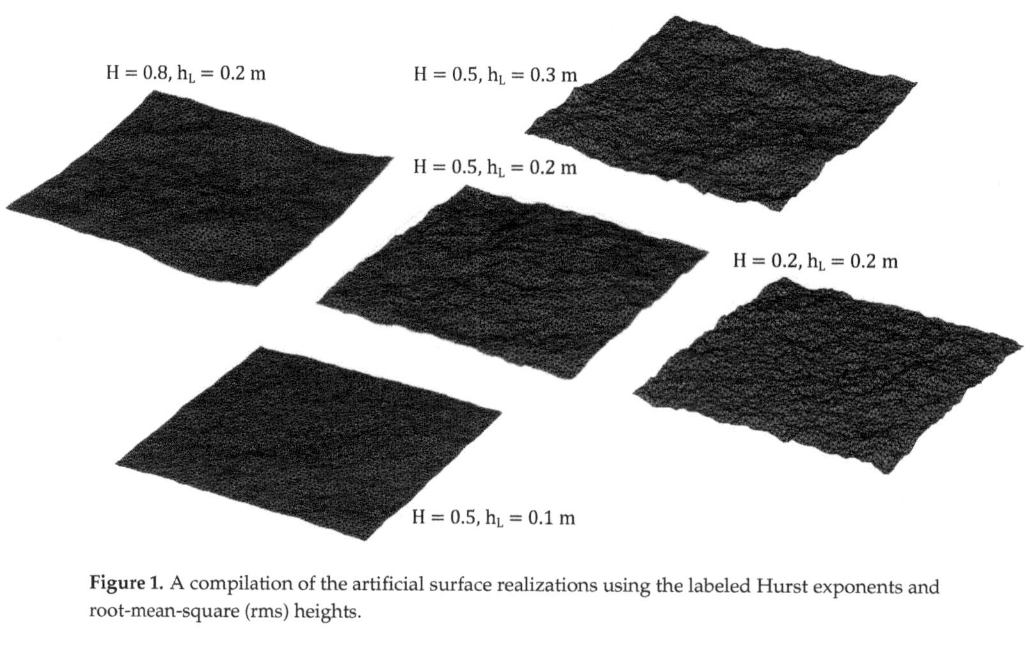

Figure 1. A compilation of the artificial surface realizations using the labeled Hurst exponents and root-mean-square (rms) heights.

2.2. Scattering Properties

For the scattering properties, two different algorithms were used: (1) a simplified but fast algorithm to simply sum together the reflected power by all illuminated and visible triangles but shadowing omitted, available online under the name "surface-scattering", and (2) a full ray-tracer that keeps account of shadowing, multiple bounces, and the polarization of the rays, available online as Ray-optics for self-affine fractal surfaces (ROSAS). Both codes are open source and available online; see the Data Availability Statement.

In the first algorithm, the relative Fresnel reflection powers of each triangular element of the shape model were summed together, but each element's local incidence angle, $\theta_{0,i}$ (the triangles facing away from the incident direction were thus fully excluded), and the relative power to the corresponding phase angle (α) bin, was recorded so that $\alpha = 2\theta_0$. Then, I rotated the incident direction by 2° around the y-axis while keeping the topography model's mean-height plane parallel to the xy-plane and repeated the process up to $\theta_0 = 80°$. The algorithm does not take into account elements that shadow other elements or multiple scattering but considers that if a ray reflects from an element near the edge downward outside the area of the synthetic surface, the reflection is excluded. In other words, only the reflected vectors with a positive z-component in the shape model reference frame are included.

In ROSAS, I used a ray-tracing algorithm to track which triangle a ray intersects first and whether it intersects another triangle after the reflection, and if so, how many times, and in which direction the ray eventually emits. The algorithm reads or generates the surface topography model, generates rays that hit a random (but illuminated) location at the surface, and tracks their propagation during the interaction with the surface, keeping account of the number of reflections that take place and the polarization of the rays. The user can select the incidence angle, the number of rays, the number of azimuth angles for ensemble averaging, the number of phase angle bins, the refractive properties, and the wavelength. Here, the reflectivity was investigated as a function of incidence angle using 30,000 rays for five different surface realizations (i.e., 150,000 rays in total) for incidence angles from 0° to 60° with 10° intervals. The manual for the ROSAS code is included in the online repository for the code and includes a more detailed description of its functionality. The effect of multiple scattering is small and will be discussed in more detail in Section 3.

In both algorithms, the total scattering efficiency was obtained by normalizing the total scattered power by the incident power, in this case, represented by the number of rays. The phase function (the first element of the scattering phase matrix as a function of the phase angle) was normalized so that the integral of the phase function over the full space angle is equal to 4π. Mathematically dressed:

$$\int_0^{2\pi}\int_0^{\pi} P_{1,1}(\alpha,\theta_0)\sin\alpha \, d\alpha \, d\phi = 4\pi. \tag{3}$$

The BSC function was then derived by multiplying the phase function value at the phase angle of 0° by the scattering efficiency (q_{sca}) for each incidence angle:

$$\sigma_0(\theta_0) = q_{\text{sca}}(\theta_0) P_{1,1}(0°,\theta_0)\cos\theta_0. \tag{4}$$

The scattering efficiency increases as a function of incidence angle as the average of the Fresnel reflection coefficients increases. At normal incidence,

$$R_F = \left[\frac{\sqrt{\epsilon}-1}{\sqrt{\epsilon}+1}\right]^2. \tag{5}$$

Finally, the scattering laws were fit to the computed BSC functions using SciPy's curve_fit function, and the results were recorded and evaluated. For each law, the fit parameter C was assumed s^{-2} and compared to the rms slope ($s = \tan\theta_s$) so that the slope angle is obtained from the vertical component of unit normal vectors of the surface elements: $\theta_s = \arccos(\hat{n}_z)$.

2.3. Statistical Properties of the Surfaces

The statistical properties include the rms height h, rms deviation, ν, and its square, the rms variance, the rms slope, and the autocorrelation function. The definitions follow those presented in [24]. Let us begin with the square of the rms height as a function of the sample length, or rather, the standard deviation of the heights around the mean height:

$$h^2(L) = \frac{1}{N-1}\sum_{i=1}^{N}[z(r_i) - z_{av,L}]^2, \tag{6}$$

where $z_{av,L}$ is the average of all N height points $z(r_i)$ included in the sample length L. The square of the rms height as a function of step size, Δr, is:

$$h^2(\Delta r) = \frac{1}{2N}\sum_{i=1}^{N} z(r_i)^2 + z(r_i + \Delta r)^2, \tag{7}$$

The rms height is constant as a function of step size but increases as a function of sample length, L, whereas the rms deviation and rms slope are constant as functions of sample length but change as a function of step size, as shown, for example, in [25]. The rms variance as a function of step size is:

$$\nu^2(\Delta r) = \frac{1}{N}\sum_{i=1}^{N}[z(r_i) - z(r_i + \Delta r)]^2 \tag{8}$$

where N is the number of pairs of points separated by a lag, Δr. The rms slope is:

$$s(\Delta r) = \frac{\nu(\Delta r)}{\Delta r}, \quad \Delta r > 0. \tag{9}$$

The rms slope for a two-dimensional (2D) surface profile, $z(x)$, is denoted using a subscript p. The adirectional rms slope for a 3D surface, $z(x, y)$, denoted using a subscript A, is related to the 2D rms slope so that $s_A = \sqrt{2}s_p$. Here, the 2D rms slope is calculated as shown above for each individual line and column of the synthetic surfaces, and the 3D rms slope using the vertical component of the normals, $\hat{n}_z$, of the topography model's individual elements: $s_A = \langle\tan(\arccos(\hat{n}_z))\rangle$, where the chevrons indicate an ensemble average. The numerical results were compared at different values of H for possible variations from $\sqrt{2}$ but the ratio was always consistently $\sqrt{2}$.

The rms-height-normalized autocorrelation function is defined as:

$$\rho(\Delta r) = \frac{\langle z(r_i)z(r_i + \Delta r)\rangle}{\langle z^2(r_i)\rangle}. \tag{10}$$

The autocorrelation functions are typically given as:

$$\rho(\Delta r) = \exp\left[-\left(\frac{\Delta r}{l}\right)^b\right], \tag{11}$$

so that $b = 1$ for the exponential and 2 for the Gaussian autocorrelation function and l is the autocorrelation length. Also, other options such as x-exponential and x-power correlation functions have been presented (e.g., [21]) but are significantly less used in planetary science applications. For stationary surfaces, the rms slope can be derived from the second derivative of the autocorrelation function ($s^2 = -h^2\rho''(0)$), i.e., at a small enough step, the local slope can be found. On the contrary, in fractal surfaces, the roughness depends on the scale as defined by the fractal dimension (or the related Hurst exponent), so the given relation is not valid. Here, the synthetic surfaces are generated using the power spectrum (the Fourier transform of the autocorrelation function), and autocorrelation functions are derived from the generated surfaces to understand how the autocorrelation at the wavelength-scale sampling is related to the Hurst exponent. Examples of the obtained autocorrelation functions are further discussed and visualized in Section 3.

For stationary surface profiles, the autocorrelation function is related to the rms variance so that

$$\nu^2(\Delta r) = \langle [z(r_i) - z(r_i + \Delta r)]^2 \rangle = 2h^2 \langle 1 - \rho(\Delta r) \rangle, \quad (12)$$

but this relation may depend on the self-affinity of the surface [18], which will be considered in the analysis.

The roughness characteristics can also be derived from the power spectrum:

$$W(q) = W_0 \left(\frac{q}{q_0} \right)^{-2(H+1)} \quad (13)$$

Here, W is in units of m^4 and the wavevectors q and q_0 are in units of m^{-1} for three-dimensional (hereafter 3D) surfaces. Note that units of m^3 and a power of $-(2H+1)$ would be used for the power spectrum of a two-dimensional (hereafter 2D) surface profile. The values of q_0 and q_1 depend on the choice of the integration limits and coordinate system. For an analytical solution that ensures isotropic characteristics with respect to any reference point, integration in a polar coordinate system is more convenient. If $q_0 = \pi L^{-1}$ for a 2D surface profile, in a 3D case, the surface area remains equal only if $L^2 = \pi L_p^2$, where L_p is the radius of the circular sample area. Therefore, if $L_p = 0.5L_e$, $q_0 = 2\sqrt{\pi} L_e^{-1}$. The upper limit of the wavevectors, q_1, is constrained by the sampling interval (i.e., the element size) so that $q_1 = q_0 m^{-1} = 2\sqrt{\pi} d_e^{-1}$.

Using the power spectrum, the square of the rms height over an isotropic self-affine fractal surface area with an edge length of L_e and a sampling interval of d_e is

$$h^2(q_1, q_0) = 2\pi \int_{q_0}^{q_1} q W(q) \mathrm{d}q, \quad (14)$$

and the square of the rms slope is given as

$$s^2(q_1, q_0) = 2\pi \int_{q_0}^{q_1} q^3 W(q) \mathrm{d}q \quad (15)$$

as shown in [26]. When $0 < H < 1$, the rms slope as a function of the rms height and the Hurst exponent is then (as shown in more detail in Appendix B):

$$s(q_1, q_0) = h_L q_0 \sqrt{\frac{H}{1-H}} \sqrt{\frac{(\frac{q_1}{q_0})^{2(1-H)} - 1}{1 - (\frac{q_1}{q_0})^{-2H}}}. \quad (16)$$

Here, $h_L = h(d_e, L_e)$. When $q_1 \gg q_0$, the wavelength-scale rms slope can be approximated as:

$$s(q_1, q_0) = h_L q_0 \left(\frac{q_1}{q_0} \right)^{1-H} \sqrt{\frac{H}{1-H}}. \quad (17)$$

For the end case of $H = 0$:

$$s(q_1, q_0) = \frac{h_L q_1}{\sqrt{2 \ln \left(\frac{q_1}{q_0} \right)}}, \quad q_1 \gg q_0, \quad (18)$$

and for $H = 1$:

$$s(q_1, q_0) = h_L q_0 \sqrt{2\ln\left(\frac{q_1}{q_0}\right)}, \quad q_1 \gg q_0. \tag{19}$$

The equations above assume that the power spectrum follows the $q^{-2(H+1)}$ relation through all included wavevectors from q_0 to q_1 ($L_r = 0.5L_e$). If $L_r < 0.5L_e$, the low-frequency components of the surface profile are cut, and the power spectrum has a constant value of $W(q) \propto (\pi q_r/L_e)^{-2(H+1)}$ when $q < q_0$. In this case, the roll-off wavevector index has to be set to $2 < q_r < m$, and s requires an additional factor of $(0.5q_r)^H$.

Because of the discretization, small differences in the analytical solutions may occur in the numerical surface model realizations. A confidence test of the Hurst exponent was conducted by computing the adirectional rms slopes for synthetic surfaces generated using two different values of m and L_e but equal d_e and h_L for different values of H. A selection of $m = 64$ and 128, and, respectively, $L_e = 8.0$ and 16.0, keeps $d_e = 0.125$. In this case, the ratio of the rms slopes should only depend on the Hurst exponent and the ratio of either m or L_e so that $s_A(L_e = 8)/s_A(L_e = 16) = 2^H$. The derived H was found to be about $0.1 + 0.8H_{in}$, where H_{in} denotes the value used for generating the surface. Therefore, some error in the derived self-affinity can be expected for the synthetic surfaces that are generated using values of H not equal to 0.5 with decreasing confidence toward 0 and 1; whereas surfaces with $H = 0.5$, have a high level of confidence.

To verify the power-spectrum reference value q_0, I generated synthetic surfaces of $L_e = 8.0$ and $m = 64$ ($d_e = 0.125$) and $L_e = 16.0$ and $m = 128$ using $h_L \in [0.01, 0.3]$ and $H = 0$, 0.3, 0.5, 0.8, or 1, and recorded the adirectional rms slopes, then solved for q_0 (assuming that $q_1 = q_0 m$). For values of $H \in [0.3, 0.8]$, I found $q_0 L_e$ to be in the range from 3.33 to 3.67, with a positive correlation with H and small differences due to the discretization. This range for intermediate values of H is consistent with $\sqrt{4\pi} \approx 3.54$, which is consistent with the analytical solution as described above. For $H = 0$ and $H = 1$, I found $q_0 L_e \approx 3.0$, and ~ 4.3 for the synthetic surfaces, respectively; however, the confidence for the statistical properties when using the end values of H is low because of the bias towards $H = 0.5$.

For surface profiles, the corresponding wavevectors would be $q_0 \approx \sqrt{2\pi}L_e^{-1}$ and $q_1 \approx \sqrt{2\pi}d_e^{-1}$. The discretization plays a small role, which will be addressed in more detail in Sections 3 and 4. When using the limiting wavevectors given above, Equation (17) takes the form

$$s(d_e, L_e) = \frac{2h_L}{L_e} m^{1-H} \sqrt{\frac{\pi H}{1-H}} = \frac{2h_L}{d_e} m^{-H} \sqrt{\frac{\pi H}{1-H}}. \tag{20}$$

Note that here, $h_L = h_\lambda m^H$, when using Equation (1), and the step size is equal to the wavelength: $d_e = \lambda$. Because $s_\lambda \lambda = \nu_\lambda$,

$$\nu_\lambda = h_\lambda \sqrt{\frac{4\pi H}{1-H}} \tag{21}$$

If written as a ratio of the rms variance to the square of the same-scale rms height, a simple form can be found:

$$\frac{\nu_\lambda^2}{h^2(L = \lambda)} \approx \frac{4\pi H}{1-H} \tag{22}$$

for surface areas, whereas for surface profiles:

$$\frac{\nu_\lambda^2}{h^2(L = \lambda)} \approx \frac{2\pi H}{1-H}. \tag{23}$$

2.4. The Radar Scattering Laws

The BSC function, also known as a differential or area-normalized radar cross section, is defined as $\sigma_0(\theta_0) = \frac{d\sigma(\theta_0)}{dA(\theta_0)}$. It quantifies the backscattering reflectivity of a surface as a function of incidence angle, i.e., the angle between the source of illumination and the normal of a surface element. The presented laws are only valid for the backscattered power of the quasi-specular reflection, which is the same-sense polarization as the transmitted signal when using linear polarization, typically denoted as HH for horizontal-to-horizontal or VV for vertical-to-vertical, or the opposite-circular (OC) when using circular polarization.

The Gaussian law is mathematically the most convenient one because it is based on a Gaussian height distribution and a Gaussian autocorrelation function; the fit parameters directly describe the statistical properties of the surface so that $R_G = R_F$. It is defined as

$$\sigma_G(\theta_0) = \frac{R_G C_G}{\cos^4\theta_0}\exp(-C_G \tan^2\theta_0), \quad C_G = \frac{2h_0^2}{l^2}, \tag{24}$$

where l is the autocorrelation length, and h_0 is the rms height. The Gaussian law can also be derived using only the slope probability density function (p.d.f.), assuming that it follows a Gaussian form. For example, Simpson and Tyler [27] presented a compilation of some of the legacy radar-scattering laws and how they were derived.

Hagfors' law [3] assumes the height distribution to be Gaussian and the autocorrelation function to be exponential. It is traditionally presented in the form:

$$\sigma_H(\theta_0) = \frac{R_H C_H}{2}(\cos^4\theta_0 + C_H \sin^2\theta_0)^{-3/2}, \quad C_H = \frac{\lambda^2 l^2}{16\pi^2 h_0^4}, \tag{25}$$

where R_H is described as the Fresnel power reflection coefficient at normal incidence, and C_H is a (wavelength-scale) roughness parameter, and λ is the wavelength. It is useful to highlight the difference in the definitions of C_H and C_G, as both are often simply C in the literature and are described to be related to the "slope" or "surface roughness"; however, the definition of C differs between the laws.

Mitchell et al. [6] chose three different scattering laws for evaluating the radar scattering profiles of the dwarf planets (1) Ceres, (2) Pallas, and (4) Vesta: the Gaussian law, the cosine law, and the flat law.It should be noted that the scattering profiles in this study were Doppler-only echo spectra rather than BSC functions. The flat law is omitted here as it was found to be a bad fit for most radar profiles and is thus not used widely. The cosine law is defined as

$$\sigma_C(\theta_0) = R_c(C_c + 1)\cos^{2C_c}\theta_0. \tag{26}$$

This law was selected as part of the study because it is the default radar scattering law in the shape-modeling software Shape (version 2; for more details, see, e.g., [10,28]). The users of the software typically fix R_c to an arbitrary value and let C_c float through the multitude of iterations that the numerical sculpting of the model requires. As the shape model reaches its final shape, R_c can be allowed to vary with the purpose of constraining it using the Doppler-echo-power spectra; however, this step is often skipped and the scattering-law fit parameters left unpublished, or only C_c is included. Therefore, I selected the cosine law here as one scattering law for our analysis, so that a more robust interpretation for the fit parameters can be provided using the better-established laws as a reference, and thus help inform Shape users. Analytically, $\cos x \approx \exp(-\tan x)$ when x is small, so for small slopes, the cosine law can be expected to be comparable to the Gaussian law.

Shepard and Campbell [14] presented near-nadir scattering laws for self-affine fractal surfaces that are defined using the Hurst exponent (H) and a slope (s) that depends on the scale (so that the Hurst exponent defines the scaling, as described in detail in [24]). They derived the BSC function in the near-nadir regime for cases $H = 0.5$, $H = 1.0$, and a general case using the wavelength-scale slope (s_λ). I adopt the case of $H = 0.5$ as one of our scattering laws to fit by using their ρ, which they define as the Fresnel reflection coefficient, and s_λ as the fitting parameters R_S and s_S, respectively:

$$\sigma_S(\theta_0, H = 0.5) = \frac{R_S \pi s_S^4 \cos^4\theta_0}{16(\pi^2 s_S^4 \cos^4\theta_0 + \sin^2\theta_0)^3}, \tag{27}$$

They note that the term πs_S^2 appears analogous to $1/\sqrt{C_H}$ when compared to the Hagfors' scattering law but do not attach any further physical interpretation to the parameters. When $H = 1.0$, they define the BSC function as:

$$\sigma_S(\theta_0, H = 1.0) = \frac{\rho}{4\pi s_\lambda^4 \cos^4 \theta_0} \exp(-2s_\lambda^{-2} \tan^2 \theta_0). \tag{28}$$

This form using $H = 1.0$ is evidently similar to the form of the Gaussian law (Equation (24)) with a difference of the (incorrect) factor of 4π. The general form is omitted here because it is complicated to use for fitting, and is incorrect, as will be discussed in Section 4.

Another widely used set of more comprehensive models for microwave scattering is the integral equation method (IEM) [29,30] and the improved integral equation method (I^2EM) [21]. Similar to the models presented above, IEM and I^2EM are based on a given autocorrelation function, an autocorrelation length, and a stationary rms height. The benefit of these models is that the reflection coefficients take into account both horizontal and vertical polarizations and I^2EM also includes diffuse scattering, which improves the modeling of the polarization properties [21], but makes inverse-problem solutions more challenging. The comparison to these models is not provided in this paper due to the complexity of the fitting in a large parameter space that is not analytically derived for self-affine fractal surfaces.

Figure 2 demonstrates the forms of the selected radar scattering laws when using a fixed rms slope of 15° and assuming that $C_c = C_G = C_H = s_S^{-2}$. Similarly, all R-parameters as equal to the normal-incidence Fresnel reflection coefficient of a plane interface with an electric permittivity of 5.0. Several other radar scattering models exist in the literature; however, as stated, only those in wider use and not restricted to rms slopes below 10° are included. As is evident in the graph, the Gaussian and cosine laws are very similar to each other, whereas Hagfors' and Shepard's laws have $\sigma_0(0°)$ approximately one-half of that of the Gaussian and cosine laws. Also, Hagfors' law has a distinct tail, whereas the other functions decrease to zero more steeply.

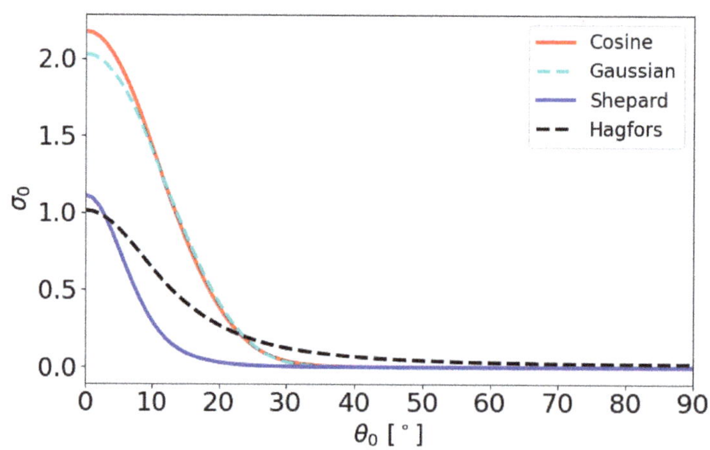

Figure 2. An example illustration of the radar scattering laws selected for the analysis when using $\theta_s = 15°$, which is equal to $s = 0.26$ and $C = 14.8$ when $C = s^{-2}$.

In addition, one empirical law was selected to compare analytical laws to measured data with surface statistics and radar-scattering properties validated numerically. Campbell [22] modeled the backscattering coefficients measured from the Hawaii lava flows empirically, and found that $\sigma_{HH} = \sigma_{VV}$, and that the left-to-right (LR), or more generally the OC, polarization component follows a similar form but with a difference of a constant factor:

$$\sigma_{HH} = 1.11\sigma_{OC} = 0.16[1 - \exp(-70.372 s_\lambda^2 \exp(-0.0644\theta_0))], \quad (29)$$

the ratio of the cross-polarization component σ_{HV} to σ_{VV} is

$$\frac{\sigma_{HV}}{\sigma_{VV}} = \frac{\theta_0}{270}[1 - \exp(-4.5 s_\lambda)], \quad (30)$$

and the ratio of σ_{HV} to the left-to-left (LL) polarized component, more commonly notated as the same-circular (SC) polarization, is

$$\frac{\sigma_{HV}}{\sigma_{LL}} = 0.3 + 0.2[1 - \exp(-4.5 s_\lambda)]. \quad (31)$$

The total BSC function computed for the synthetic surfaces is $\sigma_{HH} + \sigma_{HV}$ using the linear polarization components and $\sigma_{LR} + \sigma_{LL}$ using the circular polarization components. From these equations, also the circular-polarization ratio, i.e., the ratio of the echo power in the LL or SC polarization to that in the LR or OC polarization can be easily derived. The measurements were conducted using incidence angles from 35° to 60°, so the model can be considered valid only within this range of incidence angles. The surface statistics and the measured backscattering coefficients were averaged over surface profiles covering 50–100 m per side as shown in [22].

I selected three cases out of the ten sample surfaces presented by Campbell [22,25], hereafter referred to as Test cases.

1. $h_1 = 0.017$, $s_{\lambda,p} = 0.076$, and $H = 0.63$;
2. $h_1 = 0.046$, $s_{\lambda,p} = 0.225$, and $H = 0.49$;
3. $h_1 = 0.120$, $s_{\lambda,p} = 0.705$, and $H = 0.26$.

The corresponding rms slope angles in degrees are 4.3°, 12.7°, and 35.2°. For the surface models, the edge length is assumed to be $L_e = 9$ m, because the spatial resolution of the images used for deriving the empirical equations was within the range of 5–10 m (Bruce Campbell, personal communication), and using a number in the high end of the range is beneficial for statistical reasons due to the greater number of elements per sample surface. Furthermore, when using 25 cm as the sampling interval, following the reported surface statistics, the discretization number is $m = 36$. The wavelength of the BSC measurements was 24 cm (L-band). For using the values above in the model, two other issues must be considered: first, the rms heights and slopes were measured for surface profiles, and second, the Hurst exponent's dependence on the size scale either due to different surface formation mechanisms [25], which can be considered mathematically as the roll-off length (as described earlier). As described in more detail in [25], the Hurst exponent of these sample surfaces does not remain constant at all measured length scales, which could be a sign of the proximity of the roll-off length to the measurement scales. Also concerning Test Case 3, the wavelength-scale rms height significantly exceeds the Rayleigh limit of 3.1 cm at normal incidence; this fact is included in the discussion.

In Section 2.3, an equation for the slope was shown for the case, where the length scale extended from the wavelength scale to the roll-off length. Here, assuming that the roll-off length is a lot greater than 1 m, but the rms heights are reported only up to a length scale of 1 m, the result of integration from the wavelength scale to an arbitrary integer factor of the wavelength, $x = m_x \lambda$ ($\lambda < x < L_r$) is (as described in more detail in Appendix B):

$$s(d_e = \lambda) = h_1 q_x \sqrt{\frac{H}{1 - H}} \sqrt{\frac{m_x^2 - m_x^{2H}}{m_x^{2H} - 1}}. \quad (32)$$

In this case, $x = 1$ and $m_x = 4$ because the length scale for the rms height is four times the wavelength. Also here, a factor of $\sqrt{2}$ is accounted for in $q_x = \sqrt{2}\pi x^{-1}$ (profiles) vs.

$q_x = \sqrt{4\pi}x^{-1}$ (areas). The same equation can also be used to derive the rms height at any value of x (within the limits stated above) so that

$$h(x) = \frac{s_\lambda}{q_x}\sqrt{\frac{1-H}{H}}\sqrt{\frac{m_x^{2H}-1}{m_x^2 - m_x^{2H}}}, \tag{33}$$

which gives $h(9) = 0.06$ m for Test Case 1, 0.13 m for Test Case 2, and 0.28 m for Test Case 3. The values for Test Cases 1 and 2 are comparable to using Equation (1), in this case, $h(9) = h_1 9^H$, which gives, respectively, 0.07 and 0.14. For Test Case 3, Equation (1) gives $h(9) = 0.21$ m—significantly below 0.28 m—possibly because of a roll-off length $L_r < 4.5$ m, while Equation (1) would require a much greater value of L_r to apply, as the next section discusses in more detail.

3. Results

The results are presented in four parts: first, how consistent the surface statistics of the synthetic surfaces are with the theory, which is a fundamental part of validating the further results; second, the evaluation of the consistency of the selected analytical scattering laws with the computed BSC functions; third, the evaluation of the consistency of an empirical scattering law that models the measured BSC functions of Hawaii lava flows with the analytical scattering laws validated numerically; and fourth, the contribution of multiple scattering, which quantifies the validity of single-scattering assumptions of analytical scattering laws. The contribution of multiple scattering by a three-dimensional undulating surface is challenging to derive analytically and is thus better to compute numerically.

3.1. Surface Statistics

First, the statistical properties of the surfaces are shown so that sample profiles are extracted from the individual rows and columns of several surface realizations, and the properties are averaged. The autocorrelation function, the rms heights, and the rms deviations and slopes are shown. The code for computing the surface statistics is included in the same online repository as the codes introduced in Section 2; see the Data Availability Statement.

The autocorrelation function at a specific length scale for the synthetic surfaces was derived to quantify how it is related to the Hurst parameter (at low values of lag):

$$\rho(\Delta r) = \exp\left[-K_c\left(\frac{\Delta r}{L_r}\right)^{y(H)}\right], \tag{34}$$

where a linear relation of $y(H)$ can be found for the exponent (to be discussed below) and $K_c \approx 3.1 \pm 0.5$ when $H \in [0.3, 0.9]$, but 4.0 ± 0.2 at $H = 0.1$. Figure 3 illustrates the change in the shape of the autocorrelation functions when $H = 0.5$, $h_L = 0.1$ m, $L_e = 16$ m, $m = 64$, and $q_r = 8$ ($L_r = 2$ m). The fitting is limited to the values of lag up to 1 m.

Figure 4 shows the rms height as a function of profile length ($2d_e < L < L_e$) based on Equation (6), with a linear fit to the data points at $\Delta r < L_r$, and the rms height as a function of step size based on Equation (7). The fit to the average of ten surface realizations gave a linear relation $h(L) = 0.07(LL_r^{-1})^{0.51}$, which is nearly exactly as expected for the given input characteristics for the sample surfaces. The power of 0.51 is consistent with the value of H used for generating the surface, i.e., 0.5. The factor 0.07 is, here, the value of $h(L_r)$ and thus depends on q_r, which defines L_r. Beyond L_r, the power was found to be up to about 60% of the input H for $H > 0.5$, comparable at $H \approx 0.3$, but greater than zero even for $H = 0$.

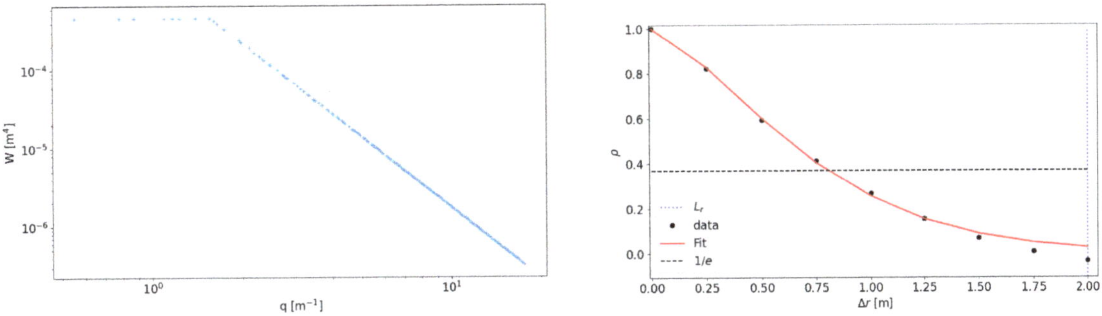

Figure 3. The power spectrum (on the **left**) and the autocorrelation function (on the **right**) when $H = 0.5$, $h_L = 0.1$ m, $L_e = 16$ m, $m = 64$, and $q_r = 8$ ($L_r = 2$ m, depicted as a vertical blue dotted line). In the power spectrum, the transition point is at $q_r \pi L_e^{-1} = 0.5\pi$, and the power-law gradient is proportional to $q^{-2H-2} = q^{-3}$. The black horizontal line depicts $1/e$, i.e., the correlation length is here the value of lag where the line crosses the autocorrelation function (at $\Delta r \approx 0.8$ m).

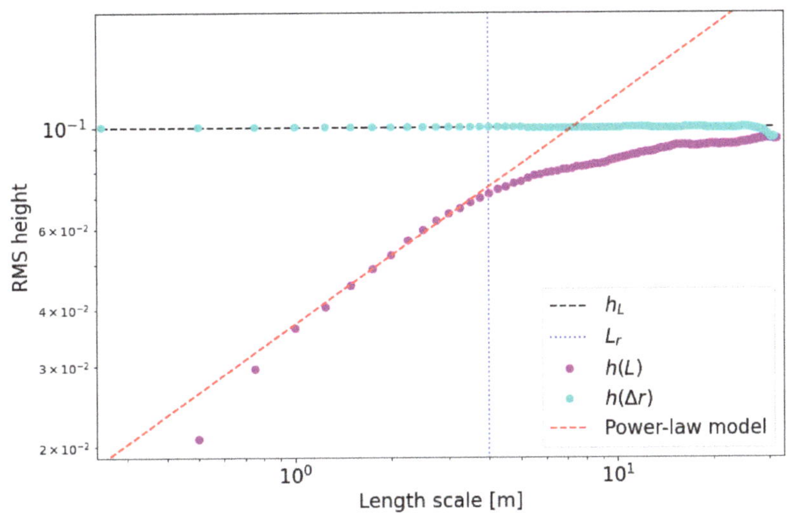

Figure 4. The root-mean-square (rms) height as a function of step size (cyan) and profile length (magenta) when $H = 0.5$, $h_L = 0.1$ m, $L_e = 32$ m, $m = 128$, and $q_r = 8$ ($L_r = 4$ m, depicted as a vertical blue dotted line), where each data set is an average of 10 surface realizations. The fit to the rms height as a function of profile length (L) is shown as a red dashed line, and the horizontal black dashed line depicts h_L given as input to define the surface.

Figure 5 illustrates the rms deviation and slope of a surface profile as functions of lag. Here, the roll-off length appears as a plateau in the function $\nu(\Delta r)$ at $\Delta r > L_r$. Consequently, the Allan (rms) deviation follows

$$\nu(\Delta r) \approx \sqrt{2} h_L [1 - \exp(-K_\nu \left(\frac{\Delta r}{L_r}\right)^{y(H)})], \tag{35}$$

where K_ν and $y(H)$ were found to be nearly equal to those for the autocorrelation function above: $K_\nu \approx 3.2 \pm 0.3$ when $H \in [0.3, 0.9]$ and $K_\nu \approx 4.4 \pm 0.3$ at $H = 0.1$, and $y(H)$ is discussed below mutually for all cases. The respective rms slope as a function of lag is $s = \nu(\Delta r)^{-1}$, for which $K_s \approx 3.3 \pm 0.3$ when $H \in [0.3, 0.9]$ and $K_s \approx 4.6 \pm 0.3$ at $H = 0.1$.

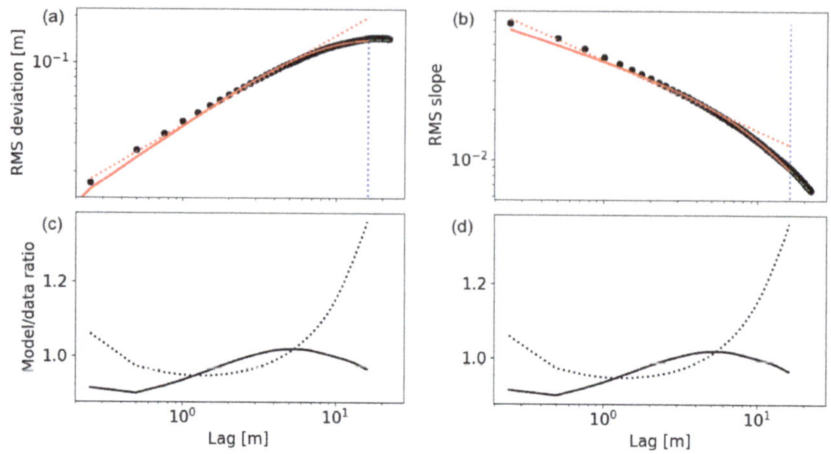

Figure 5. The surface properties as functions of lag when $H = 0.6$, $h_L = 0.1$ m, $L_e = 64$ m, $m = 256$, and $q_r = 4$ ($L_r = 16$ m, depicted as a vertical blue dotted line). (**a**) The root-mean-square (rms) deviation for surface profiles sampled from synthetic surfaces (black circles) and a fitted model (Equation (35); red solid line at $\Delta r < L_r$), and $\nu(\Delta r > L_r) = \sqrt{2}h_L$ (horizontal green dashed line at $\Delta r \geq L_r$). The power-law model (Equation (2)) fitted to only small values of Δr is added for reference (red dotted lines). (**b**) The rms slope (black circles) and the fitted model (with the same line styles as in panel (**a**)). (**c**) The ratio of the rms deviation model's values to the data values as a function of lag (the solid and dotted line corresponding to the red lines of the same style). (**d**) The ratio of the rms slope model's values to the data values as a function of lag (with the same line styles as in panel (**c**)).

Similar to the rms height, when the rms deviation increases exponentially, it appears linear in the log scale at low values of lag, i.e., if measurements are conducted only at low values of lag with respect to the roll-off length, only this part may be visible. However, when approaching the roll-off length, the gradient of the deviogram decreases. When the lag is longer than the roll-off length, the rms deviation remains approximately stationary at $\sqrt{2}h_L$, whereas the slope continues to decrease as $(\Delta r)^{-1}$. In the numerical approach, when $q_r \leq 2$, the function $\nu(\Delta r)$ has excessive downward curving at higher values of lag, so for modeling purposes, using $q_r > 2$, provides more realistic results when the lag extends beyond L_r. Likely due to the same reason, the equation for $\nu(\Delta r)$, as shown above, was not as good a fit for $q_r \leq 2$ as for greater values of q_r. Equation (17) was found to be a good approximation for $\sqrt{2}s(d_e)$ for $H \in [0.3, 0.9]$ (the square root added due to the difference in dimensions—3D for the analytical equation but 2D for the surface-statistics data computed here).

As for $y(H)$, the values from fits to the autocorrelation function, rms deviation, and rms slope as functions of lag produced similar trends: $y(H) \approx 0.83(H+1)$. The value of $y(H = 0.5)$ was consistently about $1.25 \approx \sqrt{0.5\pi}$ and at $y(H = 1), \approx 1.65$. In none of the tested cases $y(H = 1) > 1.7$, likely due to the bias in the self-affinity, as was discussed earlier. Nevertheless, the form of the slope p.d.f. of the surface is similar to a Rayleigh distribution (consistent with [18]) regardless of the wavelength-scale autocorrelation function when the rms $\theta_s \in [2°, 30°]$ ($s \in [0.03, 0.6]$), whereas the slope angle p.d.f. is nearly Gaussian at moderate slopes. For significantly greater rms slopes, the Gaussian distribution cannot model the skew toward greater angles (see Figure A1 in Appendix D for histograms of the slope angle p.d.f.); however, a combination of two or more Gaussian distributions could be more successful.

The surface statistics theory presented so far can be tested on the surface-profile statistics presented by Campbell [22]. Figure 6 depicts the 2D wavelength-scale rms slopes for surface profiles obtained for the Hawaii lava flows versus the rms slope model following

Equation (32). The surface profile with the greatest slope has the highest residual value possibly due to a shorter roll-off length than the other cases. The power spectral density function for the specific case in question is presented in [25] (Figure 3.13). The spectrum shows more variation in the slope of the function for the lowest Hurst exponent than those of the other two profiles with greater Hurst exponents, which could be indicative of a roll-off length near the measurement scale. A lower Hurst exponent would provide a better fit to the model for the case in question.

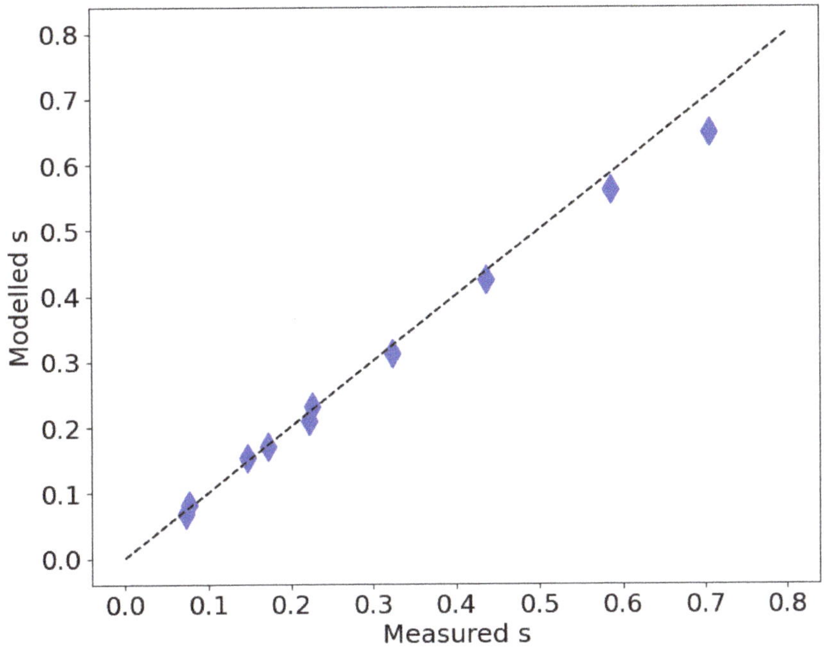

Figure 6. The root-mean-square (rms) slopes measured for Hawaii lava flow surface profiles [22] compared to a model following Equation (32) using $m_x = 4$. The black dashed line has been added as a visual reference of an optimal fit. The values are listed in Table A1 in Appendix C.

Campbell [22] found the following empirical model for the rms deviation for a step size Δr and rms height of profile length (Δr):

$$\nu(\Delta r) = \sqrt{2} h_{\Delta r} \exp(2H), \tag{36}$$

which is intriguingly similar to Equation (21) when $\Delta r = \lambda$ and $H \in [0.26, 0.63]$, that is, for all their sample surfaces. Also, computing $\nu^2 h^{-2}$ for the surface statistics data shown in Figures 4 and 5 suggests that Equation (23) is a good analytical solution for the ratio, although there was a lot of variation for individual surfaces at $H = 0.5$ (from 3 to 15, while the expected value was 2π). The evaluation of the consistency of the synthetic surfaces' statistical properties with the analytical solutions for 3D surfaces (e.g., Equation (17)) was found, in general, to be challenging due to the significant variance.

3.2. Evaluation of Radar Backscatter Functions

First, the laws based on autocorrelation functions, the Gaussian law, Hagfors' law, and the cosine law are evaluated. In the examples below, the wavelength-scale element discretization is kept at 25 cm, while varying h_L and H. The Rayleigh limit for the wavelength-scale h_λ is then 0.031 m at zero incidence. The electric permittivity is 5.0.

Figures 7 and 8 demonstrate the quality of the different laws in modeling the BSC functions of the synthetic surfaces computed using the simplified surface-scattering algorithm. Each curve in Figure 7 is an ensemble average of 40 surfaces and 4 azimuth angles. Despite the ensemble- and orientation-averaged results, there is a significant deviation in the phase function at low phase angles when $h_L = 0.5$ m due to the steep slopes of individual elements. The error in the first phase angle bin is mitigated by using the mean of the first three phase-angle bins for the BSC functions. For $h_L = 0.1$ m, $h_\lambda = 0.008$ m, which is well below the Rayleigh limit. For $h_L = 0.3$ m, $h_\lambda = 0.031$, which is at the Rayleigh limit at the zero incidence, but not through all local incidence angles of the individual elements. For $h_L = 0.5$ m, $h_\lambda = 0.040$ m, which is significantly above the Rayleigh limit, and therefore the result should be considered with caution. Figure 7 also shows the change in the shape of the phase function as the incidence angle increases. The phase functions are useful for verifying that the distributions of the surface reflections are correct, but also have practical use when modeling scenarios where the source of illumination and the observer are not at the same location.

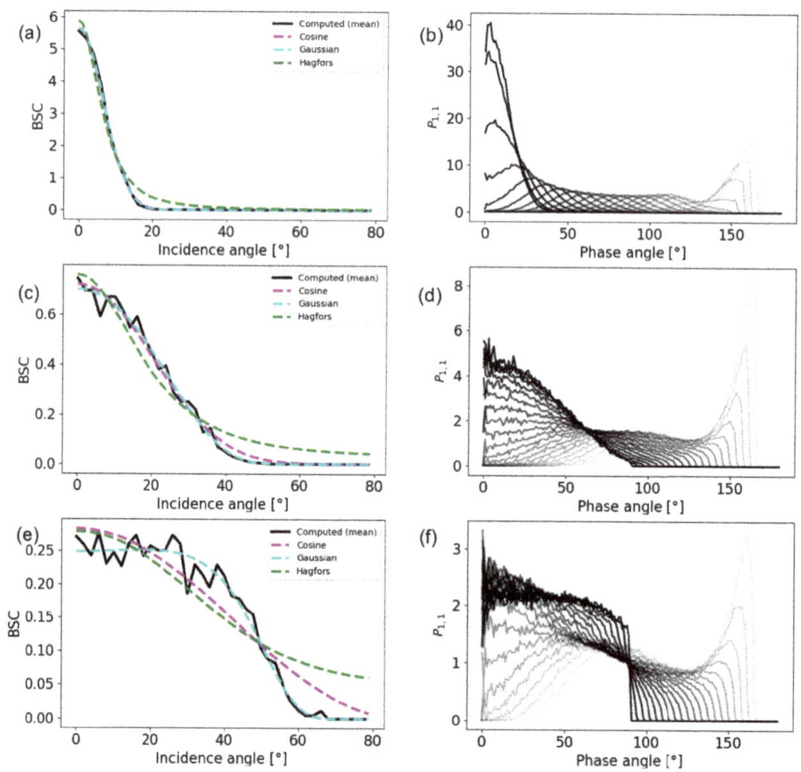

Figure 7. The backscattering coefficients as a function of incidence angle (the left-side panels (**a**,**c**,**e**)) and the scattering phase matrix $P_{1,1}$ element as a function of the phase angle (the right-side panels (**b**,**d**,**f**)) for $H = 0.8$, $L_e = 9$ m, $m = 36$, and $q_r = 2$, when $h_L = 0.1$ m (**a**,**b**), $h_L = 0.3$ m (**c**,**d**), and $h_L = 0.5$ m (**e**,**f**). The scattering models based on the cosine (magenta), Gaussian (cyan), and Hagfors' laws (green) were fitted to the computed BSC functions at incidence angles from 0 to 80°. In the phase curves on the right, the normal incidence is shown as a black line, and each consecutive incidence angle at 4° intervals in a lightening shade of gray.

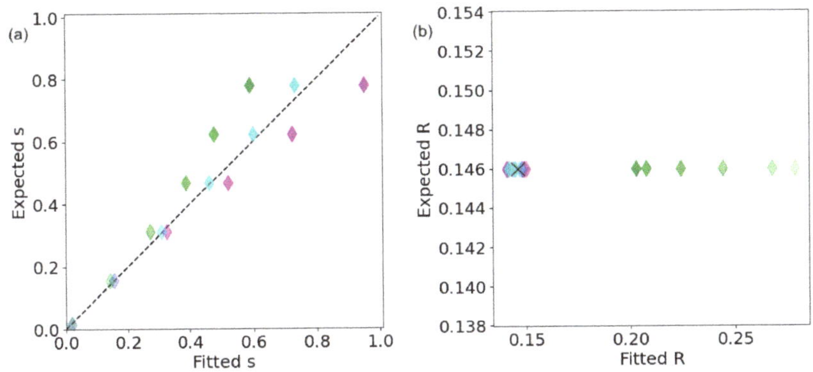

Figure 8. (**a**) The comparison of the model-predicted (fitted) root-mean-square (rms) slope versus the true (expected) rms slope of the synthetic surface when using the cosine (magenta), the Gaussian (cyan), and the Hagfors' law (green) depicted with the reference for an ideal fit (the dashed line); and (**b**) the same comparison for the Fresnel normal-incidence reflectivity coefficient with the ideal fit at 0.146 (the black cross). Here, $H = 0.8$ and h_L varies from 0.01 m to 0.5 m with steps of 0.1 m (respectively, the markers from the lower left to the upper right corner in panel (**a**), and the corresponding colors on the right).

In Figure 8, the rms slopes derived from the BSC functions using the Gaussian law are the most consistent with the rms slopes computed geometrically using the vertical component of the normal of each element. The rms slopes derived using the cosine law were increasingly over the true rms slope as the rms slope increased (up to a factor of 1.19 at $s = 0.62$), while the Fresnel reflection coefficient was close to the expected value. The rms slopes derived using Hagfors' law were increasingly below the true rms slope (down to a factor of 0.77 at $s = 0.62$), while the Fresnel reflection coefficient was consistently above the true value by a factor of 1.42–1.92 with a negative correlation to the rms slope. Here, I assumed that $C = s^{-2}$, which is not analytically correct for an exponential correlation function that the Hagfors' law is based on, but this result can be useful information for estimating the uncertainty of results that have been previously derived using Hagfors' law. For the Gaussian and cosine scattering laws (Equations (24) and (26)), the accuracy of the fit parameter R was 5% up to an rms slope of 0.467 ($\theta_s = 25°$). It should be noted that although the autocorrelation function at the wavelength scale was not found to have a Gaussian form at any value of H within the range $[0, 1]$, the Gaussian scattering law still provides the best fit to the synthetic BSC functions. The accuracy of the Gaussian law deteriorated below rms slope angles of 3° as well as above 40°. One reason is the slope angle p.d.f: at very low and very high rms slopes, the slope angle p.d.f. skews from a Gaussian shape, as the slopes of the elements increase but cannot reach 90° (Appendix D). This is an important note for the assumption of a gently undulating surface. As the slopes approach 0°, the reflections are more accurately described by simple Fresnel equations for planar surfaces.

The cosine law was performed to equal accuracy only up to slopes of about 15°, and increasingly overestimated the steeper rms slopes. Hagfors' law performed in a similar quality to the cosine law but underestimated the rms slopes close to a comparable extent to what the cosine law overestimated them. The distinct "tail" at high incidence angles was not observed to occur in any of the simulations regardless of whether multiple scattering was included or not.

The Shepard's model for $H = 0.5$ fits the computed BSC functions at higher incidence angles better than the cosine law but no better than the Gaussian law. Despite the similarity to Hagfors' law, there is no distinct tail visible at high incidence angles. The simulations showed that R_S is not equal to R_F as claimed in [14], but a correction factor that depends on the slope and the Hurst exponent is required. Shepard's model for $H = 0.5$ was able to predict the rms slope with any value of H when $s \approx 2.7 s_S^{2.2}$. The Fresnel reflection coefficient also requires a non-trivial term that depends on the rms slope to correctly predict the reflection coefficient.

One reason for the discrepancies is a misplaced factor of 4π. Also, the integration to infinity used in many analytical approaches can be questioned in physical applications. It is mathematically convenient in the analytical derivation of the models; however, the statistical properties are, in reality, limited to specific length scales. When modeling radar imaging data, the integration is ideally done through the area covered by each pixel or a set of pixels, i.e., the area that contributes to the pixel. Because the rms height depends on the horizontal scale, as given by the Hurst exponent, the measured, finite area might play a role that should not be omitted.

3.3. *Backscatter Functions of Hawaii Lava Flows*

The consistency of the analytical and numerical solutions with measured data is a crucial part of evaluating their validity. Campbell [22,25] reported surface statistics for ten surface profiles of lava flows at Kilauea, Hawaii. The measured rms slopes are available in both the wavelength scale of 25 cm (s_λ) and a unit scale of 1 m, the rms heights are measured over 1-m profile lengths (h_1), the Hurst exponents are available in the unit scale of 1 m for all ten profiles, and some additional information available for a few profiles (Hurst exponent when using a different step size, deviograms, and power spectral density plots).

Figure 9 illustrates the scattering law fits for the three test cases (Section 2.4), with the empirical model for the data presented in Campbell [22] included at the incidence angle range from 35° to 60°. The shown BSC functions for the synthetic surfaces are ensemble-averaged for 30 sample surfaces and orientation-averaged for 4 different azimuth orientations. The electric permittivity was set here to $4.5 + 0.042i$. For Test Cases 1 and 2, $q_r = 2$, and for Test Case 3, $qr = 3$. The fits include the Gaussian, cosine, and Hagfors' laws. The surface model was able to reproduce a comparable rms slope using the rms height and Hurst exponent for all test cases to an accuracy of 97% (with the dimension factor of $\sqrt{2}$ considered).

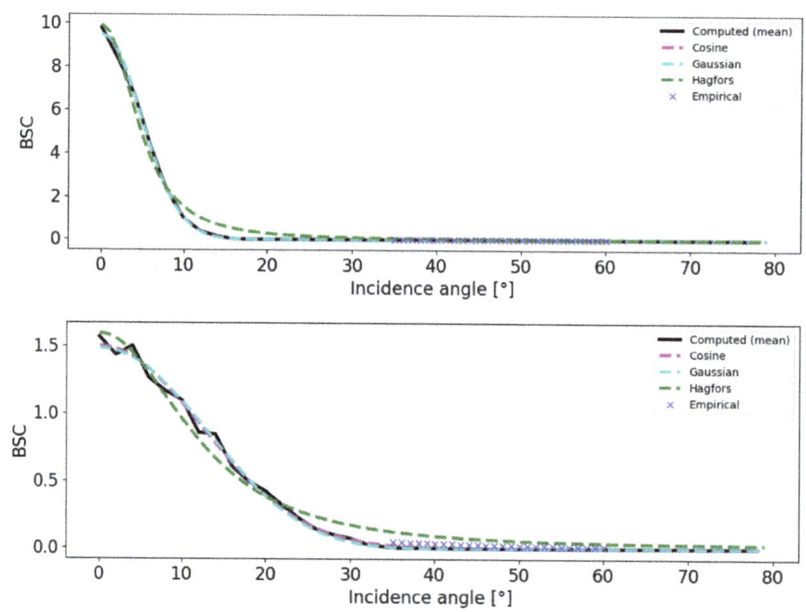

Figure 9. *Cont.*

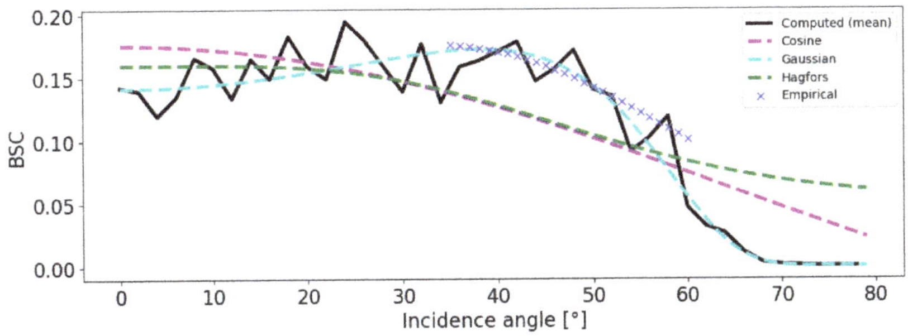

Figure 9. The backscattering coefficient (BSC) functions for Test Case 1 (**top**), Test Case 2 (**middle**), and Test Case 3 (**bottom**). The scattering models based on the cosine, Gaussian, and Hagfors' laws were fit to the computed BSC functions at incidence angles from 0° to 78°. Campbell's empirical model (the sum of Equations (29) and (30)) for backscattering coefficients is depicted in blue "x" markers for the wavelength-scale root-mean-square (rms) slope for the synthetic surface.

3.4. Contribution of Multiple Scattering

In order to quantify the contribution of multiple scattering in the synthetic surfaces selected for the study, I used the more elaborate ray-tracing algorithm ROSAS described in Section 2. For the computations presented here, the number of rays was 20,000, and the refractive index was set to $2.12 + 0.01i$ (electric permittivity of $4.5 + 0.042i$). The scattering properties were averaged over four azimuth angles for a synthetic surface with $L_e = 9$ m and $m = 36$.

For Test Case 1, no multiple scattering took place. For Test Case 2, multiple scattering took place at a level of 1–2 per 600 rays, but the contribution of multiple scattering is negligible in the total echo power, particularly at backscattering. For Test Case 3, it was not clear what the value of h_L should be for an edge length of 9 m because, as discussed earlier, there are indications that the roll-off length is less than 4.5 m, and the same wavelength-scale rms slope could be reproduced with different combinations of h_L and q_r. Also, it is interesting to look at two different examples, where multiple bounces take place: one where the peak contribution is visible for double and triple bounces at about $2\theta_0$ (Figure 10 panels a and b, $\theta_s = 36.7°$, $H = 0.26$, $h_L = 0.212$ m, and $q_r = 2.5$), and a second one where the phase function for the double and triple bounces decreases (panels c and d; $\theta_s = 45.8°$, $H = 0.26$, $h_L = 0.27$ m, and $q_r = 3.0$). In the first example, the power contribution of the double bounce was, at most, about 10% of the power contribution of single scattering and that of the triple bounce, at most, about 1%, the maxima for the double bounce occurring at about $2\theta_0$ at both incidence angles and also for the triple bounce at $\theta_0 = 40°$. In the second example, the maxima for all numbers of bounces are at zero phase angle for both incidence angles, 0° and 40°, and the power contribution of double bounces was up to $\sim$ 20% of the total at backscattering while remaining at 1% for triple bounces.

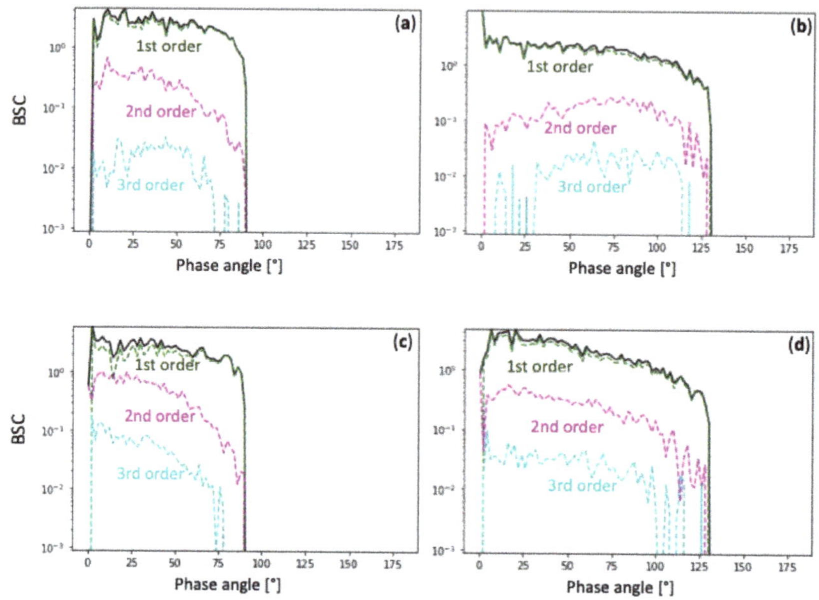

Figure 10. The contribution of multiple scattering. The phase function with a breakdown of the contribution of first-order reflection in green, double bounce depicted in magenta, and that of triple reflections in cyan, when the incidence angle is (**a**) 0° and (**b**) 40°, when $s = 0.745$ ($\theta_s = 36.7°$), and (**c**) 0° and (**d**) 40°, when $s = 1.027$ ($\theta_s = 45.8°$).

Figure 11 shows that for Test Case 3, the backscattered echo power decreases in both polarization states as the incidence angle increases from 0° to 60°. The BSC function is (as given in Equation (4) $\sigma_0 = q_{\text{sca}} P_{1,1}(0°, \theta_0) \cos\theta_0$, whereas $\sigma_{OC} = 0.5\sigma_0(1 - P_{4,4}P_{1,1}^{-1})$ and $\sigma_{SC} = 0.5\sigma_0(1 + P_{4,4}P_{1,1}^{-1})$. The BSC function for the total and the OC polarization follows the empirical model for the lava flows well. The difference between the simplified surface-scattering code and the ROSAS code can be attributed to the improved tracking of multiple scattering in ROSAS. The electric permittivity of the lava flows was not well constrained in [22], which is one factor in the difference to the empirical model. I used an electric permittivity of 4.5 (real part); although a lower value down to 4.0 could further improve the fit at an incidence angles of 30°–40°, it could produce a worse fit at $\theta_0 \geq 50°$. The SC polarization of the BSC function decreases as a function of incidence angle more steeply as the empirical function, as is visualized in Figure 11; this finding is contrary to observations, which demonstrates the well-known inadequacy of ray-optics models in simulating depolarization effects even when second- and third-order scattering are included. As Fa et al. [31] showed, simulating the SC polarization component so that it is consistent with measurements is difficult even when volume scattering by wavelength-scale ellipsoids is included in the model. On the other hand, in their surface model, which is based on IEM and uses an exponential autocorrelation function for the surface statistics, the circular-polarization ratio increases as a function of incidence angle, which suggests that the ROSAS code could be developed further by incorporating relevant field coefficients [29,30].

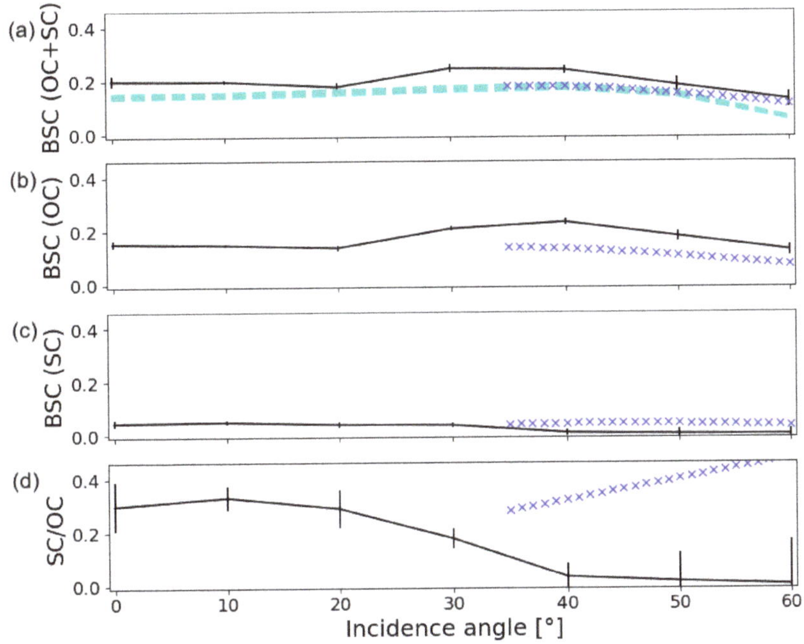

Figure 11. The backscattering coefficients and circular polarization components as a function of incidence angle for Test Case 3 computed using ROSAS (here, $H = 0.26$, $h_L = 0.27$, $q_r = 3.0$). (**a**) The total BSC function, (**b**) the opposite-circular-polarization component, (**c**) the same-circular-polarization component, and (**d**) the circular-polarization ratio. The errorbars were derived using the one-sigma uncertainty of the zero-phase-angle intercept for a line fit to the phase angles from 0° to 20° of the $P_{1,1}$ and $P_{4,4}$ elements using a linear regression fit (the functions are sufficiently linear in this range of angles). The blue crosses depict the empirical models for the measurements by Campbell [22] (using Equations (29)–(31) and their combinations), and the cyan dashed line depicts the Gaussian fit using the simplified surface-scattering algorithm.

4. Discussion

First, it is important to discuss how well the artificial surface generation algorithm performed with respect to the existing analytically derived statistical properties for self-affine surfaces. The discretization of the artificial surface is a good analog to empirical measurements, where the topographic height distribution is derived using sampling at regular intervals. However, differences to mathematically ideal surfaces are possible due to the discretization, and therefore the analytical solution could be, to some extent, different from the generated surface despite ensemble averaging. The algorithm was created based on the theory presented in [23], and the numerical evaluation revealed that the surface properties were quite consistent with the theory. The consistency was the best near $H = 0.5$ and degraded towards the end values $H = 0$ and $H = 1$. Specifically, the low values of H caused issues. For artificial surfaces that were generated using Hurst exponents near the end values, the algorithm tended to generate surfaces with their self-affinity biased toward $H = 0.5$. Considering applications, Campbell's measurements [22,25] showed that natural rocky and grain surfaces have Hurst exponents typically below $H = 0.65$ but above $H = 0.15$. Therefore, the surface generation algorithm can be considered a good analog when the Hurst exponent of the surface is measured at $H \in [0.3, 0.8]$, which applies to the majority of natural surfaces. The issues related to the low values of H could be related to the fact that the surface slopes should be larger than the discretization can reproduce.

The modeling of the measured Hawaii lava flows was successful; the synthetic surfaces were able to reproduce the measured rms slopes when generated using the measured rms height and Hurst exponent as input parameters. For Test Case 3, the fit was the most challenging one but surprisingly good considering that no diffuse component was included while simultaneously the rms height within a wavelength scale was greater than the Rayleigh limit.

The presented exponential form for the rms deviation (Equation (35)) is analytically sensible and considers the roll-off length better than a simple power-law function. The downward curvature is consistent with the empirical deviograms presented in [24] Figure 3. An increased consideration of the roll-off length was also a key to being able to reproduce the rms slope of Test Case 3.

The main scope of the paper is to evaluate how well the various radar scattering laws are able to model the rms slope. The surface slope p.d.f. is the main factor of a rough surface that determines how the scattered power distributes in space. The simulations showed that although the synthetic surfaces did not have a Gaussian autocorrelation function at the wavelength scale nor the slope or the slope angle p.d.f were precisely Gaussian (a combination of two Gaussian laws could be a better fit), the Gaussian law performed the best at modeling the rms slopes at 95% accuracy at all adirectional wavelength-scale rms slopes from 2° to 40°. At $s < 2^\circ$, the surface can be assumed planar, and regular Fresnel reflection coefficients are a decent approximation.

The cosine law performed to equal accuracy with the Gaussian law only up to slopes of about 15°, and increasingly overestimated the steeper rms slopes, when the slope was derived from the scattering profile. Hagfors' law (Equation (25)) also fits well at low slopes but was found to underestimate the rms slopes close to an equal amount to what cosine law overestimates them. Also, R_H was inconsistent with the true Fresnel reflection coefficient, whereas R_G and R_C provided better fits. The distinct diffuse scattering "tail" of Hagfors' law at high incidence angles did not fit well using a ray-optics model even when multiple scattering was included in the model. In fact, the tail was not observed to occur in any of the simulations regardless of the inclusion of multiple scattering. Therefore, it is questionable whether such a feature can be produced when only Fresnel scattering is included, which is consistent with Hagfors' own suspicion that "an entirely different type of reflection mechanism should be sought to describe the tail of the echo" due to an observation of depolarization of the radar echo [3]. For example, wavelength-scale particles on the surface could reproduce the tail, because there is always a non-zero backscattering intensity component and a non-zero depolarization component for non-spheroidal, asymmetric particles in the resonance-scattering regime (e.g., [32]).

In conclusion, the Gaussian law can be recommended as the most reliable radar-scattering law for all natural surfaces for the inverse modeling of the rms slopes and the Fresnel reflection coefficient from the quasi-specular backscattering peak in disk functions. If diffuse scattering is observed, a separate law that considers the physical properties of wavelength-scale scatterers rather than rms slopes could be a more realistic approach. For large objects such as planets and moons, also considering a realistic distribution of roughness parameters could be an improvement to assuming homogeneous roughness characteristics for the whole object.

Shadowing models for self-affine surfaces have been comprehensively investigated by Shepard and Campbell [20]. The shadowing produced by the artificial surface model presented here was found to be consistent with the numerical shadowing models for up to moderate slopes. This result is expected for the shadowing models that presume a Gaussian slope p.d.f. For the test case with the steepest slopes, the analytical model of Smith [33] was valid with an uncertainty of 85%, which is a relatively good performance as well for such complex topographies. Also here, it is useful to remember that the model is derived for an undulating surface without wavelength-scale particles on the surface. The shadowing function for a surface covered with wavelength-scale particles would increasingly differ as a function of the incidence angle, because the resonance-regime scattering functions of particles with various shapes, sizes, and orientations are much more complicated than the Fresnel scattering functions in the ray-optics regime, and their relative contribution to the total backscattering is greater at higher incidence angles.

Campbell [34] showed that double reflections by dihedral corners may increase the SC polarization component significantly; however, the probability density function of the favorable angles with respect to the full measured area is a complex question, and only 2D surface profiles were considered. In this paper, the ray-tracing computations for multiple scattering revealed that for very rough surfaces (rms slope angle above 30°), the contribution of each increasing order of scattering is about one order of magnitude less than the previous, i.e., the second and third-order contributions are, respectively, about one-tenth and one-hundredth of the contribution of the first-order reflection (Figure 10). The reflected double-bounce phase function's maximum is centered at $2\theta_0$, whereas for a single reflection, the phase function remains at a stationary level up to $2\theta_0$ before decreasing. The SC polarization component was up to 33% of the OC polarization component when multiple scattering was included and the roughest Test Case 3 was used as the surface model. The circular-polarization ratio decreased as a function of incidence angle, which is contrary to what measurements suggest. Ray-optics approximation is known to be insufficient for modeling polarization, which is highly sensitive to the scatterer's morphology and may also require the wave-interference effects to be taken into account. In Test Case 3, also the element-scale rms height significantly exceeding the Rayleigh limit likely plays a role. Finer discretization would improve the accuracy of the true morphology, but element sizes smaller than the wavelength make the use of ray-optics approximation questionable.

A discrete dipole approximation approach such as that presented in Riskilä et al. [35] (and the references therein) would allow including sub-wavelength-scale roughness, so it was also evaluated as a modeling approach, but it did not produce results that would be more consistent; in fact, the backscattering functions were much less consistent with the lava flow measurements due to a strong diffraction peak (visible in Figures 8–20 in [35]). Separating the diffraction peak from the incoherent scattering field would require additional computations that are out of the scope of this paper but, as future work, could potentially help resolve discrepancies. The same issue prevents the use of other scattering software based on volumetric discretization of the medium.

Applications to Planetary Surfaces

The Hawaii lava flows are a good example of natural surfaces for which the Hurst exponents, rms heights, and slopes have been measured. I found a relatively good consistency between the synthetic surfaces in this study and the radar measurements for the flows. The consistency validates the synthetic surfaces as realistic and provides a further interpretation of the lava flows; for example, the electric permittivity based on the modeling in this paper was found to be 4.5 (or slightly less) in contrast to the estimate of 6 by Campbell [22]. The porosity of the lava flows could thus be higher than originally expected. There was, however, some discrepancy in the rate of decrease of the BSC function for rough surfaces, which the ray-optics models were not able to model. For the smoother surfaces, the simulations provided an excellent fit to the measurements, but there was measurement data available only at incidence angles where BSC is close to zero, so the most relevant information on the specular enhancement is lacking.

Furthermore, the surfaces of the atmosphereless solar-system objects (SSOs) are mostly composed of regolith. Depending on the object, the size-frequency distribution of regolith varies from fine-grained micrometer-scale particles to multi-meter-scale boulders. The surfaces of the atmosphereless terrestrial planets and their moons are dominated by the fine-grained regolith as the Moon landings and numerous spacecraft missions have shown. Helfenstein and Shepard [36] reported an average Hurst exponent of about $H = 0.69 \pm 0.06$ for the Moon, which was considered consistent with other literature. In contrast, while fine-grained regolith is common on asteroids as well, some asteroids have been found to be lacking the fine-grained regolith and are composed primarily of centimeter-scale and larger particles, possibly deep into the subsurface and even into the core, for example, the Origins, Spectral Interpretation, Resource Identification, Security, Regolith Explorer (OSIRIS-REx) mission revealed for asteroid (101955) Bennu [37], and Hayabusa2 for asteroid (162173) Ryugu [38]. Harmon and Ostro [39] estimated that 25% of the lunar echo power is contributed by diffuse scattering by decimeter-scale particles and an even greater fraction of more than 40% for Mars. When a surface is dominantly composed of wavelength-scale particles, ray optics is not a sufficient scattering mechanism. Also, rms slopes above 40° are unphysical for a regolith surface because regolith particles would likely not retain such high-angled slopes over large areas for a long time due to dynamical and weathering processes. The scattering by wavelength-scale particles using particle shapes characteristic to the surface in question is thus crucial to realistically simulate the SC polarization component [32,40]. So far, the radar-scattering simulations have focused on particles in free space or near a smooth, Fresnel-scattering interface. Recently, there have also been useful developments in understanding the scattering in dense random media of irregular particles (e.g., [41]). The next key step is understanding the scattering interactions by irregular, wavelength-scale particles and the rough surface of a medium of fine-grained regolith.

Although this paper focuses on applications at microwave wavelengths, the results can be useful at visual and near-infrared wavelengths as well. For planetary surfaces composed of larger grains (greater than a millimeter), the scattering by the particles can be modeled more precisely when a rough surface is assumed instead of an ideal ellipsoid. Statistical treatment of the surface roughness could enhance the effectiveness of simulations. Furthermore, there are significant opportunities for other light-scattering applications using synthetic surfaces. There are both easy and challenging opportunities for software development; although the focus is here on incidence angles, the presented codes also compute phase functions, which can be used for modeling visual observations, when ray optics can be considered a valid approximation. An example of a more challenging task would be to add a diffuse scattering component that is not simply empirical but has physical meaning. Moreover, the observed self-affinity bias towards $H = 0.5$ requires further consideration.

5. Conclusions

In summary, this paper presents a set of open-source Python codes for generating synthetic self-affine fractal surfaces and computing their phase and BSC functions, which were used to evaluate radar scattering laws. The codes can provide a useful tool for the scientific communities from various applications of physics from remote sensing and modeling of electromagnetic radiation to material sciences. The synthetic surfaces were shown to be able to reproduce measured rms slopes for a real-life example, lava flows, which gives significant confidence to the validity of the model. Some challenges were also encountered, the most significant of which is the self-affinity bias of the generated synthetic surfaces towards $H = 0.5$. This issue is fortunately not a major constraint for most natural surfaces. Moreover, the roll-off length (i.e., the frequency limits of the undulations) of the self-affine fractal surfaces has been given little consideration in the existing analytical solutions for the surface-roughness characteristics, although it plays a clear role. Here, the wavelength-scale rms slope was derived using the power spectrum with different options of roll-off length (low-frequency limit; Equations (17) and (32)).

The Gaussian law was confirmed to be the optimal scattering model for surfaces for predicting rms slope angles up to 40° when the volumes below and above the interface are homogeneous (free from scatterers larger than the Rayleigh limit), which is consistent with the literature (e.g., [18]). The Gaussian law reproduced measured rms slopes well, even when the rms height within a wavelength-scale sample length was above the Rayleigh limit; however, the produced BSC function could be less accurate. A Gaussian autocorrelation function or a Gaussian slope p.d.f. are not required for the Gaussian law to be valid for self-affine fractal surfaces. The evaluation of other widely used radar scattering laws showed that the cosine law can predict the rms slope when the slope is below 15° but overestimates the larger slopes. Neither the Gaussian nor cosine models should be used when the rms slopes are below 1°, but Fresnel reflection functions for a planar surface can be used as a valid approximation. Hagfors' model has been shown to be an empirically well-fitting model for regolith surfaces; however, the model underestimates large rms slopes and overestimates the Fresnel reflectivity. Hagfors' law is distinct from the other laws due to the diffuse-scattering tail at high incidence angles. The tail was not visible in the computed BSC functions at any value of the Hurst exponent or rms height, but a diffuse scattering component such as wavelength-scale structures or particles on the surface or as volume scatterers below the surface would be required to reproduce it. The scattering models presented in Shepard and Campbell [14] were incorrect; the fitted parameters require additional correction functions.

The ROSAS code was used to quantify multiple scattering, which is complex to quantify analytically. When the rms slope of the 3D surface was 0.745 (0.527 for a surface profile), the double-bounce contribution of the backscattered power was up to 10% of the total at normal incidence and the peak of the phase function appeared near $2\theta_0$ for any incidence angle, whereas for a greater rms slope of 1.027 (0.726 for a profile), the backscattered power was up to 20% of the total and the peak of the phase function was at 0°. Deriving further empirical or analytical solutions for the multiple-scattering contributions is not in the scope of the paper, but the code provides a concrete tool for the estimation of the multiple scattering for any relevant applications and paves the way for further work.

Further modeling work is also needed to understand the interactions of wavelength-scale particles in contact with the self-affine fractal surfaces. A large number of elaborate scattering modeling applications exist, some are optimized for modeling volume scattering by buried inclusions, and some for individual objects of finite size with respect to the wavelength, such as applications based on the discrete-dipole approximation or the T-matrix method. A key challenge that arises in the modeling of planetary surfaces is combining a wavelength-scale object with a surface much larger than the wavelength; the scenario environment is too large for the discretization-based full-wave approaches, but ray tracing is not a sufficient approximation for a wavelength-scale object. A solution to this challenge requires future work.

Funding: This research was funded by the Academy (Research Council) of Finland grant number 347627.

Data Availability Statement: The numerical data used in the article was generated using the presented codes available through https://bitbucket.org/planetarysystemresearch/workspace/projects/SAS (accessed on 20 December 2023).

Acknowledgments: I would like to thank Bruce Campbell for providing additional information regarding the Hawaii lava flow measurements, Antti Penttilä for providing the scattering computations using the discrete dipole approximation, Juan F. Alcaide for the language check of the manuscript, and the anonymous reviewers for comments that improved the manuscript.

Conflicts of Interest: The author declares no conflicts of interest. The funders had no role in the design of the study; in the collection, analyses, or interpretation of data; in the writing of the manuscript; or in the decision to publish the results.

Abbreviations

The following abbreviations are used in this manuscript:

BSC	Backscattering coefficient
HH	Horizontal to horizontal
HV	Horizontal to vertical
LL	Left to left
LR	Left to right
OC	Opposite circular
P.d.f.	Probability density function
Rms	Root mean square

ROSAS	Ray-optics for self-affine fractal surfaces
SAR	Synthetic aperture radar
SC	Same circular
SSO	Solar-system object
VNIR	Visual and near-infrared
VV	Vertical to vertical

Appendix A

The artificial-surface algorithm, based on the theory presented in [23], generates and discretizes self-affine surfaces. The key steps of the algorithm are summarized here. First, an array of wavevectors is generated so that for each array element j:

$$q_j = \frac{2\pi j}{m}, \tag{A1}$$

a corresponding array q_k is generated for the orthogonal dimension. For each dimension, the array is shifted and scaled by the pixel width, d_e, so that the center of the array is at 0, and the wavevector ranges from $-2\pi m(2md_e)^{-1} = -\pi(d_e)^{-1}$ to $\pi(m-2)(md_e)^{-1}$. The two arrays are used for generating a two-dimensional array $\rho(\mathbf{q})$ where an element $\rho(j,k) = \sqrt{q_j^2 + q_k^2}$. This array is then used to generate a two-dimensional array $w(\mathbf{q}) = (\rho(\mathbf{q}))^{-2(H+1)}$ for all elements $\rho(j,k) \geq q_r\pi/L_e$, and $w(\mathbf{q}) = (q_r\pi/L_e)^{-2(H+1)}$ for all elements $\rho(j,k) < q_r\pi/L_e$. The array $w(\mathbf{q})$ is then summed; let us define this sum as w_{sum}. The normalized power spectrum array is then:

$$W(\mathbf{q}) = \frac{h_L^2 L_e^2 w(\mathbf{q})}{4\pi^2 w_{\text{sum}}}. \tag{A2}$$

The array $W(\mathbf{q})$ can be used to derive the one-dimensional power-spectral-density function to check that the gradient of the function is $-2(H+1)$ at all wavevectors that are greater than the roll-off wavevector and constant at the wavevectors below the roll-off limit.

Next, the Fourier-transformed height array $h(\mathbf{q}) = B(\mathbf{q})\exp(i\phi(\mathbf{q}))$ is derived so that the amplitudes are

$$B(\mathbf{q}) = \frac{2\pi m}{d_e}\sqrt{W(\mathbf{q})} \tag{A3}$$

and the phases, $\phi(\mathbf{q})$, are (uniformly) random in the range from $-\pi$ to π. The Fourier-transformed height array has a conjugate symmetry so that $W(q_j, q_k) = W(q_j, -q_k) = W(-q_j, q_k) = W(-q_j, -q_k)$. To ensure the symmetry, $B(\mathbf{q})$ and $\phi(\mathbf{q})$ of the corners of the first quarter, i.e., elements (1,1), (1,$m/2$), ($m/2$,1), and ($m/2$, $m/2$) are set to 0 (1 referring here to the first element). Then, the values along the rows and columns in both the amplitude and phase arrays are mirrored across to the other side so that the second half of each row (columns extending from $m/2+1$ to m) are copied and flipped (i.e., in reverse order) to replace the first half of the row (columns extending from 2 to $m/2-1$). For columns 1 and $m/2$, the arrays are copied from row 2 to $m/2-1$ and flipped to replace the rows from $m/2+1$ to m. The mirrored phase arrays are also multiplied by -1.

Finally, the topography is generated by the inverse fast-Fourier transformation of an array $Z(\mathbf{q}) = B(\mathbf{q})[\cos(\phi(\mathbf{q})) + i\sin(\phi(\mathbf{q}))]$ (where i is the imaginary unit) after shifting the origin of the array $Z(\mathbf{q})$ from the middle ($q \in [-0.5q_0, 0.5q_0]$) back to the beginning ($q \in [0, q_0]$).

Appendix B

In this section, Equations (17) and (32) are derived. The power spectrum for a 3D surface is defined as

$$W(q) = W_0\left(\frac{q}{q_0}\right)^{-2(H+1)}.$$

The square of the rms height over an isotropic self-affine surface area is then

$$h^2(q_1, q_0) = 2\pi \int_{q_0}^{q_1} qW_0 \left(\frac{q}{q_0}\right)^{-2(H+1)} dq,$$

which gives

$$h^2(q_1, q_0) = \frac{2\pi W_0 q_0^2}{2H} \left[1 - \left(\frac{q_1}{q_0}\right)^{-2H}\right],$$

where $q_1 q_0^{-1} = m$. If $m \gg 1$,

$$h(q_1, q_0) = q_0 \sqrt{\frac{W_0 \pi}{H}}.$$

The square of the rms slope is given as

$$s^2(q_1, q_0) = 2\pi \int_{q_0}^{q_1} q^3 W_0 \left(\frac{q}{q_0}\right)^{-2(H+1)} dq,$$

which gives

$$s^2(q_1, q_0) = \frac{2\pi W_0 q_0^4}{2(1-H)} \left[\left(\frac{q_1}{q_0}\right)^{2(1-H)} - 1\right],$$

where, with a similar assumption for m as above,

$$s(q_1, q_0) = q_0^2 m^{1-H} \sqrt{\frac{W_0 \pi}{1-H}}.$$

The ratio of the two products is then

$$\frac{s}{h} = q_0 m^{1-H} \sqrt{\frac{H}{1-H}} = q_1 m^{-H} \sqrt{\frac{H}{1-H}}.$$

Here, the power spectrum is integrated from the roll-off wavevector q_0 to the single-element wavevector q_1. To solve the rms height and the wavelength-scale rms slope for an arbitrary length (e.g., the spatial resolution in a radar image) defined by a wavevector q_x, the power spectrum must be integrated from q_1 to q_x, while the roll-off wavevector is still q_0 to obtain:

$$h^2(q_1, q_x, q_0) = \frac{\pi W_0 q_0^2}{H} \left[\left(\frac{q_0}{q_x}\right)^{2H} - \left(\frac{q_0}{q_1}\right)^{2H}\right]$$

and

$$s^2(q_1, q_x, q_0) = \frac{\pi W_0 q_0^2 q_x^2}{1-H} \left[\left(\frac{q_1}{q_x}\right)^2 \left(\frac{q_0}{q_1}\right)^{2H} - \left(\frac{q_0}{q_x}\right)^{2H}\right].$$

In the ratio of the rms slope and rms height, q_0 is eliminated and the ratio is expressed as a function of only $q_1 q_x^{-1} = m_x$ and q_x, that is, Equation (32). To clarify, this solution is different from the rms slope for a lag of $\Delta r = x$ by solving the wavelength-scale rms slope.

Appendix C

Table A1 lists the statistical roughness properties of the Hawaii lava flows presented in [22,25]. The Rayleigh roughness limit is 0.031 m (at normal incidence) for the rms height in the wavelength-scale sample length, which can be calculated by $h_{0.25} = h_1(0.25)^H$. All sites with $h_1 > 0.06$ m are above the Rayleigh limit at a sample length of 0.25 m.

Table A1. Roughness statistics for the Hawaii lava flows presented in [22,25] (sorted by roughness, i.e., root-mean-square (rms) slope and rms height, but with the original site numbers shown in the first column) and used in Figure 6. Test Cases 1, 2, and 3 are, respectively, sites 10, 8, and 5.

Site	H	s_λ (25 cm)	s_1 (1 m)	h_1 (1 m) [m]	$h_{0.25}$ (25 cm) [m]
1	0.62	0.073	0.039	0.014	0.006
10	0.63	0.076	0.046	0.017	0.007
9	0.64	0.147	0.088	0.032	0.013
2	0.68	0.172	0.104	0.036	0.014
6	0.44	0.222	0.099	0.041	0.022
8	0.49	0.225	0.108	0.046	0.023
4	0.48	0.324	0.155	0.062	0.032
3	0.51	0.436	0.215	0.085	0.042
7	0.29	0.586	0.217	0.105	0.070
5	0.26	0.705	0.240	0.120	0.084

Appendix D

Figure A1 shows four different slope angle p.d.f. histograms so that h_L varies from 0.01 m to 0.3 m, and other parameters are kept constant ($H = 0.3$, $L_e = 16$ m, $m = 128$, and $q_r = 2$). The slope angle is obtained from the vertical component of unit normal vectors of the surface elements so that $\theta_s = \arccos(\hat{n}_z)$. A Gaussian p.d.f. fit is added for reference. The slope ($\tan\theta_s$) distributions are best fit by the Rayleigh p.d.f. as shown by Shepard et al. [18].

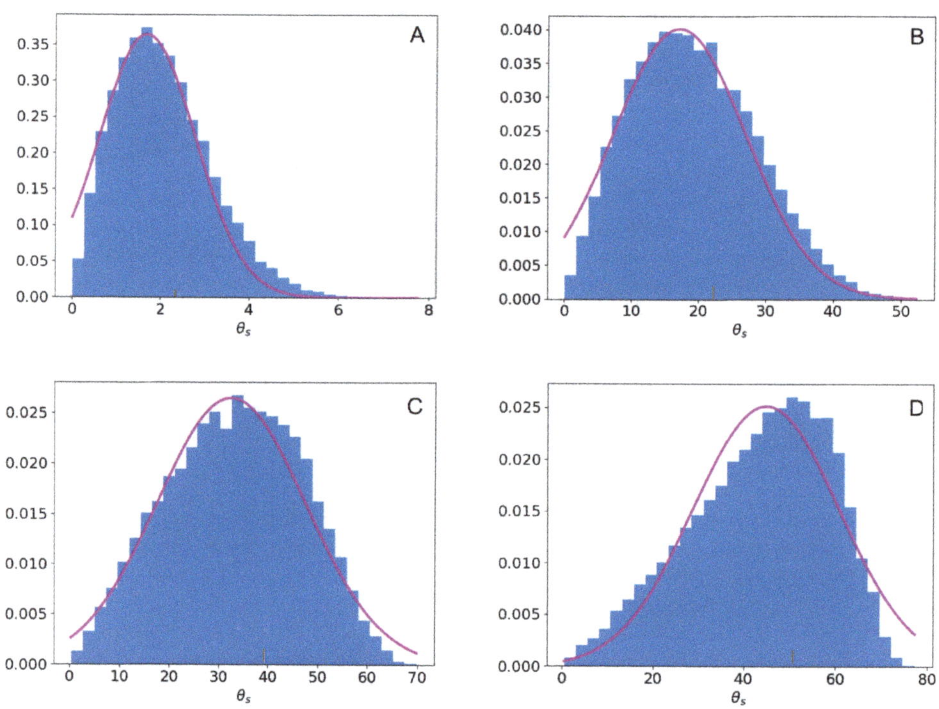

Figure A1. The slope angle p.d.f. for four different cases of h_L: **(A)** $h_L = 0.01$ m, **(B)** $h_L = 0.1$ m, **(C)** $h_L = 0.2$ m, and **(D)** $h_L = 0.3$ m, when $H = 0.3$, $L_e = 16$ m, $m = 128$, and $q_r = 2$. The magenta curve is a Gaussian p.d.f. fit.

References

1. Patterson, G.W.; Stickle, A.M.; Turner, F.S.; Jensen, J.R.; Bussey, D.B.J.; Spudis, P.; Espiritu, R.C.; Schulze, R.C.; Yocky, D.A.; Wahl, D.E.; et al. Bistatic radar observations of the Moon using Mini-RF on LRO and the Arecibo Observatory. *Icarus* **2017**, *283*, 2–19. [CrossRef]
2. Bhiravarasu, S.S.; Chakraborty, T.; Putrevu, D.; Pandey, D.K.; Das, A.K.; Ramanujam, V.M.; Mehra, R.; Parasher, P.; Agrawal, K.M.; Gupta, S.; et al. Chandrayaan-2 Dual-frequency Synthetic Aperture Radar (DFSAR): Performance Characterization and Initial Results. *Planet. Sci. J.* **2021**, *2*, 134. [CrossRef]
3. Hagfors, T. Backscattering from an undulating surface with applications to radar returns from the Moon. *J. Geophys. Res.* **1964**, *69*, 3779–3784. [CrossRef]
4. Muhleman, D.O. Symposium on Radar and Radiometric Observations of Venus during the 1962 Conjunction: Radar scattering from Venus and the Moon. *Astron. J.* **1964**, *69*, 34. [CrossRef]
5. Rea, D.G.; Hetherington, N.; Miffin, R. The Analysis of Radar Echoes from the Moon. *J. Geophys. Res.* **1964**, *69*, 5217–5223. [CrossRef]
6. Mitchell, D.L.; Ostro, S.J.; Hudson, R.S.; Rosema, K.D.; Campbell, D.B.; Velez, R.; Chandler, J.F.; Shapiro, I.I.; Giorgini, J.D.; Yeomans, D.K. Radar Observations of Asteroids 1 Ceres, 2 Pallas, and 4 Vesta. *Icarus* **1996**, *124*, 113–133. [CrossRef]
7. Jurgens, R.F.; Goldstein, R.M. Radar observations at 3.5 and 12.6 CM wavelength of asteroid 433 Eros. *Icarus* **1976**, *28*, 1–15. [CrossRef]
8. Ostro, S.J.; Campbell, D.B.; Shapiro, I.I. Radar observations of asteroid 1685 Toro. *Astron. J.* **1983**, *88*, 565–576. [CrossRef]
9. McCollom, T.M.; Jakosky, B.M. Interpretation of planetary radar observations: The relationship between actual and inferred slope distributions. *J. Geophys. Res.* **1993**, *98*, 1173–1184. [CrossRef]
10. Nolan, M.C.; Magri, C.; Howell, E.S.; Benner, L.A.M.; Giorgini, J.D.; Hergenrother, C.W.; Hudson, R.S.; Lauretta, D.S.; Margot, J.L.; Ostro, S.J.; et al. Shape model and surface properties of the OSIRIS-REx target Asteroid (101955) Bennu from radar and lightcurve observations. *Icarus* **2013**, *226*, 629–640. [CrossRef]
11. Marshall, S.E.; Howell, E.S.; Magri, C.; Vervack, R.J.; Campbell, D.B.; Fernández, Y.R.; Nolan, M.C.; Crowell, J.L.; Hicks, M.D.; Lawrence, K.J.; et al. Thermal properties and an improved shape model for near-Earth asteroid (162421) 2000 ET70. *Icarus* **2017**, *292*, 22–35. [CrossRef]
12. Rożek, A.; Lowry, S.C.; Nolan, M.C.; Taylor, P.A.; Benner, L.A.M.; Fitzsimmons, A.; Zegmott, T.J.; Weissman, P.R.; Green, S.F.; Rozitis, B.; et al. Shape model and spin-state analysis of PHA contact binary (85990) 1999 JV6 from combined radar and optical observations. *Astron. Astrophys.* **2019**, *631*, A149. [CrossRef]
13. McGlasson, R.A.; Marshall, S.E.; Venditti, F.C.F.; Naidu, S.P.; Benner, L.A.M.; Brozović, M.; Giorgini, J.D.; Taylor, P.A.; Aponte, B.; Virkki, A.K.; et al. Radar and Lightcurve Observations and a Physical Model of Potentially Hazardous Asteroid 1981 Midas. *Planet. Sci. J.* **2022**, *3*, 35. [CrossRef]
14. Shepard, M.K.; Campbell, B.A. Radar Scattering from a Self-Affine Fractal Surface: Near-Nadir Regime. *Icarus* **1999**, *141*, 156–171. [CrossRef]
15. Franceschetti, G.; Iodice, A.; Migliaccio, M.; Riccio, D. Scattering from natural rough surfaces modeled by fractional Brownian motion two-dimensional processes. *IEEE Trans. Antennas Propag.* **1999**, *47*, 1405–1415. [CrossRef]
16. Biccari, D.; Picardi, G.; Seu, R.; Corradini, A.; Orosei, R. The Mars High Resolution Advanced Radar for 2005 Space Mission. In Proceedings of the Conference on the Geophysical Detection of Subsurface Water on Mars, Houston, TX, USA, 6–10 August 2001; p. 7029.
17. Turcotte, D.L. *Fractals and Chaos in Geology and Geophysics*; Cornell University: New York, NY, USA, 1997.
18. Shepard, M.K.; Brackett, R.A.; Arvidson, R.E. Self-affine (fractal) topography: Surface parameterization and radar scattering. *J. Geophys. Res. Planets* **1995**, *100*, 11709–11718. [CrossRef]
19. Austin, R.T.; England, A.W.; Wakefield, G.H. Special problems in the estimation of power-law spectra as applied to topographical modeling. *IEEE Trans. Geosci. Remote Sens.* **1994**, *32*, 928–939. [CrossRef]
20. Shepard, M.K.; Campbell, B.A. Shadows on a Planetary Surface and Implications for Photometric Roughness. *Icarus* **1998**, *134*, 279–291. [CrossRef]
21. Ulaby, F.T.; Long, D.G. *Microwave Radar and Radiometric Remote Sensing*; University of Michigan Press: Ann Arbor, MI, USA, 2014.
22. Campbell, B.A. Scale-Dependent Surface Roughness Behavior and ItsImpact on Empirical Models for Radar Backscatter. *IEEE Trans. Geosci. Remote Sens.* **2009**, *47*, 3480–3488. [CrossRef]
23. Persson, B.N.J.; Albohr, O.; Tartaglino, U.; Volokitin, A.I.; Tosatti, E. TOPICAL REVIEW: On the nature of surface roughness with application to contact mechanics, sealing, rubber friction and adhesion. *J. Phys. Condens. Matter* **2005**, *17*, R1–R62. [CrossRef]
24. Shepard, M.K.; Campbell, B.A.; Bulmer, M.H.; Farr, T.G.; Gaddis, L.R.; Plaut, J.J. The roughness of natural terrain: A planetary and remote sensing perspective. *J. Geophys. Res.* **2001**, *106*, 32777–32796. [CrossRef]
25. Campbell, B.A. *Radar Remote Sensing of Planetary Surfaces*; Cambridge University Press: Cambridge, UK, 2002.
26. Persson, B.N.J. On the Fractal Dimension of Rough Surfaces. *Tribol. Lett.* **2014**, *54*, 99–106. [CrossRef]
27. Simpson, R.A.; Tyler, G.L. Radar scattering laws for the lunar surface. *IEEE Trans. Antennas Propag.* **1982**, *30*, 438–449. [CrossRef]
28. Magri, C.; Ostro, S.J.; Scheeres, D.J.; Nolan, M.C.; Giorgini, J.D.; Benner, L.A.M.; Margot, J.L. Radar observations and a physical model of Asteroid 1580 Betulia. *Icarus* **2007**, *186*, 152–177. [CrossRef]
29. Fung, A.K. *Microwave Scattering and Emission Models and Their Applications*; Artech House: Norwood, MA, USA, 1994.

30. Fung, A.K.; Li, Z.; Chen, K.S. Backscattering from a randomly rough dielectric surface. *IEEE Trans. Geosci. Remote Sens.* **1992**, *30*, 356–369. [CrossRef]
31. Fa, W.; Wieczorek, M.A.; Heggy, E. Modeling polarimetric radar scattering from the lunar surface: Study on the effect of physical properties of the regolith layer. *J. Geophys. Res. (Planets)* **2011**, *116*, E03005. [CrossRef]
32. Virkki, A.K.; Bhiravarasu, S.S. Modeling Radar Albedos of Laboratory-Characterized Particles: Application to the Lunar Surface. *J. Geophys. Res. (Planets)* **2019**, *124*, 3025–3040. [CrossRef]
33. Smith, B.G. Lunar Surface Roughness: Shadowing and Thermal Emission. *J. Geophys. Res.* **1967**, *72*, 4059. [CrossRef]
34. Campbell, B.A. High circular polarization ratios in radar scattering from geologic targets. *J. Geophys. Res. (Planets)* **2012**, *117*, E06008. [CrossRef]
35. Riskilä, E.; Lindqvist, H.; Muinonen, K. Light scattering by fractal roughness elements on ice crystal surfaces. *J. Quant. Spectrosc. Radiat. Transf.* **2021**, *267*, 107561. [CrossRef]
36. Helfenstein, P.; Shepard, M.K. Submillimeter-Scale Topography of the Lunar Regolith. *Icarus* **1999**, *141*, 107–131. [CrossRef]
37. Lauretta, D.S.; Dellagiustina, D.N.; Bennett, C.A.; Golish, D.R.; Becker, K.J.; Balram-Knutson, S.S.; Barnouin, O.S.; Becker, T.L.; Bottke, W.F.; Boynton, W.V.; et al. The unexpected surface of asteroid (101955) Bennu. *Nature* **2019**, *568*, 55–60. [CrossRef] [PubMed]
38. Michikami, T.; Honda, C.; Miyamoto, H.; Hirabayashi, M.; Hagermann, A.; Irie, T.; Nomura, K.; Ernst, C.M.; Kawamura, M.; Sugimoto, K.; et al. Boulder size and shape distributions on asteroid Ryugu. *Icarus* **2019**, *331*, 179–191. [CrossRef]
39. Harmon, J.K.; Ostro, S.J. Mars: Dual-polarization radar observations with extended coverage. *Icarus* **1985**, *62*, 110–128. [CrossRef]
40. Virkki, A.; Muinonen, K. Radar scattering by planetary surfaces modeled with laboratory-characterized particles. *Icarus* **2016**, *269*, 38–49. [CrossRef]
41. Markkanen, J.; Väisänen, T.; Penttilä, A.; Muinonen, K. Scattering and absorption in dense discrete random media of irregular particles. *Opt. Lett.* **2018**, *43*, 2925. [CrossRef]

MDPI AG
Grosspeteranlage 5
4052 Basel
Switzerland
Tel.: +41 61 683 77 34

Remote Sensing Editorial Office
E-mail: remotesensing@mdpi.com
www.mdpi.com/journal/remotesensing

www.ingramcontent.com/pod-product-compliance
Lightning Source LLC
Chambersburg PA
CBHW061835210926
51850CB00006B/899